METHODS IN MOLECULAR BIOLOGY

Series Editor
John M. Walker
School of Life and Medical Sciences
University of Hertfordshire
Hatfield, Hertfordshire, AL10 9AB, UK

For further volumes:
http://www.springer.com/series/7651

DNA Repair

Methods and Protocols

Edited by

Lata Balakrishnan

Indiana University – Purdue University, Indianapolis, IN, USA

Jason A. Stewart

Department of Biological Sciences, University of South Carolina, Columbia, SC, USA

Humana Press

Editors
Lata Balakrishnan
Indiana University – Purdue University
Indianapolis, IN, USA

Jason A. Stewart
Department of Biological Sciences
University of South Carolina
Columbia, SC, USA

ISSN 1064-3745 ISSN 1940-6029 (electronic)
Methods in Molecular Biology
ISBN 978-1-4939-9499-1 ISBN 978-1-4939-9500-4 (eBook)
https://doi.org/10.1007/978-1-4939-9500-4

This Humana imprint is published by the registered company Springer Science+Business Media, LLC, part of Springer Nature.
The registered company address is: 233 Spring Street, New York, NY 10013, U.S.A.

Preface

Maintenance of genome integrity is important for all aspects of cellular physiology and development. Alterations in our genome due to endogenous and exogenous factors threaten the integrity of our genome and therein have a direct consequence to human health. Evolution of DNA repair mechanisms has ensured that the majority of the DNA damage acquired is often quickly and efficiently reversed to restore genome stability. Based on the type of damage encountered, different repair proteins are mobilized to the site of damage for DNA repair. Over the past five decades, our understanding of these repair pathways has increased by convergent knowledge from chemists, biochemists, geneticists, molecular biologists, radiobiologists, and oncologists, among many others. At the heart of understanding the molecular details of these complex processes is the wealth of information that has been gained by the development of a wide range of experimental techniques.

This edition of *Methods in Molecular Biology* will outline techniques that will serve as a guide to assist researchers in their experimental design as they study the underlying mechanisms of these important DNA repair pathways. The chapters in this book have been divided into five general parts, each focused on specific aspects of repair biology.

In Part I, two timely reviews are provided on different aspects of DNA repair. Chapter 1 discusses the role of post-translational modifications on homology-directed repair (HDR) and recent advances in the treatment of HDR-defective cancers. The second review focuses on DNA damage caused by defective telomere replication, with a focus on telomere fragility in cells utilizing the recombination-based alternative lengthening of telomeres (ALT) pathway.

Part II, titled "Cellular Assays to Detect and Measure DNA Damage and Damage Response," describes methods in the detection and measurement of DNA breaks and proteins involved in the DNA damage response (DDR). Chapter 3 describes in detail the use of micro-irradiation to create DSBs followed by the use of live-cell imaging or immunofluorescence (IF) to quantify the dynamic association and dissociation of proteins involved in both the recognition and repair of DSBs. In Chapter 4, Pond and Ellis provide a detailed description of the use of pulse-field gel electrophoresis (PFGE) to detect and measure DSBs in genomic DNA. Chapter 5 then shifts focus to the detection of tri-nucleotide repeat (TNR) expansions and deletions caused by DNA base lesions. In this method, Lai et al. describe how to measure the distribution and location of these lesions and their affect TNR stability. Finally, Chapter 6 presents a qPCR method to measure cellular nucleotide pools, which, when not properly maintained, leads to DNA damage and defective DNA repair.

Part III, titled "Cellular and Cell Extract-Based Assays to Measure DNA Repair," encompasses methods designed to measure DNA repair efficiency and characterize factors involved in the repair process. Firstly, Chapter 7 focuses on a much-needed method to measure HDR following genome-editing (i.e., CRISPR, TALEN) through a fluorescence conversion assay, which allows for flow cytometry-based detection of converted cells followed by DNA sequencing to characterize the conversion tracts. Chapter 8 follows up with another reporter-based assay to directly measure base-excision repair (BER) in human cells. The next two chapters make use of *Xenopus* egg extracts, as a cell-free method to detect

DNA repair. Chapter 9 describes a newly developed assay, which creates site-specific ssDNA breaks on plasmid DNA. This plasmid can then be incubated with *Xenopus* egg extracts to measure ssDNA repair efficiency as well as protein recruitment and the DDR. Chapter 10 makes use of *Xenopus* extracts to perform chromatin-immunoprecipitation (ChIP) to measure dynamic protein associations and displacement in a synchronized, cell-free system, which has distinct advantages over cell-based assays. Chapter 11 then describes the use of cell viability assays, IF, and metaphase spread analysis to understand the functional importance of DNA helicases during DNA repair. The final two chapters of this part describe methods to screen for genetic complementation following DNA damage by the colony formation assay (Chapter 12) and the identification of small molecules that inhibit DNA repair proteins (Chapter 13).

Part IV, titled "In Vitro Biochemical and Biophysical Methods to Study DNA Repair," outlines different in vitro assays using purified recombinant proteins to study DNA repair enzyme mechanisms. Chapter 14 outlines the expression and purification of three proteins involved in DSB repair and use of a single-molecule DNA curtain assay to temporally and spatially assess the mechanisms of DNA resection end resection at DSBs. In Chapter 15, Sausen et al. describe two techniques to study the ATPase activity of recombinant helicase/ATPase enzymes. The first outlines a direct method for assaying ATPase activity using radiolabeled ATP and thin layer chromatography and the second uses a high-throughput 96-well plate assay that couples ATPase activity with nicotinamide adenine dinucleotide (NADH) oxidation for real-time monitoring. Chapter 16 details biochemical methods to assess the efficiency of repair enzymes during different DNA transactions, namely, DNA unwinding and annealing. Strand exchange mechanisms are frequently observed during homologous recombination and repair and Chapters 16 and 17 outline methodologies to study strand exchange reactions with purified recombination proteins. With a similar focus on the HDR pathway, Chapters 18 and 19 describe methodologies to assess branch migration and nucleoprotein filament formation, respectively.

The tail end of the book focuses on methods to study DNA damage and repair within the telomeric regions and/or arising during DNA replication. Part V is titled "Methods to Assess Telomeric and Replication-Induced DNA Damage and Repair." Fouquerel et al. in Chapter 20 present highly sensitive protocols, which allow for measurement of UV photoproducts that induce telomere damage in human cells and subsequent assessment of their removal and repair. Genome stability is directly impacted by the dynamics of replication fork movement during the duplication process. DNA fiber analysis allows for direct assessment of fork movement and replication dynamics. Assays describing DNA fiber methods to measure genome-wide and telomere replication are presented in Chapters 21 and 22, respectively. The presence of guanine-rich sequences at telomere ends promote the formation of G-quadraplex structures, which both protect the ends of the chromosomes as well as impede repair and replication. Chapter 23 outlines an in vivo method to detect these structures and their colocalization to telomeric regions using IF. The final technique (Chapter 24) describes the use of FISH to detect DNA damage on metaphase chromosomes, using telomeric FISH probes.

Many of the techniques outlined in this book can be easily adapted or modified by researchers interested in studying DNA transactions to assess their specific protein or repair pathway of interest. Our hope is that these methods will be useful for both those new and well-established in the field and help advance our understanding of DNA repair. We would like to specifically acknowledge and thank all the authors who contributed to this book with their expert descriptions and detailed outlines of various protocols, which will undoubtedly be an excellent resource to many in the field. We would also like to thank the series editor, John Walker, for assistance in the development and editing of this book.

Indianapolis, IN, USA *Lata Balakrishnan*
Columbia, SC, USA *Jason A. Stewart*

Contents

Contributors

SANKET AWATE • *Laboratory of Molecular Gerontology, National Institute on Aging, NIH, NIH Biomedical Research Center, Baltimore, MD, USA*

TARASWI BANERJEE • *DSFederal, Rockville, MD, USA*

RYAN P. BARNES • *Department of Environmental and Occupational Health, UPMC Hillman Cancer Center, University of Pittsburgh, Pittsburgh, PA, USA*

MATTHEW L. BOCHMAN • *Department of Molecular and Cellular Biochemistry, Indiana University, Bloomington, IN, USA*

ROBERT M. BROSH JR • *Laboratory of Molecular Gerontology, National Institute on Aging, NIH, NIH Biomedical Research Center, Baltimore, MD, USA*

ARINDAM DATTA • *Laboratory of Molecular Gerontology, National Institute on Aging, NIH, NIH Biomedical Research Center, Baltimore, MD, USA*

SRIJITA DHAR • *Laboratory of Molecular Gerontology, National Institute on Aging, NIH, NIH Biomedical Research Center, Baltimore, MD, USA*

NATHAN A. ELLIS • *Department of Cellular and Molecular Medicine, University of Arizona Cancer Center, University of Arizona, Tuscon, AZ, USA*

ILYA J. FINKELSTEIN • *Department of Molecular Biosciences and Institute for Cellular and Molecular Biology, The University of Texas at Austin, Austin, TX, USA; Center for Systems and Synthetic Biology, The University of Texas at Austin, Austin, TX, USA*

RACHEL LITMAN FLYNN • *Department of Pharmacology and Experimental Therapeutics, Boston University School of Medicine, Boston, MA, USA; Department of Medicine, Cancer Center, Boston University School of Medicine, Boston, MA, USA*

ELISE FOUQUEREL • *Department of Biochemistry and Molecular Biology, Thomas Jefferson University, Philadelphia, PA, USA; Department of Environmental and Occupational Health, UPMC Hillman Cancer Center, University of Pittsburgh, Pittsburgh, PA, USA*

HIMABINDU GALI • *Department of Pharmacology and Experimental Therapeutics, Boston University School of Medicine, Boston, MA, USA; Department of Medicine, Cancer Center, Boston University School of Medicine, Boston, MA, USA*

FADE GONG • *Department of Molecular Biosciences, Institute for Cellular and Molecular Biology, The University of Texas at Austin, Austin, TX, USA; Department of Biochemistry and Molecular Biology, Baylor College of Medicine, Houston, TX, USA*

STEVEN D. GOODSON • *Department of Genetics and Biochemistry, Clemson University, Clemson, SC, USA; Eukaryotic Pathogens Innovation Center, Clemson University, Clemson, SC, USA*

ANH HA • *Department of Biological Sciences, University of North Carolina at Charlotte, Charlotte, NC, USA*

RUSSELL B. HAWES • *Department of Genetics and Biochemistry, Clemson University, Clemson, SC, USA*

ERIC A. HENDRICKSON • *BMBB Department, University of Minnesota Medical School, Minneapolis, MN, USA*

YOONCHAN HWANG • *Department of Biochemistry, Jacobs School of Medicine and Biomedical Sciences, State University of New York at Buffalo, Buffalo, NY, USA*

LAUREN JOHENGEN • *Department of Biochemistry, Jacobs School of Medicine and Biomedical Sciences, State University of New York at Buffalo, Buffalo, NY, USA*

YINAN KAN • *eGenesis, Cambridge, MA, USA*
ANDREW A. KELSO • *Department of Genetics and Biochemistry, Clemson University, Clemson, SC, USA; Eukaryotic Pathogens Innovation Center, Clemson University, Clemson, SC, USA*
JAE JIN KIM • *Department of Molecular Biosciences, Institute for Cellular and Molecular Biology, The University of Texas at Austin, Austin, TX, USA*
RAMHARI KUMBHAR • *Department of Molecular Biosciences, Institute for Cellular and Molecular Biology, The University of Texas at Austin, Austin, TX, USA*
YANHAO LAI • *Department of Chemistry and Biochemistry, Florida International University, Miami, FL, USA*
NATALIE A. LAMB • *Department of Biochemistry, Jacobs School of Medicine and Biomedical Sciences, State University of New York at Buffalo, Buffalo, NY, USA*
RUIPENG LEI • *Biochemistry Ph.D. Program, Florida International University, Miami, FL, USA*
YUNFENG LIN • *Department of Biological Sciences, University of North Carolina at Charlotte, Charlotte, NC, USA*
YUAN LIU • *Department of Chemistry and Biochemistry, Florida International University, Miami, FL, USA; Biochemistry Ph.D. Program, Florida International University, Miami, FL, USA; Biomolecular Sciences Institute, Florida International University, Miami, FL, USA*
RUI LIU • *Department of Genetics, School of Basic Medical Sciences, Tianjin Medical University, Tianjin, People's Republic of China*
DAVID T. LONG • *Department of Biochemistry and Molecular Biology, Medical University of South Carolina, Charleston, SC, USA*
EMILY MASON-OSANN • *Department of Pharmacology and Experimental Therapeutics, Boston University School of Medicine, Boston, MA, USA; Department of Medicine, Cancer Center, Boston University School of Medicine, Boston, MA, USA*
KYLE M. MILLER • *Department of Molecular Biosciences, Institute for Cellular and Molecular Biology, The University of Texas at Austin, Austin, TX, USA*
LOGAN R. MYLER • *Department of Molecular Biosciences and Institute for Cellular and Molecular Biology, The University of Texas at Austin, Austin, TX, USA*
PATRICIA L. OPRESKO • *Department of Environmental and Occupational Health, UPMC Hillman Cancer Center, University of Pittsburgh, Pittsburgh, PA, USA*
DOROTA PIEKNA-PRZYBYLSKA • *Department of Microbiology and Immunology, School of Medicine and Dentistry, University of Rochester, Rochester, NY, USA*
DOUGLAS L. PITTMAN • *Department of Drug Discovery and Biomedical Sciences, College of Pharmacy, University of South Carolina, Columbia, SC, USA*
KELVIN W. POND • *Department of Cellular and Molecular Medicine, University of Arizona Cancer Center, University of Arizona, Tuscon, AZ, USA*
NICOLE M. REILLY • *Fondazione Piemontese per la Ricerca sul Cancro ONLUS, Candiolo, Italy*
YAOU REN • *Biochemistry Ph.D. Program, Florida International University, Miami, FL, USA*
CODY M. ROGERS • *Department of Molecular and Cellular Biochemistry, Indiana University, Bloomington, IN, USA*
CHRISTOPHER W. SAUSEN • *Department of Molecular and Cellular Biochemistry, Indiana University, Bloomington, IN, USA*

P. LOGAN SCHUCK • *Department of Biological Sciences, University of South Carolina, Columbia, SC, USA*

MICHAEL G. SEHORN • *Department of Genetics and Biochemistry, Clemson University, Clemson, SC, USA; Eukaryotic Pathogens Innovation Center, Clemson University, Clemson, SC, USA; Clemson University School of Health Research, Clemson, SC, USA; Center for Optical Materials Science and Engineering Technologies, Clemson University, Clemson, SC, USA*

JOSHUA A. SOMMERS • *Laboratory of Molecular Gerontology, National Institute on Aging, NIH, NIH Biomedical Research Center, Baltimore, MD, USA*

MICHAEL M. SONIAT • *Department of Molecular Biosciences and Institute for Cellular and Molecular Biology, The University of Texas at Austin, Austin, TX, USA*

JASON A. STEWART • *Department of Biological Sciences, University of South Carolina, Columbia, SC, USA*

RADHA SUBRAMANIAM • *Genetics, Genomics and Bioinformatics Program, Jacobs School of Medicine and Biomedical Sciences, State University of New York at Buffalo, Buffalo, NY, USA*

JENNIFER A. SURTEES • *Genetics, Genomics and Bioinformatics Program, Jacobs School of Medicine and Biomedical Sciences, State University of New York at Buffalo, Buffalo, NY, USA; Department of Biochemistry, Jacobs School of Medicine and Biomedical Sciences, State University of New York at Buffalo, Buffalo, NY, USA*

JOHN J. TURCHI • *Department of Medicine, Indiana University School of Medicine, Indianapolis, IN, USA; Department of Biochemistry and Molecular Biology, Indiana University School of Medicine, Indianapolis, IN, USA*

PAMELA S. VANDERVERE-CAROZZA • *Department of Medicine, Indiana University School of Medicine, Indianapolis, IN, USA*

SARAH M. WALDVOGEL • *Department of Genetics and Biochemistry, Clemson University, Clemson, SC, USA*

HONG WANG • *Department of Physics, North Carolina State University, Raleigh, NC, USA; Center for Human Health and the Environment, North Carolina State University, Raleigh, NC, USA*

FENG WANG • *Department of Genetics, School of Basic Medical Sciences, Tianjin Medical University, Tianjin, People's Republic of China*

KELLY B. WOLFE • *Department of Biochemistry and Molecular Biology, Medical University of South Carolina, Charleston, SC, USA*

SHAN YAN • *Department of Biological Sciences, University of North Carolina at Charlotte, Charlotte, NC, USA*

BRIAN D. YARD • *Department of Translational Hematology and Oncology Research, Lerner Research Institute, Cleveland Clinic, Cleveland, OH, USA*

MIAOMIAO ZHANG • *Department of Genetics, School of Basic Medical Sciences, Tianjin Medical University, Tianjin, People's Republic of China*

Part I

Reviews

Chapter 1

Homologous Recombination-Mediated DNA Repair and Implications for Clinical Treatment of Repair Defective Cancers

Nicole M. Reilly, Brian D. Yard, and Douglas L. Pittman

Abstract

Double-strand DNA breaks (DSBs) are generated by ionizing radiation and as intermediates during the processing of DNA, such as repair of interstrand cross-links and collapsed replication forks. These potentially deleterious DSBs are repaired primarily by the homologous recombination (HR) and nonhomologous end joining (NHEJ) DNA repair pathways. HR utilizes a homologous template to accurately restore damaged DNA, whereas NHEJ utilizes microhomology to join breaks in close proximity. The pathway available for DSB repair is dependent upon the cell cycle stage; for example, HR primarily functions during the S/G2 stages while NHEJ can repair DSBs at any cell cycle stage. Posttranslational modifications (PTMs) promote activity of specific pathways and subpathways through enzyme activation and precisely timed protein recruitment and degradation. This chapter provides an overview of PTMs occurring during DSB repair. In addition, clinical phenotypes associated with HR-defective cancers, such as mutational signatures used to predict response to poly(ADP-ribose) polymerase inhibitors, are discussed. Understanding these processes will provide insight into mechanisms of genome maintenance and likely identify targets and new avenues for therapeutic interventions.

Key words Homologous recombination, Double-strand break repair, Ubiquitination, RAD51, Chromosome integrity, Nonhomologous end joining, Alternate end joining, Break induced replication, Synthesis-dependent strand annealing, PARP inhibitors

Abbreviations

DSBs	Double-strand breaks
dsDNA	Double-stranded DNA
HR	Homologous recombination
ICLs	Interstrand cross-links
NHEJ	Nonhomologous end joining
PARP	Poly(ADP)-ribose polymerase
PTMs	Posttranslational modifications
SSBs	Single-strand breaks
ssDNA	Single-stranded DNA

Lata Balakrishnan and Jason A. Stewart (eds.), *DNA Repair: Methods and Protocols*, Methods in Molecular Biology, vol. 1999, https://doi.org/10.1007/978-1-4939-9500-4_1, © Springer Science+Business Media, LLC, part of Springer Nature 2019

1 Introduction

Near the turn of the twentieth century, an insightful summary by Boveri [1][1] made the connection that "most chromosome combinations that deviate from the norm lead to the death of the cell, but there are some that allow the cell to survive but not to function normally." When new analytical tools became available during the 1960s and 1970s [2], Dr. Loeb's "Mutator Phenotype Hypothesis" further stated that "mutations occur randomly throughout the genome, and among these would be mutations in genes that guarantee the fidelity of DNA replication" (summarized in [3]). Drs. Hanahan and Weinberg in the Cancer Hallmark series later proposed "genome instability" as an enabling characteristic, stating that accumulation of diverse mutations is one of the critical steps leading to cancer development, and concluding that accurate repair of DNA damage is essential for maintaining genomic integrity [4]. Based upon these fundamental concepts, a number of pathways and subpathways to repair DNA damage and maintain genome integrity are now known. Understanding how DNA repair factors interact, integrate, modify, and influence the kinetics of each step is the next major challenge.

Both endogenous and exogenous insults can lead to the accumulation of mutations and chromosome instability. DNA double-strand breaks (DSBs), generated by ionizing radiation (IR) or through the processing of stalled replication forks, methylation DNA damage, or interstrand cross-links (ICLs), represent particularly challenging structures to repair [5–7]. Mammalian cells have evolved two central pathways for DSB repair: nonhomologous end joining (NHEJ) and homologous recombination (HR) that employ distinct repair mechanisms [8]. NHEJ facilitates direct ligation of unprocessed DNA ends, an error-prone process that can result in loss of genetic material at the DSB site [9]. HR is more accurate and restores the fidelity of the DNA using a homologous template, typically an intact sister chromatid.

DNA damage repair pathways and subpathways are integrated and regulated by posttranslational modifications (PTMs) [10]. For example, ubiquitination events during HR-mediated DSB repair are essential for pathway progression and accurate repair of the damage. Ubiquitin modification can activate proteins, initiate the recruitment and binding of downstream proteins to a damage site, or signal proteasomal-mediated protein degradation. Ubiquitin may be added to single or multiple lysine residues along a target protein to produce mono- or multiple single (multimono) modifi-

[1] Boveri's original manuscript was published in 1914, and the 2008 reference is an English translation by Henry Harris.

cations as well as ubiquitin chains along a single lysine residue [11]. This chapter outlines the current knowledge of PTMs, such as phosphorylation, PARP, and ubiquitination of proteins during HR-mediated DNA damage response. An overview of the clinical relevance of HR defective cancers, specifically mutational signatures and therapy sensitivity associated with loss of HR proteins, is discussed. Together, this information may be used toward understanding cancer stage as well as for combination therapies to guide clinical treatment.

2 Homologous Recombination Mediated Repair

2.1 Ionizing Radiation Induced DNA Strand Breaks

Genotoxic damage induced by ionizing radiation (IR) occurs when radiative energy directly produces repulsive energy along the phosphodiester bonds, generating both single- and double-stranded DNA breaks [12]. Most frequently, the radiation energy is absorbed by water particles within the cell leading to the release of reactive oxygen species (ROS) [13]. Typically, ROS induces SSBs that impair replication, resulting in stalled replication forks and ultimately DSBs [14].

DSBs are resolved by the NHEJ or HR pathway. NHEJ functions throughout the cell cycle and DSBs are recognized by the Ku70/Ku80 heterodimer, which binds the ends of the DNA on both sides of the break. Formation of this heterodimer leads to activation of the catalytic subunit of DNA-dependent protein kinase (DNA-PK), a member of the ATM family of kinases. DNA-PK binds to Ku70 and Ku80 and stabilizes the ends of the DNA. After stabilization, the DNA ligase IV/XRCC4 complex binds and joins the ends of the DNA together, resulting in repair of the DSB [15, 16]. Pathway choice between NHEJ and HR during S/G2 is influenced by the extent of end processing at the DSB site [17]. NHEJ does not have a prerequisite for DNA end processing, but HR requires 5′ end resection at the break [18], a process promoted during the S and G2 phases of the cell cycle but not during the G0/G1 stages [19].

At least two distinct homologous recombination (HR) subpathways have been identified for conservative repair: "classical HR" and the synthesis-dependent strand annealing (SDSA) pathway [20, 21] (Fig. 1; left and center panels). In both pathways, HR-mediated repair begins with end processing in the 5′–3′ direction to yield 3′ single-stranded DNA (ssDNA) overhangs. During S phase, SSBs and DSBs are initially detected by the nuclear enzyme poly(ADP-ribose) polymerase 1 (PARP1), an enzyme that catalyzes poly-ADP-ribosylation of proteins in response to genotoxic stress [22]. PARP1 activity triggers recruitment of the MRN complex, comprised of the meiotic recombination 11 (MRE11), RAD50, and Nijmegen Breakage Syndrome

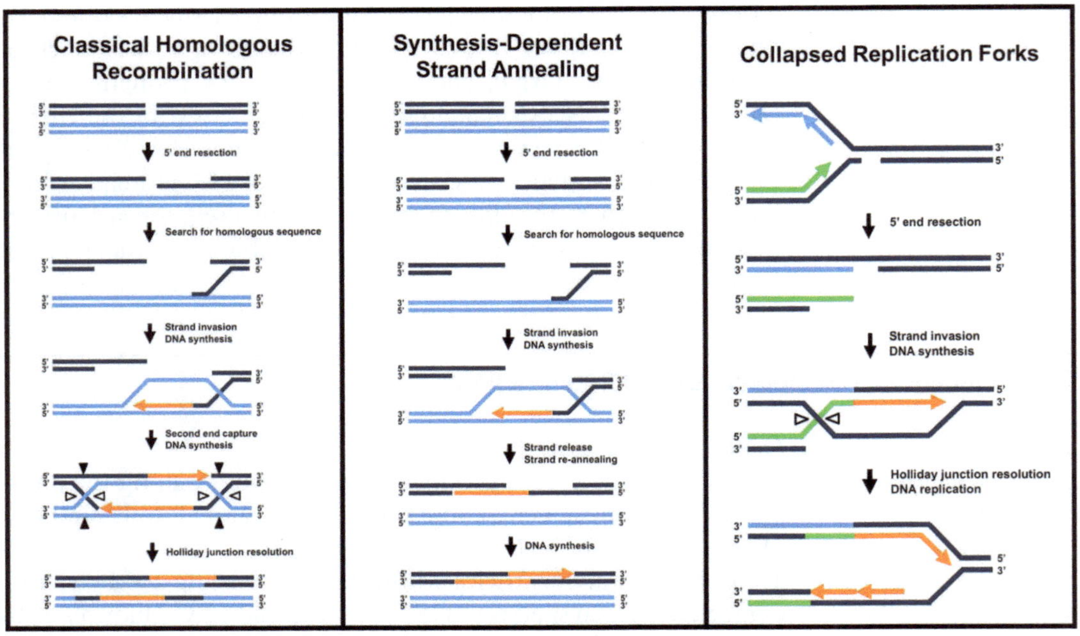

Fig. 1 *Left panel:* Homologous recombination-mediated repair of double strand breaks by double-strand break repair subpathways. In this model, navy blue and light blue lines represent sister chromatids. The DSB is indicated by a gap in both of the navy blue lines. After DSB formation, the DNA ends at the break site are processed by endonucleases to expose 3′ ssDNA overhangs. One of the 3′ ends invades the sister chromatid and primes DNA synthesis using homologous sequence as a template. The other 3′ end is captured by the sister chromatid forming a second Holliday junction. Following DNA synthesis, the double Holliday junctions are resolved in the same or opposite planes creating either noncrossover or crossover products respectively. In this illustration, the junctions are cleaved at the white arrowheads generating noncrossover products. Newly synthesized DNA is highlighted in orange with arrows indicating the direction of synthesis. *Center panel:* HR-mediated repair of DSBs by the SDSA (Synthesis Dependent Strand Annealing) subpathway. In this model, navy blue and light blue lines represent sister chromatids. The DSB is indicated by a gap in both of the navy blue lines. Following the formation of a DSB, the DNA ends at the break site are processed by endonucleases to expose 3′ ssDNA overhangs. One of the overhangs invades the sister chromatid and binds to homologous sequence. The invading strand primes DNA synthesis beyond the break site. The newly replicated strand then reanneals to the damaged sister chromatid. DNA synthesis then fills in the remaining gap on the damaged sister chromatid. Only noncrossover products result from SDSA. Newly synthesized DNA is highlighted in orange with arrows indicating the direction of synthesis. *Right panel:* Model for HR-mediated repair of collapsed replication forks. Navy blue lines represent parental DNA. Green and light blue lines represent the leading and lagging strands respectively. A replication fork collapses after encountering a gap on a parental strand, resulting in a one-ended DSB. The exposed DNA end undergoes 5′ to 3′ resection to yield a 3′ ssDNA overhang. For ease of viewing the top strands have been flipped. The 3′ end invades the homologous DNA duplex to prime new DNA synthesis. The Holliday junction is resolved to repair the DNA replication fork and enable new DNA replication. Newly synthesized DNA is illustrated in orange

1 (NBS1) proteins, to initiate HR-mediated repair [23]. The MRN complex has both exonucleolytic and endonucleolytic capabilities, and produces short ssDNAs at the break site [24, 25]. CtIP (C-terminal binding protein 1) is recruited to the break site and, in conjunction with the MRN complex, mediates 5′ to 3′ end

resection [19, 23, 26]. The 3′ overhang facilitates search and invasion of a DNA duplex with homologous sequence, forming a Holliday junction. Following MRN/CtIP-mediated nucleotide excision, the single-strand binding protein replication protein A (RPA) stabilizes the newly produced ssDNA overhangs [27]. The primary role of RPA during classical HR is to stabilize ssDNA and prevent the formation of secondary structures [28]. A central key step in the HR pathway is localization of RAD51 to the damage site, which is mediated by BRCA1, BRCA2, as well as the RAD51 paralogs [29–32]. BRCA2 directly interacts with RAD51 and targets RAD51 to ssDNA 3′ overhangs by displacing RPA [33–35]. RAD51 then forms a nucleoprotein filament along the 3′ ssDNA overhang and initiates the search for a homologous template, usually a sister chromatid, by binding the dsDNA duplex [20, 36, 37]. The invading strand is then utilized as a primer and replicated beyond the original DSB site (represented by the orange line). Although conventionally thought to only be found during S phase when a sister chromatid is present, the formal possibility remains that the template may also be found in the homologous chromosome at any cell cycle stage. After a homologous sequence is identified, RAD54 integrates into the RAD51 filament, displacing RAD51, and allowing DNA polymerases to access the DNA and fill in the gap [38]. This replication activity during classical HR produces junctions, a cruciform structure that contains the four DNA strands joined together [39]. After the DNA has been extended, the Mus81-Mms4 resolvase facilitates crossover or noncrossover events that resolve the Holliday junctions and produce two intact DNA strands [40].

During classical HR, the 3′ ssDNA overhang not involved in initial strand invasion forms a second Holliday junction with the homologous chromatid (second end capture). The two Holliday junctions are then resolved in the same or opposite directions creating either noncrossover or crossover products, respectively [41]. During SDSA, the initial steps are identical except the second DNA end does not form a Holliday junction. In contrast, following the invading 3′ end being replicated beyond the original break site it is subsequently annealed to the 3′ overhang of the damaged chromatid [41–43]. In mammals, SDSA is the dominant HR pathway for the repair of two-ended DSBs, which yields noncrossover products [44].

In addition to two-ended DSBs, HR is responsible for the repair of collapsed replication forks [45] (Fig. 1; right panel). Replication forks that encounter gaps in single-stranded DNA or forks that have been processed by endonucleases following replication arrest will collapse into one-ended DSBs [7, 46–48]. When a replication fork collapses into a one-ended DSB, a free DNA end becomes available for 5′–3′ processing. The 3′ overhang (green) can then invade the newly replicated sister chromatid, form a single

Holliday junction, and initiate new DNA synthesis. Resolution of the Holliday junction restores the replication fork and allows for reactivation of DNA replication.

2.2 DNA Interstrand Cross-Link Induced Damage

DNA interstrand cross-links (ICLs) are caused by endogenous reactive aldehydes or exogenous chemicals, such as platinum-based chemotherapy drugs. Platinum-induced ICLs occurring at 5′-CG-3′ sequences distort the DNA helix, shifting the platinum atom into the minor groove and bending the DNA strand and the cytosine nucleotide protruding from the helical plane [49]. These distortions increase DNA structure flexibility, allowing for more thermodynamically favorable binding of damaged DNA into the active sites of repair enzymes [50].

To remove an ICL lesion that inhibits DNA replication, a core complex comprised of Fanconi Anemia (FA) proteins (FANCA, -B, -C, -E, -F, -G, -L, -M) binds the DNA strands encompassing the lesion. The complex promotes recruitment of FANCP/SLX4 and the endonucleases XPF, MUS81/ERCC1, and SLX1. Together, these proteins catalyze the incision of one DNA strand on either side of the ICL lesion, producing a double-stranded break and leaving the cross-link on the opposite strand [51–54] (Fig. 2). Following DSB formation, CtIP excises the DNA surrounding the break to produce a 3′ single-strand overhang [55]. A second incision occurs at the 5′ end of the ICL, unhooking the lesion. Translesion synthesis proteins fill in the missing sequence and the cross-link removed by nucleotide excision repair [56, 57]. RAD51-dependent HR is then carried out similarly to replication fork repair (Fig. 1; right panel), and Holliday junction (or junctions if no SDSA) resolution enables re-activation of replication [58].

3 RAD51 Paralogs Promote HR-Mediated DNA Double-Strand Break Repair

Structural analysis using cryo-electron microscopy demonstrated that RAD51 nucleofilaments form a helical structure around ssDNA to promote HR-mediated DSB repair [20, 30, 31, 36, 37]. Although RAD51 oligomers have a higher affinity for ssDNA, the ability to bind dsDNA is essential for promoting homology search and strand exchange during repair [59]. When bound to dsDNA, RAD51 filaments extend the DNA strand creating tension along the helix [60]. This tension exposes Watson-Crick base pairings and brief interactions of the nonhomologous 3′ tail to dsDNA during the homology search [59].

In response to IR, RAD51 activity is regulated by phosphorylation. During S phase, phosphorylation along RAD51 is mediated by c-Abl at Tyr54 and Tyr315 [61–64]. These modifications inhibit RAD51 oligomerization and can enhance its strand exchange activity [63, 65]. BRCA2, a known RAD51 binding partner, also

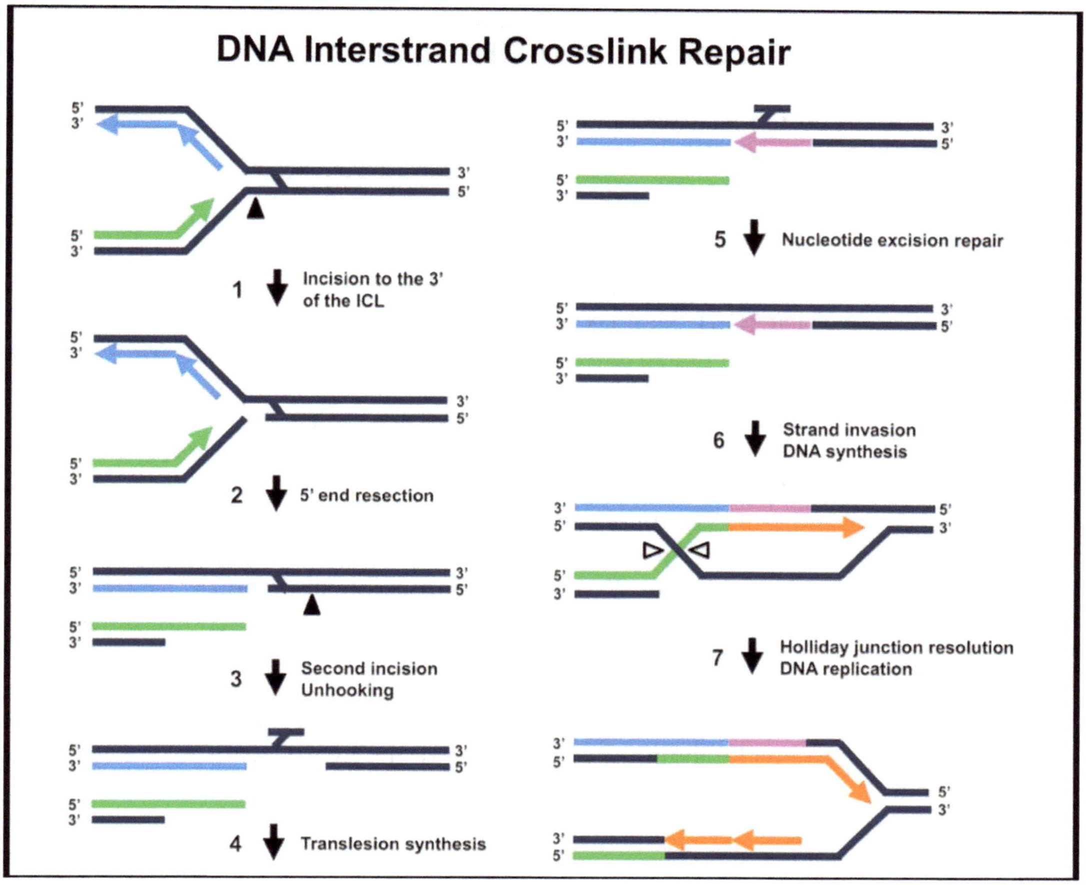

Fig. 2 Model for ICL repair mediated by homologous recombination. Navy blue lines represent parental DNA. The ICL is indicated by the short diagonal line between the two parental strands. Green and light blue lines represent the leading and lagging strands respectively. (*1*) A replication fork stalled at an ICL collapses after the ICL is incised on the 3′ end. (*2*) The exposed DNA end is subject to 5′ end resection. (*3*) The ICL is incised of the 5′, becomes unhooked, and swings away from the helix. In the model illustrated by Helleday and colleagues, the second incision is triggered by the arrival of another replication fork that is not depicted here [56]. (*4*) Translesion synthesis fills in the gap to restore an intact template. (*5*) The remaining lesion is removed by nucleotide incision repair. It has been suggested that the lesion may be removed following HR since it should not hinder the HR machinery. For ease of viewing the top strands have been flipped. (*6*) The 3′ end invades the DNA duplex and facilitates DNA synthesis. (*7*) The Holliday junction is resolved to repair the replication fork. Alternatively, if a second replication fork is present as depicted by Helleday et al., ICL repair may proceed by SDSA [56]. Sequence resulting from translesion synthesis is highlighted in pink, while other newly synthesized DNA is illustrated in orange

coordinates the phosphorylation of RAD51 at Ser14 by polo-like kinase 1 (PLK1) in response to DNA damage at the G2/M junction of the cell cycle [66]. Phosphorylation at Ser14 in turn initiates phosphorylation at Thr13 by casein kinase 2 (CK2), a modification that promotes RAD51 interaction with NBS1, a core component of the MRN complex [67]. Interaction between RAD51 and NBS1 is important for RAD51 recruitment and binding to DNA damage sites [66].

Five additional mammalian RAD51-like proteins—RAD51B, RAD51C, RAD51D, XRCC2, and XRCC3—were identified by homology searches [68–70] and by cloning genes that complemented hamster cell line mutations that conferred hypersensitive phenotypes to DNA damaging agents [71, 72]. As key components during HR, cells deficient for any of the RAD51 family members exhibited increased sensitivity to DNA cross-linking agents, gross chromosomal rearrangements, and defective RAD51 foci formation [72–77]. Yeast-two-hybrid protein interaction studies found the RAD51 paralogs interact with one another in different combinations: (1) RAD51B with RAD51C, (2) RAD51C with RAD51D, (3) RAD51D with XRCC2, and (4) RAD51C with XRCC3 [69, 72, 78, 79]. Immunoprecipitation experiments further identified two distinct complex formations: RAD51B-RAD51C-RAD51D-XRCC2 (BCDX2) and RAD51C-XRCC3 (CX3) [80, 81]. The BCDX2 complex preferentially binds to two distinct DNA structures: Y-shaped DNA and DNA Holliday junctions [82], and is required for RAD51 foci formation in response to IR-induced DSBs and for proper resolution of Holliday junctions [83–85].

Each of the RAD51 paralogs has the conserved Walker Box A and B ATPase motifs, multimer (BRC) interface, and helix–hairpin–helix region (Fig. 3) [86]. The Walker Box A and B motifs are ATP binding sites that catalyze the hydrolysis of ATP to promote

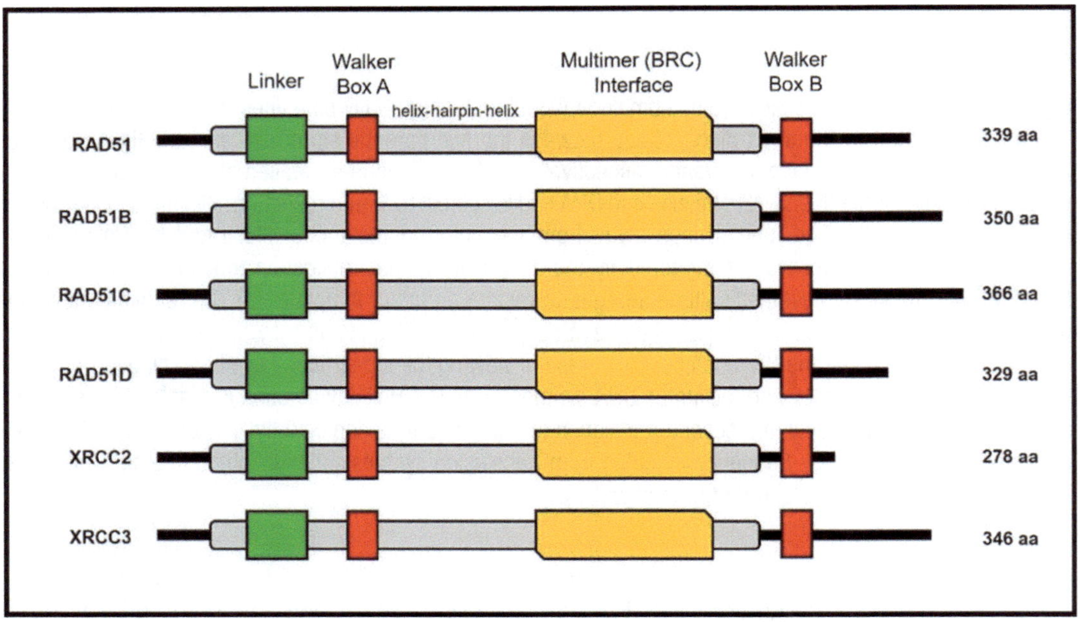

Fig. 3 RAD51 and its paralogs indicating known domains. The RAD51 family share approximately 20–30% identity and have several conserved domains, including a linker region (green) and a helix–hairpin–helix structure (grey). Two Walker Box ATPases motifs (A and B; red) are present in all paralogs and a multimer (BRC) interface domain (yellow) mediates interaction between RAD51 and BCRA2 and is predicted to mediate interactions between the paralogs

ssDNA binding activity of the paralogs [78, 86, 87], and mutations in the Walker Box motifs were shown to decrease ATPase activity and confer cellular sensitivity to DNA damaging agents [88]. The BRC interface is a region of homology between the RAD51 paralogs and the breast cancer associated 2 (BRCA2) protein [89, 90]. Interestingly, RAD51 interacts with BRCA2 through this interface, but none of the other paralogs have been shown to bind BRCA2 [90]. Of note, peptide fragments from the BRC region of BRCA2 are sufficient to act as inhibitors of RAD51 binding to BRCA2, and prevent RAD51 nucleoprotein filament formation [91].

4 Mechanisms of Ubiquitination

Ubiquitin modification of proteins is a major regulatory component of HR [92, 93]. The process of ubiquitination can be illustrated as three-steps (Fig. 4a, b) [94, 95]. First, it is initiated by an ATP-dependent mechanism in which ubiquitin becomes covalently attached to an ubiquitin-activating enzyme (E1). In a second step, ubiquitin is transferred from the E1 to an ubiquitin-conjugating enzyme (E2) in a transthioesterification reaction, also known as "E2 charging." Lastly, an E3 ubiquitin ligase associates with the E2 and facilitates conjugation of ubiquitin to a lysine residue on the target protein. This is accomplished through the formation of an isopeptide bond with the ubiquitin carboxy-terminal glycine residue. The E3 ligase also provides specificity for the reaction through direct interaction with the target substrate [94, 96].

The human genome is predicted to encode over 600 E3 ligases [97]. Based on the mechanism in which ubiquitin is transferred to the target substrate, E3 ligases are divided into two superfamilies [94, 95, 97]. HECT (homologous to E6-AP carboxy terminal) domain E3 ligases form thioester intermediates with ubiquitin prior to the transfer of ubiquitin to the target substrate. In contrast, RING (really interesting new gene) domain E3 ligases do not directly interact with ubiquitin but instead catalyze transfer of ubiquitin directly from the E2 to the targeted protein [94, 95]. As discussed below, RING finger E3 ligases are intricately involved in the regulation of HR mechanisms.

Approximately 95% of E3 ligases are RING finger proteins and are anticipated to have intrinsic E3 ubiquitin ligase activity or function as subunits in E3 ligase complexes [97, 98]. The RING domain is typically 35–45 amino acid residues in length and has the consensus sequence of Cys-X_2-Cys-$X_{(9-19)}$-Cys-$X_{(1-3)}$-His-$X_{(2-3)}$-Cys-X_2-Cys-$X_{(10-18)}$-Cys-X_2-Cys in which X represents any amino acid [99–101]. The cysteine and histidine amino acid residues coordinate interaction with two zinc ions and are essential for proper folding of the RING domain. Association between the RING domain and the E2 is necessary for E3 ubiquitin ligase

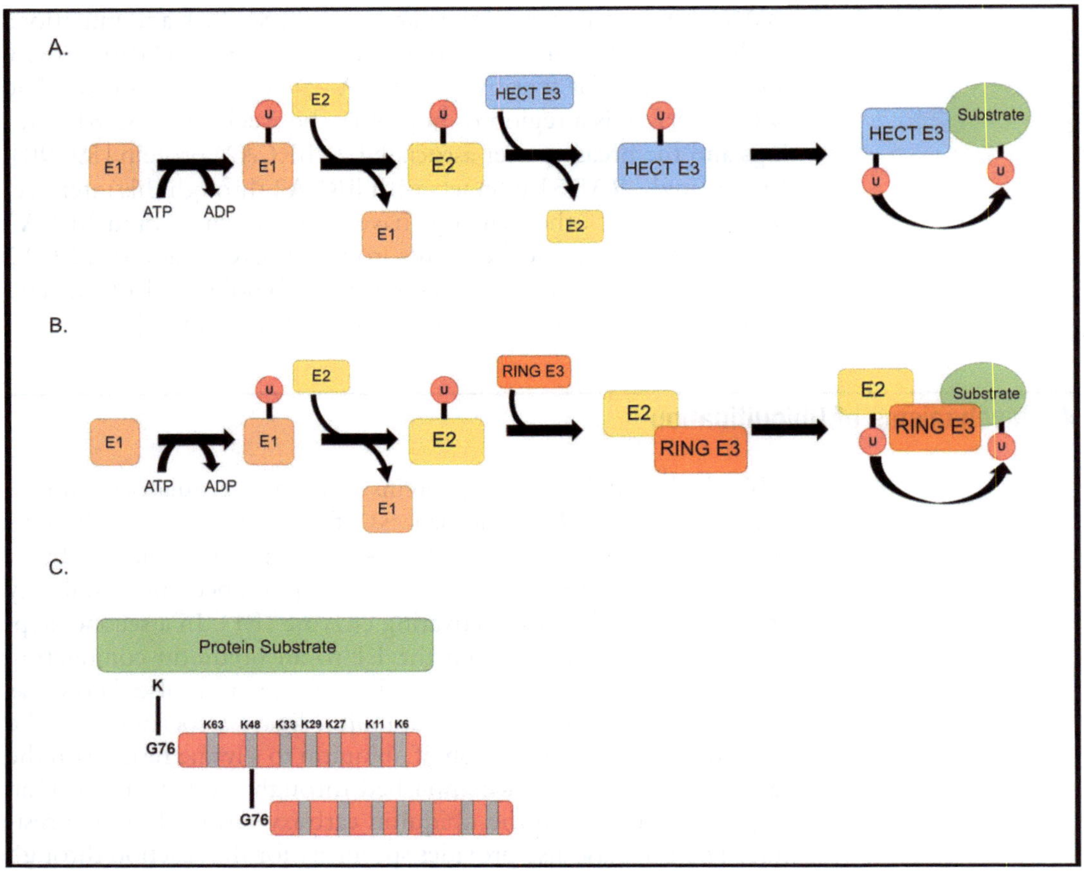

Fig. 4 Ubiquitin modification of target proteins. (**a**) Three-step enzymatic addition of a ubiquitin molecule to a substrate protein. The ubiquitin is activated and ligated to an E1 enzyme in a reaction that consumes ATP. The activated ubiquitin is transferred to an E2 enzyme that interacts directly with a RING E3 ligase. The RING E3 binds the target protein and promotes the transfer of the ubiquitin molecule from the E2 to the substrate. (**b**) The activated ubiquitin molecule is transferred from the E2 enzyme directly to the HECT E3 ligase. The HECT interacts with the substrate and facilitates the transfer of the ubiquitin to the protein. (**c**) Polyubiquitin chains are formed between the terminal glycine of one ubiquitin and one of seven lysine residues (indicated by the grey box) of another ubiquitin molecule

activity. Amino acid substitutions in the RING domain that disrupt E2 interaction also abolish catalytic activity [98, 102, 103].

While E3 ligases target specific proteins for modification, the lysine residues on the target proteins appear to be ubiquitinated without specificity [94, 98]. Lysine residues targeted for ubiquitination tend to be: (1) along the protein surface, (2) in the vicinity of basic amino acid residues, (3) present in ordered secondary structures, particularly α-helices, and (4) overlap with acetylation sites [104]. Thus, lysine residues targeted for ubiquitination are regulated by accessibility rather than recognition of a consensus sequence.

The ubiquitin protein contains seven lysine residues (K6, K11, K27, K29, K33, K48, and K63), and all seven are capable of supporting polyubiquitination in vivo [105]. The effect of ubiquitination on a protein's cellular function depends not only on the pattern of ubiquitination (the lysine(s) residues targeted for modification) but also the topology of the ubiquitin chains (how the ubiquitin moieties are linked together) if the substrate is polyubiquitinated (Fig. 4c). K63-linked chains are nondegradative and modify protein function, serve as docking sites for other proteins, or alter cellular localization [106–109]. In solution, di-ubiquitin K63 chains adopt both "open" and compact "closed" conformations that are recognized by ubiquitin interacting motifs of DNA damage response proteins [110, 111]. In contrast, K48-linked chains that are four or more subunits in length signal for proteasome-mediated degradation of the target protein. K48-linked chains predominantly adopt a "closed" conformation specifically recognized by the 19S subunit of the proteasome and target a protein for proteasomal degradation [112–115].

5 Ubiquitin Signaling Regulates HR-Mediated DNA Repair

RING finger E3 ligases mediate assembly of HR proteins at DSB sites [116] (Fig. 5). The initiating event is the phosphorylation of the histone variant H2AX on serine 139 (known as γH2AX) by ATM [117], which is observed up to two megabases from DSBs. γH2AX provides a DNA damage beacon for the recruitment of HR signaling proteins [118] such as the mediator of DNA checkpoint 1 (MDC1), which binds γ-H2AX and is subsequently phosphorylated by ATM [119–121]. Phosphorylation of MDC1 marks the transition from a phosphorylation cascade to an ubiquitination cascade.

The first RING finger E3 ligase to function in this response is RNF8 (RING finger protein 8). RNF8 binds to the ATM-phosphorylated motifs on MDC1 [122, 123] and, along with the E2 protein UBC13, facilitates the formation of K63-linked polyubiquitin chains on H2A and H2AX histones neighboring the DSB site [124]. These RNF8/UBC13-mediated ubiquitination events further promote the recruitment of downstream HR signaling proteins. One of the HR signaling proteins that binds the K63-linked polyubiquitin chains is RNF168 (RING finger protein 168). RNF168 contains two adjacent motif interacting with ubiquitin (MIU) domains that specifically recognize K63-linked polyubiquitin chains [125]. Similar to RNF8, RNF168 interacts with UBC13 and stimulates the formation of K63-linked polyubiquitin chains on surrounding histones, thus amplifying the original ubiquitin signal initiated by RNF8 [126–128]. In addition, a third E3 ligase HERC2 (HECT Domain

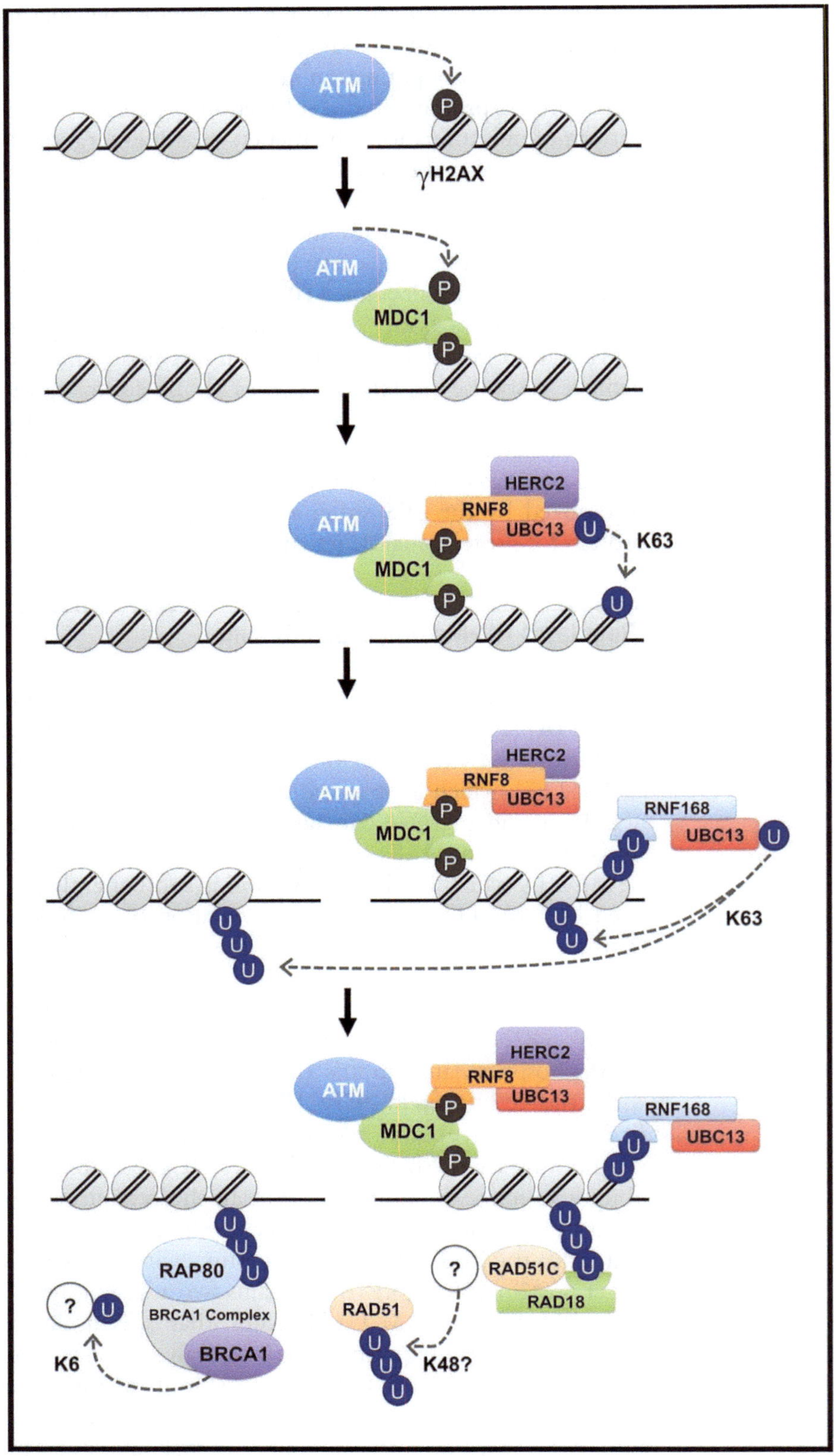

Fig. 5 Ubiquitin signaling regulates the assembly of HR proteins at DSBs. MDC1 binds γ-H2AX and is phosphorylated by ATM. Phosphorylated MDC1 is recognized by the RING finger protein RNF8. RNF8:UBC13

and RLD2), accumulates at DSBs through direct interaction with RNF8. HERC2 appears to support the interaction between RNF8 and UBC13 [129, 130], and has been shown to target the nucleotide excision repair protein, XPA, for ubiquitination [131].

The mechanisms by which RNF8 and RNF168-mediated ubiquitination events regulate HR proteins are only beginning to be elucidated. However, it is apparent that the formation of K63-linked polyubiquitin chains is essential for recruiting and retaining downstream HR repair proteins that recognize these ubiquitin moieties through specialized ubiquitin biding domains. Several ubiquitin binding domains have been identified in proteins, and each domain has specificity for certain ubiquitin modifications. Ubiquitin interacting motifs (UIMs) have specificity for longer ubiquitin chains, such as K63 chains generated along histones, while MIUs preferentially bind K63-linked chains [125, 132]. Additionally, internal zinc finger motifs function as an ubiquitin-binding domain, and are found in repair proteins such as RAD18 [129, 133]. Interestingly, RAD18 directly interacts with RAD51C and is responsible for the accumulation of RAD51C at DNA damage sites [133]. Thus, RAD18 functions as an adapter between K63-linked polyubiquitin chains at DSBs and RAD51C.

BRCA1 also accumulates at DSBs in a manner dependent upon RNF8 and RNF168-mediated ubiquitination events. While BRCA1 does not contain any ubiquitin binding motifs, components of the BRCA1 associated complex including RAP80 (receptor associated protein 80) have the ability to associate with K63-linked polyubiquitin chains [110, 111]. RAP80 contains two UIMs critical for BRCA1 localization to HR repair sites [134–136]. In fact, mutations in the RAP80 UIM have been discovered in BRCA1/BRCA2 negative breast cancers patients, thus highlighting the importance of ubiquitin association [137]. In addition, BRCA1 is itself a RING finger protein. Approximately 20% of the clinically relevant BRCA1 mutations are expected to diminish BRCA1 E3 ligase activity [138]. DNA damage also induces association between BRCA1 and the E2 protein UBCH5C that facilitates K6-linked ubiquitination events [139].

Fig. 5 (continued) with support from HERC2 catalyzes the formation of K63-linked polyubiquitin chains on H2A and H2AX, which is further amplified by RNF168. Formation of K63-linked polyubiquitin chains is essential for recruiting downstream HR repair proteins. For example, both RAD18 and RAP80 recognize K63-linked polyubiquitin chains trough specialized ubiquitin binding domains. RAD18 and RAP80 facilitate assembly of RAD51C and BRCA1, respectively, at the break site by serving as molecular adapters. RAD51C also regulates the ubiquitination and proteasomal degradation of RAD51 through an undetermined mechanism. In addition, BRCA1 facilitates formation of K6-linked ubiquitin polymers on unidentified protein targets at the break site. The black line with a gap indicates a DSB, while the shaded circles represent histones. Abbreviations: *P* phosphorylation, *U* ubiquitin, *K63* lysine 63-linked, *K48* lysine 48-linked

Lysine 6 (K6)-linked polyubiquitin chains are present at DSBs, but their significance and protein targets are topics of active investigation [92, 140]. A recent study identified a novel role for K27-linked ubiquitin chains as early signals of DSB damage. These modifications are generated along histone H2A by RNF168 and are necessary to activate DNA damage response. Several repair proteins, including 53BP1 and the RAP80 subunit of the BRCA1 E3 ligase complex, recognize these chains and loss of this modification along chromatin disrupts repair activity [141, 142].

In addition to ubiquitination events directly mediated by RNF8 and RNF168, RAD51 is ubiquitinated and subject to proteasome-mediated degradation following exposure to IR [143]. Interestingly, ubiquitination of RAD51 is regulated by RAD51C. In the absence of RAD51C, RAD51 is ubiquitinated independent of DSB formation [143], suggesting that RAD51C prevents the ubiquitination of RAD51. Thus, the ubiquitin landscape at DSBs is complex and likely to require involvement of additional RING finger proteins, such as the recently identified RNF138 [144–146] and RNF169 that functions as a negative regulator of DNA DSB repair [147–149].

Ubiquitination also serves as a mechanism for pathway choice at the DSB site. In response to IR, RNF138, a RING E3 ubiquitin ligase, interacts with Ku70/Ku80 and initiates ubiquitination of the Ku80 protein. This modification leads to the degradation of Ku80, an event that disrupts NHEJ and enhances HR activity [145]. Additionally, in conjunction with the E2 ligase UBE2D, RNF138 ubiquitinates CtIP to promote end resection during the early stages of HR [146]. Recently, the role of deubiquitinating enzymes (DUBs) in regulating HR has become apparent. Ubiquitin specific protease 21 (USP21) removes ubiquitin moieties from BRCA2 as a means of stabilizing the protein and promoting BRCA2 activity [150]. BRCA2 stability facilitates RAD51 loading onto ssDNA allowing HR to progress efficiently. Depletion of USP21 was shown to decrease HR efficiency, increase genomic instability, and inhibit tumor cell growth, further demonstrating that deubiquitination is an additional mechanism for the regulation of DNA repair pathways [151]. USP7, another DUB, was implicated in regulating DSB repair through its interaction with RNF169. Loss of USP7 results in decreased RNF169 protein levels, suggesting that USP7 deubiquitinates RNF169 to stabilize the protein and promote RNF169 loading at DSB sites. Furthermore, loss of USP7 confers cellular sensitivity to poly(ADP)-ribose polymerase inhibitors, suggesting that its activity is required for DNA break repair [149].

Identifying modifications that occur directly along HR proteins will help to understand repair mechanisms. However, at this time, only limited experiments to identify modifications under DNA damage conditions have been performed. Proteomic analyses of

Table 1
Predicted ubiquitin modification sites along key homologous recombination and nonhomologous end joining proteins

Protein	Ubiquitination site
RAD51	K58, K64, K107, K156
RAD51C	K26, K42, K48, K54, K76, K83, K328
BRCA1	K970, K1667
BRCA2	K268, K343, K457, K757, K799, K811, K817, K829, K838, K862, K907, K985, K1984, K1139, K1153, K1202, K1226, K1323, K1381, K1406, K1419, K1449, K1474, K1549, K1631, K1678, K1729, K1767, K1777, K1783, K1789, K1872, K1944, K1959, K1984, K2013, K2017, K2077, K2104, K2124, K2150, K2162, K2191, K2196, K2286, K2392, K2308, K2448, K2459, K2538, K2597, K2741, K2958
CtIP	K314, K526, K782
Ku70	K31, K92, K94, K100, K114, K123, K129, K182, K238, K282, K287, K297, K317, K331, K338, K351, K357, K358, K442, K445, K451, K461, K463, K468, K516, K526, K539, K553, K556, K565, K570, K575, K582, K591, K596, K605
Ku80	K36, K125, K144, K155, K156, K171, K195, K202, K233, K265, K274, K282, K285, K307, K325, K332, K334, K338, K347, K363, K439, K443, K465, K466, K469, K481, K525, K532, K534, K543, K544, K545, K565, K566, K648, K660, K665, K702
MRE11	K384, K393, K416, K420, K434, K442, K464, K475, K673
NBS1	K665, K683
PARP1	K7, K87, K97, K108, K119, K165, K209, K221, K239, K249, K254, K262, K269, K320, K331, K337, K394, K414, K425, K434, K447, K486, K528, K548, K551, K564, K579, K600, K637, K662, K664, K667, K695, K748, K787, K796, K798, K838, K852, K940, K949, K1010
RAD50	K62, K175, K187, K211, K230, K242, K295, K461, K845, K1119, K1126, K1134, K1190, K1285

HR proteins have identified predicted ubiquitination sites summarized in Table 1 [61, 66, 67, 152–158]. One example is the E3 ubiquitin ligase RFWD3 that ubiquitinates RAD51 in response to DSBs induced by mitomycin C (MMC) as a mechanism of removing the protein from the single-strand overhangs [159]. RING mutants of RFWD3 are defective in promoting RAD51 ubiquitination, resulting in persistent RAD51 foci at the damage site, cellular sensitivity to MMC, and a Fanconi Anemia phenotype [159, 160]. Given the prevalence of phosphorylation and ubiquitin modifications in the HR pathway, it follows that investigation into these modifications (phosphorylation, sumoylation, acetylation, ubiquitination, and deubiquitination) is necessary.

6 Clinical Phenotypes of HR Defective Cancers

Throughout the life span of a cancer cell "genomic scarring" indicates a multitude of genomic alterations reflecting defects in the different DNA repair pathways [161, 162]. Cancers that exhibit unique genome instability as a result of HR defects are specifically found in ovarian, breast, and prostate as well as pancreatic and gastric cancers. Mutations most commonly detected occur in HR genes, including BRCA1, BRCA2, PALB2, and the RAD51 family, and the resulting repertoire of mutational signatures or characteristic patterns on the tumor genome can be used to guide treatment regimens and predict tumor response regarding chemotherapy, radiation, and immunotherapy [163–169].

For ovarian cancers, HR-deficiency has a clinical phenotype unique from HR-proficient ovarian cancers, being enriched for high-grade serous (HGSOC) histology, visceral relapse and slightly younger ages at diagnosis [163]. HGSOC has 300–30,000 mutations per full genome with a distinct mutational signature, identified as "Signature 3," accompanied by structural variation signatures, noted by 5–25 base pair deletions as well as amplifications and deletions of large segments, often comprising whole chromosome arms, indicating activation of the alternative end joining pathway [161, 165, 166, 170, 171]. HGSOC cells have a high degree of defects in the *TP53* gene (>90%) [172]. This is consistent with the cell culture models of the RAD51 family members requiring p53 deletion in order for the cells to proliferate under the enormous load of unstable chromosomes and telomeres [77, 173–177]. Expression profiling has also identified HR-deficiency gene expression patterns; for example, one study found 230 genes whose expression differed between control and *RAD51*, *BRCA1*, or *BRIT1* deficient cells [170, 171]. These genes were selected because each one functions in a distinct and separate step in the HR pathway.

Cancers defective in HR are highly sensitive to PARP inhibitors (PARPi). The HR deficiency signature accurately predicts sensitivity to the PARPi olaparib and rucaparib [178], and these types of analyses can be applied in the clinic to classify tumor subtypes and predict patient response to therapies, such as DNA cross-linking agents or PARPi. In some cases, PARPi resistance can be attributed to the acquisition of secondary mutations in BRCA1, BRCA2, or RAD51 restoring the HR pathway [179–181]. For example in one study, genetic analysis was performed in 12 paired samples (pretreatment and postprogression) isolated from patients treated with rucaparib. Of the 12 pretreatment samples, 6 had truncation mutations in RAD51C or RAD51D, and 5 of the 6 corresponding postprogression samples had 1 or more secondary mutations that restored the open reading frame of the affected gene and increased rucaparib resistance [182]. As another example, mutagenesis screens discovered that loss of TP53BP1, which represses HR mediated repair of DSBs in favor of NHEJ, restores HR and facilitated sensitivity to PARP inhibition [183]. More recently, a study in which expression of 134 known DNA repair genes was silenced using siRNA discovered that depletion of GPBP1 or ARID1A conferred resistance to the PARPi BMN673 (Talazoparib). Depletion of GPBP1 was subsequently found to upregulate both BRCA1 and RAD51B expression in response to PARP inhibition [184]. ARID1A is a member of the SWI/SNF chromatin remodeling complex and is recruited to DSBs in an ATR dependent manner where it has been proposed to facilitate DNA end processing [185, 186]. ARID1A also appears to promote MMR through an interaction with MSH2 [187]. While the role of ARID1A in these processes needs to be clearly defined, similar to depletion of TP53 in BRCA1 mutated patients, it is likely that depletion of ARID1A allows access to the repair sites in order to bypass HR.

In addition to driving oncogenesis, improper or deficient repair of damaged DNA can alter immune balance in the tumor microenvironment. Loss of DNA repair fidelity is thought to drive the production of neoantigens and the activation of the host immune response [188–190]. Long-term clinical benefit is most often observed in metastatic melanoma and non-small cell lung cancer patients with a higher mutational burden receiving CTLA-4 or PD-1 blockade respectively [191, 192]. However, high mutational load alone does not predict for response to immunotherapy. The key may be to discover which mutations or type of chromosome instability drive immune response. Cancers with defects in the homologous recombination pathway may also be sensitive to PD-1 blockade. BRCA1/BRCA2 mutant ovarian cancers were reported to have increased mutation burden, predicted neoantigen load, and PD-1 expression [193]. In one recent study, patients with metastatic melanoma and undergoing PD-1 blockade were more likely to respond to treatment if their tumors carried a BRCA2 mutation [194]. On the other hand, while tumors with a high mutational burden tend to respond favorably to immunotherapy, tumors with marked aneuploidy or high SCNA (both hallmarks of HR-deficient tumors) were associated with immunotherapy evasion [195]. Additional studies have linked specific mutational signatures with response to immune checkpoint blockade. Tumors containing a multitude of $C > A$ transversions, a hallmark of tobacco exposure, tend to benefit from PD-1 blockade [192]. Evidence suggests that DNA repair pathway defects, and thus PTM of DNA repair factors, can serve as biomarkers for response to immunotherapies. Sporadic colorectal tumors that carry defects in the MMR pathway exhibit a high mutational burden, microsatellite instability (MSI), and high PD-1 expression [189, 196]. Pembrolizumab has recently been approved by the FDA for treatment of colorectal patients with advanced, MSI high, MMR-deficient tumors [197]. While the association between defects in the DNA repair machinery and sensitivity to immune checkpoint inhibitors is clearly correlated, the underlying cellular mechanisms remain poorly understood and warrant further investigation. HR is clearly a useful biomarker but determining its context will be very important.

7 Concluding Remarks

At the beginning of the twentieth century, Dr. Boveri was among the first to associate the importance of genome maintenance and the potential connection with human disease states. Of course, we now know key components of the homologous recombination pathway that function to preserve genome integrity through the repair of damage affecting both DNA strands. In addition to discerning the importance of genome maintenance, Boveri had the foresight to associate genome stability with external influences, such as inflammation, proposing that "if then malignant tumors arise preferentially at sites of inflammation... then it is precisely the presence of abnormal mitoses in completely or at least relatively normal tissues that, according to my hypothesis, must be regarded as the origin of the malignant tumors." Recent advances in sequencing technology has allowed for the identification of mutational signatures present in tumors. Our current understanding of DNA repair pathways and subpathways is significantly more advanced, and now the course is set for next generation research to identify posttranslational modifications and clinical phenotypes associated with repair defects that will be exploited for companion diagnostics and combination cancer therapies.

Acknowledgments

We would like to thank Michael Wyatt, Janay Clytus, and Jason Stewart for insightful comments and suggestions during the writing of this review. The work related to the topic of this chapter was supported by a grant from the National Institute of General Medical Sciences of the National Institutes of Health Award Number R15 GM110615 and the American Cancer Society (RSG-03-158-01-GMC). The content is solely the responsibility of the authors and does not necessarily represent the official views of the National Institutes of Health. Additional support was provided to NMR from a SPARC Graduate Research Grant from the Office of the Vice President for Research at the University of South Carolina and from an Associazione Italiana per la Ricerca sul Cancro "Molini Bongiovanni" Fellowship.

References

1. Boveri T (2008) Concerning the origin of malignant tumours by Theodor Boveri. Translated and annotated by Henry Harris. J Cell Sci 121(Suppl 1):1–84
2. Nestmann ER, Hill RF (1974) Mutagenesis by mutator gene mutH1 in continuous cultures of *Escherichia coli*. J Bacteriol 119 (1):33–35
3. Loeb LA (2016) Human cancers express a mutator phenotype: hypothesis, origin, and consequences. Cancer Res 76(8):2057–2059
4. Hanahan D, Weinberg RA (2011) Hallmarks of cancer: the next generation. Cell 144 (5):646–674
5. Wyatt MD, Pittman DL (2006) Methylating agents and DNA repair responses: methylated bases and sources of strand breaks. Chem Res Toxicol 19(12):1580–1594
6. Rajesh P, Litvinchuk AV, Pittman DL, Wyatt MD (2011) The homologous recombination protein RAD51D mediates the processing of 6-thioguanine lesions downstream of mismatch repair. Mol Cancer Res 9(2):206–214
7. Li X, Heyer WD (2008) Homologous recombination in DNA repair and DNA damage tolerance. Cell Res 18(1):99–113
8. Thacker J (1999) The role of homologous recombination processes in the repair of severe forms of DNA damage in mammalian cells. Biochimie 81(1–2):77–85
9. Lieber MR, Wilson TE (2010) SnapShot: nonhomologous DNA end joining (NHEJ). Cell 142(3):496–496.e491
10. Branzei D, Foiani M (2008) Regulation of DNA repair throughout the cell cycle. Nat Rev Mol Cell Biol 9(4):297–308
11. Akutsu M, Dikic I, Bremm A (2016) Ubiquitin chain diversity at a glance. J Cell Sci 129 (5):875–880
12. Boudaiffa B, Cloutier P, Hunting D, Huels MA, Sanche L (2000) Resonant formation of DNA strand breaks by low-energy (3 to 20 eV) electrons. Science 287 (5458):1658–1660
13. Smith TA, Kirkpatrick DR, Smith S, Smith TK, Pearson T, Kailasam A, Herrmann KZ, Schubert J, Agrawal DK (2017) Radioprotective agents to prevent cellular damage due to ionizing radiation. J Transl Med 15(1):232
14. Turgeon MO, Perry NJS, Poulogiannis G (2018) DNA damage, repair, and cancer metabolism. Front Oncol 8:15
15. Radhakrishnan SK, Jette N, Lees-Miller SP (2014) Non-homologous end joining: emerging themes and unanswered questions. DNA Repair (Amst) 17:2–8
16. Karran P (2000) DNA double strand break repair in mammalian cells. Curr Opin Genet Dev 10(2):144–150
17. Ciccia A, Elledge SJ (2010) The DNA damage response: making it safe to play with knives. Mol Cell 40(2):179–204
18. Hartlerode AJ, Scully R (2009) Mechanisms of double-strand break repair in somatic mammalian cells. Biochem J 423(2):157–168
19. You Z, Bailis JM (2010) DNA damage and decisions: CtIP coordinates DNA repair and cell cycle checkpoints. Trends Cell Biol 20 (7):402–409
20. West SC (2003) Molecular views of recombination proteins and their control. Nat Rev Mol Cell Biol 4(6):435–445
21. Wyman C, Kanaar R (2006) DNA double-strand break repair: all's well that ends well. Annu Rev Genet 40:363–383
22. Jungmichel S, Rosenthal F, Altmeyer M, Lukas J, Hottiger MO, Nielsen ML (2013) Proteome-wide identification of poly(ADP-Ribosyl)ation targets in different genotoxic stress responses. Mol Cell 52(2):272–285
23. Haince JF, McDonald D, Rodrigue A, Dery U, Masson JY, Hendzel MJ, Poirier GG (2008) PARP1-dependent kinetics of recruitment of MRE11 and NBS1 proteins to multiple DNA damage sites. J Biol Chem 283(2):1197–1208
24. Jazayeri A, Balestrini A, Garner E, Haber JE, Costanzo V (2008) Mre11-Rad50-Nbs1-dependent processing of DNA breaks generates oligonucleotides that stimulate ATM activity. EMBO J 27(14):1953–1962
25. Kanaar R, Wyman C (2008) DNA repair by the MRN complex: break it to make it. Cell 135(1):14–16
26. Huen MS, Sy SM, Chen J (2010) BRCA1 and its toolbox for the maintenance of genome integrity. Nat Rev Mol Cell Biol 11 (2):138–148
27. Marechal A, Zou L (2015) RPA-coated single-stranded DNA as a platform for post-translational modifications in the DNA damage response. Cell Res 25(1):9–23
28. Wold MS (1997) Replication protein A: a heterotrimeric, single-stranded DNA-binding protein required for eukaryotic DNA metabolism. Annu Rev Biochem 66:61–92
29. Candelli A, Holthausen JT, Depken M, Brouwer I, Franker MA, Marchetti M,

Heller I, Bernard S, Garcin EB, Modesti M et al (2014) Visualization and quantification of nascent RAD51 filament formation at single-monomer resolution. Proc Natl Acad Sci U S A 111(42):15090–15095
30. Short JM, Liu Y, Chen S, Soni N, Madhusudhan MS, Shivji MK, Venkitaraman AR (2016) High-resolution structure of the presynaptic RAD51 filament on single-stranded DNA by electron cryo-microscopy. Nucleic Acids Res 44(19):9017–9030
31. Xu J, Zhao L, Xu Y, Zhao W, Sung P, Wang HW (2017) Cryo-EM structures of human RAD51 recombinase filaments during catalysis of DNA-strand exchange. Nat Struct Mol Biol 24(1):40–46
32. Jensen RB, Ozes A, Kim T, Estep A, Kowalczykowski SC (2013) BRCA2 is epistatic to the RAD51 paralogs in response to DNA damage. DNA Repair (Amst) 12(4):306–311
33. Jensen RB, Carreira A, Kowalczykowski SC (2010) Purified human BRCA2 stimulates RAD51-mediated recombination. Nature 467(7316):678–683
34. Liu J, Doty T, Gibson B, Heyer WD (2010) Human BRCA2 protein promotes RAD51 filament formation on RPA-covered single-stranded DNA. Nat Struct Mol Biol 17 (10):1260–1262
35. Thorslund T, McIlwraith MJ, Compton SA, Lekomtsev S, Petronczki M, Griffith JD, West SC (2010) The breast cancer tumor suppressor BRCA2 promotes the specific targeting of RAD51 to single-stranded DNA. Nat Struct Mol Biol 17(10):1263–1265
36. Qi Z, Redding S, Lee JY, Gibb B, Kwon Y, Niu H, Gaines WA, Sung P, Greene EC (2015) DNA sequence alignment by microhomology sampling during homologous recombination. Cell 160(5):856–869
37. Sung P (1994) Catalysis of ATP-dependent homologous DNA pairing and strand exchange by yeast RAD51 protein. Science 265(5176):1241–1243
38. Sanchez H, Kertokalio A, van Rossum-Fikkert S, Kanaar R, Wyman C (2013) Combined optical and topographic imaging reveals different arrangements of human RAD54 with presynaptic and postsynaptic RAD51-DNA filaments. Proc Natl Acad Sci U S A 110(28):11385–11390
39. Holliday R (1964) The induction of mitotic recombination by mitomycin C in Ustilago and Saccharomyces. Genetics 50:323–335
40. Wyatt HD, West SC (2014) Holliday junction resolvases. Cold Spring Harb Perspect Biol 6 (9):a023192
41. Szostak JW, Orr-Weaver TL, Rothstein RJ, Stahl FW (1983) The double-strand-break repair model for recombination. Cell 33 (1):25–35
42. Paques F, Haber JE (1999) Multiple pathways of recombination induced by double-strand breaks in *Saccharomyces cerevisiae*. Microbiol Mol Biol Rev 63(2):349–404
43. Haber JE (1999) DNA recombination: the replication connection. Trends Biochem Sci 24(7):271–275
44. Johnson RD, Jasin M (2000) Sister chromatid gene conversion is a prominent double-strand break repair pathway in mammalian cells. EMBO J 19(13):3398–3407
45. Arnaudeau C, Lundin C, Helleday T (2001) DNA double-strand breaks associated with replication forks are predominantly repaired by homologous recombination involving an exchange mechanism in mammalian cells. J Mol Biol 307(5):1235–1245
46. Saleh-Gohari N, Bryant HE, Schultz N, Parker KM, Cassel TN, Helleday T (2005) Spontaneous homologous recombination is induced by collapsed replication forks that are caused by endogenous DNA single-strand breaks. Mol Cell Biol 25(16):7158–7169
47. Hanada K, Budzowska M, Davies SL, van Drunen E, Onizawa H, Beverloo HB, Maas A, Essers J, Hickson ID, Kanaar R (2007) The structure-specific endonuclease Mus81 contributes to replication restart by generating double-strand DNA breaks. Nat Struct Mol Biol 14(11):1096–1104
48. Petermann E, Helleday T (2010) Pathways of mammalian replication fork restart. Nat Rev Mol Cell Biol 11(10):683–687
49. Malinge JM, Giraud-Panis MJ, Leng M (1999) Interstrand cross-links of cisplatin induce striking distortions in DNA. J Inorg Biochem 77(1–2):23–29
50. Lando DY, Chang CL, Fridman AS, Grigoryan IE, Galyuk EN, Hsueh YW, Hu CK (2014) Comparative thermal and thermodynamic study of DNA chemically modified with antitumor drug cisplatin and its inactive analog transplatin. J Inorg Biochem 137:85–93
51. Knipscheer P, Raschle M, Smogorzewska A, Enoiu M, Ho TV, Scharer OD, Elledge SJ, Walter JC (2009) The Fanconi anemia pathway promotes replication-dependent DNA interstrand cross-link repair. Science 326 (5960):1698–1701
52. Kottemann MC, Smogorzewska A (2013) Fanconi anaemia and the repair of Watson and Crick DNA crosslinks. Nature 493 (7432):356–363

53. Jo U, Kim H (2015) Exploiting the Fanconi anemia pathway for targeted anti-cancer therapy. Mol Cells 38(8):669–676
54. Sengerova B, Wang AT, McHugh PJ (2011) Orchestrating the nucleases involved in DNA interstrand cross-link (ICL) repair. Cell Cycle 10(23):3999–4008
55. Unno J, Itaya A, Taoka M, Sato K, Tomida J, Sakai W, Sugasawa K, Ishiai M, Ikura T, Isobe T et al (2014) FANCD2 binds CtIP and regulates DNA-end resection during DNA interstrand crosslink repair. Cell Rep 7(4):1039–1047
56. Al-Minawi AZ, Lee YF, Hakansson D, Johansson F, Lundin C, Saleh-Gohari N, Schultz N, Jenssen D, Bryant HE, Meuth M et al (2009) The ERCC1/XPF endonuclease is required for completion of homologous recombination at DNA replication forks stalled by inter-strand cross-links. Nucleic Acids Res 37(19):6400–6413
57. Bergstralh DT, Sekelsky J (2008) Interstrand crosslink repair: can XPF-ERCC1 be let off the hook? Trends Genet 24(2):70–76
58. Long DT, Raschle M, Joukov V, Walter JC (2011) Mechanism of RAD51-dependent DNA interstrand cross-link repair. Science 333(6038):84–87
59. Danilowicz C, Peacock-Villada A, Vlassakis J, Facon A, Feinstein E, Kleckner N, Prentiss M (2014) The differential extension in dsDNA bound to Rad51 filaments may play important roles in homology recognition and strand exchange. Nucleic Acids Res 42(1):526–533
60. Benson FE, Stasiak A, West SC (1994) Purification and characterization of the human Rad51 protein, an analogue of *E. coli* RecA. EMBO J 13(23):5764–5771
61. Yuan ZM, Huang Y, Ishiko T, Nakada S, Utsugisawa T, Kharbanda S, Wang R, Sung P, Shinohara A, Weichselbaum R et al (1998) Regulation of Rad51 function by c-Abl in response to DNA damage. J Biol Chem 273(7):3799–3802
62. Chen G, Yuan SS, Liu W, Xu Y, Trujillo K, Song B, Cong F, Goff SP, Wu Y, Arlinghaus R et al (1999) Radiation-induced assembly of Rad51 and Rad52 recombination complex requires ATM and c-Abl. J Biol Chem 274 (18):12748–12752
63. Subramanyam S, Ismail M, Bhattacharya I, Spies M (2016) Tyrosine phosphorylation stimulates activity of human RAD51 recombinase through altered nucleoprotein filament dynamics. Proc Natl Acad Sci U S A 113 (41):E6045–e6054
64. Conilleau S, Takizawa Y, Tachiwana H, Fleury F, Kurumizaka H, Takahashi M (2004) Location of tyrosine 315, a target for phosphorylation by cAbl tyrosine kinase, at the edge of the subunit-subunit interface of the human Rad51 filament. J Mol Biol 339 (4):797–804
65. Alligand B, Le Breton M, Marquis D, Vallette F, Fleury F (2017) Functional effects of diphosphomimetic mutations at cAbl-mediated phosphorylation sites on Rad51 recombinase activity. Biochimie 139:115–124
66. Yata K, Bleuyard JY, Nakato R, Ralf C, Katou Y, Schwab RA, Niedzwiedz W, Shirahige K, Esashi F (2014) BRCA2 coordinates the activities of cell-cycle kinases to promote genome stability. Cell Rep 7 (5):1547–1559
67. Yata K, Lloyd J, Maslen S, Bleuyard JY, Skehel M, Smerdon SJ, Esashi F (2012) Plk1 and CK2 act in concert to regulate Rad51 during DNA double strand break repair. Mol Cell 45(3):371–383
68. Cartwright R, Tambini CE, Simpson PJ, Thacker J (1998) The XRCC2 DNA repair gene from human and mouse encodes a novel member of the recA/RAD51 family. Nucleic Acids Res 26(13):3084–3089
69. Dosanjh MK, Collins DW, Fan W, Lennon GG, Albala JS, Shen Z, Schild D (1998) Isolation and characterization of RAD51C, a new human member of the RAD51 family of related genes. Nucleic Acids Res 26 (5):1179–1184
70. Pittman DL, Weinberg LR, Schimenti JC (1998) Identification, characterization, and genetic mapping of Rad51d, a new mouse and human RAD51/RecA-related gene. Genomics 49(1):103–111
71. Thacker J, Tambini CE, Simpson PJ, Tsui LC, Scherer SW (1995) Localization to chromosome 7q36.1 of the human XRCC2 gene, determining sensitivity to DNA-damaging agents. Hum Mol Genet 4(1):113–120
72. Liu N, Lamerdin JE, Tebbs RS, Schild D, Tucker JD, Shen MR, Brookman KW, Siciliano MJ, Walter CA, Fan W et al (1998) XRCC2 and XRCC3, new human Rad51-family members, promote chromosome stability and protect against DNA cross-links and other damages. Mol Cell 1(6):783–793
73. Takata M, Sasaki MS, Sonoda E, Fukushima T, Morrison C, Albala JS, Swagemakers SM, Kanaar R, Thompson LH, Takeda S (2000) The Rad51 paralog Rad51B promotes homologous recombinational repair. Mol Cell Biol 20 (17):6476–6482
74. Takata M, Sasaki MS, Tachiiri S, Fukushima T, Sonoda E, Schild D,

Thompson LH, Takeda S (2001) Chromosome instability and defective recombinational repair in knockout mutants of the five Rad51 paralogs. Mol Cell Biol 21 (8):2858–2866

75. French CA, Masson JY, Griffin CS, O'Regan P, West SC, Thacker J (2002) Role of mammalian RAD51L2 (RAD51C) in recombination and genetic stability. J Biol Chem 277(22):19322–19330
76. Godthelp BC, Wiegant WW, van Duijn-Goedhart A, Scharer OD, van Buul PP, Kanaar R, Zdzienicka MZ (2002) Mammalian Rad51C contributes to DNA cross-link resistance, sister chromatid cohesion and genomic stability. Nucleic Acids Res 30 (10):2172–2182
77. Smiraldo PG, Gruver AM, Osborn JC, Pittman DL (2005) Extensive chromosomal instability in Rad51d-deficient mouse cells. Cancer Res 65(6):2089–2096
78. Braybrooke JP, Spink KG, Thacker J, Hickson ID (2000) The RAD51 family member, RAD51L3, is a DNA-stimulated ATPase that forms a complex with XRCC2. J Biol Chem 275(37):29100–29106
79. Schild D, Lio YC, Collins DW, Tsomondo T, Chen DJ (2000) Evidence for simultaneous protein interactions between human Rad51 paralogs. J Biol Chem 275(22):16443–16449
80. Masson JY, Tarsounas MC, Stasiak AZ, Stasiak A, Shah R, McIlwraith MJ, Benson FE, West SC (2001) Identification and purification of two distinct complexes containing the five RAD51 paralogs. Genes Dev 15 (24):3296–3307
81. Rajesh C, Gruver AM, Basrur V, Pittman DL (2009) The interaction profile of homologous recombination repair proteins RAD51C, RAD51D and XRCC2 as determined by proteomic analysis. Proteomics 9 (16):4071–4086
82. Yokoyama H, Sarai N, Kagawa W, Enomoto R, Shibata T, Kurumizaka H, Yokoyama S (2004) Preferential binding to branched DNA strands and strand-annealing activity of the human Rad51B, Rad51C, Rad51D and Xrcc2 protein complex. Nucleic Acids Res 32(8):2556–2565
83. Liu Y, Masson JY, Shah R, O'Regan P, West SC (2004) RAD51C is required for Holliday junction processing in mammalian cells. Science 303(5655):243–246
84. Liu Y, Tarsounas M, O'Regan P, West SC (2007) Role of RAD51C and XRCC3 in genetic recombination and DNA repair. J Biol Chem 282(3):1973–1979
85. Chun J, Buechelmaier ES, Powell SN (2013) Rad51 paralog complexes BCDX2 and CX3 act at different stages in the BRCA1-BRCA2-dependent homologous recombination pathway. Mol Cell Biol 33(2):387–395
86. Kawabata M, Kawabata T, Nishibori M (2005) Role of recA/RAD51 family proteins in mammals. Acta Med Okayama 59(1):1–9
87. Gruver AM, Miller KA, Rajesh C, Smiraldo PG, Kaliyaperumal S, Balder R, Stiles KM, Albala JS, Pittman DL (2005) The ATPase motif in RAD51D is required for resistance to DNA interstrand crosslinking agents and interaction with RAD51C. Mutagenesis 20 (6):433–440
88. Wiese C, Hinz JM, Tebbs RS, Nham PB, Urbin SS, Collins DW, Thompson LH, Schild D (2006) Disparate requirements for the Walker a and B ATPase motifs of human RAD51D in homologous recombination. Nucleic Acids Res 34(9):2833–2843
89. Pellegrini L, Yu DS, Lo T, Anand S, Lee M, Blundell TL, Venkitaraman AR (2002) Insights into DNA recombination from the structure of a RAD51-BRCA2 complex. Nature 420(6913):287–293
90. Lo T, Pellegrini L, Venkitaraman AR, Blundell TL (2003) Sequence fingerprints in BRCA2 and RAD51: implications for DNA repair and cancer. DNA Repair (Amst) 2 (9):1015–1028
91. Nomme J, Takizawa Y, Martinez SF, Renodon-Corniere A, Fleury F, Weigel P, Yamamoto K, Kurumizaka H, Takahashi M (2008) Inhibition of filament formation of human Rad51 protein by a small peptide derived from the BRC-motif of the BRCA2 protein. Genes Cells 13(5):471–481
92. Messick TE, Greenberg RA (2009) The ubiquitin landscape at DNA double-strand breaks. J Cell Biol 187(3):319–326
93. Al-Hakim A, Escribano-Diaz C, Landry MC, O'Donnell L, Panier S, Szilard RK, Durocher D (2010) The ubiquitous role of ubiquitin in the DNA damage response. DNA Repair (Amst) 9(12):1229–1240
94. Pickart CM (2001) Ubiquitin enters the new millennium. Mol Cell 8(3):499–504
95. Weissman AM (2001) Themes and variations on ubiquitylation. Nat Rev Mol Cell Biol 2 (3):169–178
96. Berndsen CE, Wolberger C (2014) New insights into ubiquitin E3 ligase mechanism. Nat Struct Mol Biol 21(4):301–307
97. Zheng N, Shabek N (2017) Ubiquitin ligases: structure, function, and regulation. Annu Rev Biochem 86:129–157

98. Deshaies RJ, Joazeiro CA (2009) RING domain E3 ubiquitin ligases. Annu Rev Biochem 78:399–434
99. Barlow PN, Luisi B, Milner A, Elliott M, Everett R (1994) Structure of the C3HC4 domain by 1H-nuclear magnetic resonance spectroscopy. A new structural class of zinc-finger. J Mol Biol 237(2):201–211
100. Borden KL, Boddy MN, Lally J, O'Reilly NJ, Martin S, Howe K, Solomon E, Freemont PS (1995) The solution structure of the RING finger domain from the acute promyelocytic leukaemia proto-oncoprotein PML. EMBO J 14(7):1532–1541
101. Zheng N, Wang P, Jeffrey PD, Pavletich NP (2000) Structure of a c-Cbl-UbcH7 complex: RING domain function in ubiquitin-protein ligases. Cell 102(4):533–539
102. Yamada HY, Gorbsky GJ (2006) Tumor suppressor candidate TSSC5 is regulated by UbcH6 and a novel ubiquitin ligase RING105. Oncogene 25(9):1330–1339
103. Christensen DE, Brzovic PS, Klevit RE (2007) E2-BRCA1 RING interactions dictate synthesis of mono- or specific polyubiquitin chain linkages. Nat Struct Mol Biol 14 (10):941–948
104. Danielsen JM, Sylvestersen KB, Bekker-Jensen S, Szklarczyk D, Poulsen JW, Horn H, Jensen LJ, Mailand N, Nielsen ML (2011) Mass spectrometric analysis of lysine ubiquitylation reveals promiscuity at site level. Mol Cell Proteomics 10(3):M110.003590
105. Peng J, Schwartz D, Elias JE, Thoreen CC, Cheng D, Marsischky G, Roelofs J, Finley D, Gygi SP (2003) A proteomics approach to understanding protein ubiquitination. Nat Biotechnol 21(8):921–926
106. Thrower JS, Hoffman L, Rechsteiner M, Pickart CM (2000) Recognition of the polyubiquitin proteolytic signal. EMBO J 19 (1):94–102
107. Huang TT, D'Andrea AD (2006) Regulation of DNA repair by ubiquitylation. Nat Rev Mol Cell Biol 7(5):323–334
108. Newton K, Matsumoto ML, Wertz IE, Kirkpatrick DS, Lill JR, Tan J, Dugger D, Gordon N, Sidhu SS, Fellouse FA et al (2008) Ubiquitin chain editing revealed by polyubiquitin linkage-specific antibodies. Cell 134(4):668–678
109. Yan J, Jetten AM (2008) RAP80 and RNF8, key players in the recruitment of repair proteins to DNA damage sites. Cancer Lett 271 (2):179–190
110. Sato Y, Yoshikawa A, Yamashita M, Yamagata A, Fukai S (2009) Structural basis for specific recognition of Lys 63-linked polyubiquitin chains by NZF domains of TAB2 and TAB3. EMBO J 28(24):3903–3909
111. Sato Y, Yoshikawa A, Mimura H, Yamashita M, Yamagata A, Fukai S (2009) Structural basis for specific recognition of Lys 63-linked polyubiquitin chains by tandem UIMs of RAP80. EMBO J 28 (16):2461–2468
112. Chau V, Tobias JW, Bachmair A, Marriott D, Ecker DJ, Gonda DK, Varshavsky A (1989) A multiubiquitin chain is confined to specific lysine in a targeted short-lived protein. Science 243(4898):1576–1583
113. Varadan R, Assfalg M, Haririnia A, Raasi S, Pickart C, Fushman D (2004) Solution conformation of Lys63-linked di-ubiquitin chain provides clues to functional diversity of polyubiquitin signaling. J Biol Chem 279 (8):7055–7063
114. Ye Y, Blaser G, Horrocks MH, Ruedas-Rama MJ, Ibrahim S, Zhukov AA, Orte A, Klenerman D, Jackson SE, Komander D (2012) Ubiquitin chain conformation regulates recognition and activity of interacting proteins. Nature 492(7428):266–270
115. Thach TT, Shin D, Han S, Lee S (2016) New conformations of linear polyubiquitin chains from crystallographic and solution-scattering studies expand the conformational space of polyubiquitin. Acta Crystallogr D Struct Biol 72(Pt 4):524–535
116. Panier S, Durocher D (2009) Regulatory ubiquitylation in response to DNA double-strand breaks. DNA Repair (Amst) 8 (4):436–443
117. Rogakou EP, Pilch DR, Orr AH, Ivanova VS, Bonner WM (1998) DNA double-stranded breaks induce histone H2AX phosphorylation on serine 139. J Biol Chem 273 (10):5858–5868
118. Harper JW, Elledge SJ (2007) The DNA damage response: ten years after. Mol Cell 28(5):739–745
119. Stewart GS, Wang B, Bignell CR, Taylor AM, Elledge SJ (2003) MDC1 is a mediator of the mammalian DNA damage checkpoint. Nature 421(6926):961–966
120. Stucki M, Clapperton JA, Mohammad D, Yaffe MB, Smerdon SJ, Jackson SP (2005) MDC1 directly binds phosphorylated histone H2AX to regulate cellular responses to DNA double-strand breaks. Cell 123 (7):1213–1226
121. Lou Z, Minter-Dykhouse K, Franco S, Gostissa M, Rivera MA, Celeste A, Manis JP, van Deursen J, Nussenzweig A, Paull TT et al

(2006) MDC1 maintains genomic stability by participating in the amplification of ATM-dependent DNA damage signals. Mol Cell 21(2):187–200

122. Huen MS, Grant R, Manke I, Minn K, Yu X, Yaffe MB, Chen J (2007) RNF8 transduces the DNA-damage signal via histone ubiquitylation and checkpoint protein assembly. Cell 131(5):901–914

123. Kolas NK, Chapman JR, Nakada S, Ylanko J, Chahwan R, Sweeney FD, Panier S, Mendez M, Wildenhain J, Thomson TM et al (2007) Orchestration of the DNA-damage response by the RNF8 ubiquitin ligase. Science 318(5856):1637–1640

124. Mailand N, Bekker-Jensen S, Faustrup H, Melander F, Bartek J, Lukas C, Lukas J (2007) RNF8 ubiquitylates histones at DNA double-strand breaks and promotes assembly of repair proteins. Cell 131(5):887–900

125. Pinato S, Gatti M, Scandiuzzi C, Confalonieri S, Penengo L (2011) UMI, a novel RNF168 ubiquitin binding domain involved in the DNA damage signaling pathway. Mol Cell Biol 31(1):118–126

126. Doil C, Mailand N, Bekker-Jensen S, Menard P, Larsen DH, Pepperkok R, Ellenberg J, Panier S, Durocher D, Bartek J et al (2009) RNF168 binds and amplifies ubiquitin conjugates on damaged chromosomes to allow accumulation of repair proteins. Cell 136(3):435–446

127. Pinato S, Scandiuzzi C, Arnaudo N, Citterio E, Gaudino G, Penengo L (2009) RNF168, a new RING finger, MIU-containing protein that modifies chromatin by ubiquitination of histones H2A and H2AX. BMC Mol Biol 10:55

128. Stewart GS, Panier S, Townsend K, Al-Hakim AK, Kolas NK, Miller ES, Nakada S, Ylanko J, Olivarius S, Mendez M et al (2009) The RIDDLE syndrome protein mediates a ubiquitin-dependent signaling cascade at sites of DNA damage. Cell 136(3):420–434

129. Ramadan K, Dikic I (2017) Editorial: ubiquitin and ubiquitin-relative SUMO in DNA damage response. Front Genet 8:188

130. Bekker-Jensen S, Rendtlew Danielsen J, Fugger K, Gromova I, Nerstedt A, Lukas C, Bartek J, Lukas J, Mailand N (2010) HERC2 coordinates ubiquitin-dependent assembly of DNA repair factors on damaged chromosomes. Nat Cell Biol 12(1):80–86. sup pp 1–12

131. Lee TH, Park JM, Leem SH, Kang TH (2014) Coordinated regulation of XPA stability by ATR and HERC2 during nucleotide excision repair. Oncogene 33(1):19–25

132. Scott D, Oldham NJ, Strachan J, Searle MS, Layfield R (2015) Ubiquitin-binding domains: mechanisms of ubiquitin recognition and use as tools to investigate ubiquitin-modified proteomes. Proteomics 15 (5–6):844–861

133. Huang J, Huen MS, Kim H, Leung CC, Glover JN, Yu X, Chen J (2009) RAD18 transmits DNA damage signalling to elicit homologous recombination repair. Nat Cell Biol 11(5):592–603

134. Wang B, Matsuoka S, Ballif BA, Zhang D, Smogorzewska A, Gygi SP, Elledge SJ (2007) Abraxas and RAP80 form a BRCA1 protein complex required for the DNA damage response. Science 316(5828):1194–1198

135. Yan J, Kim YS, Yang XP, Li LP, Liao G, Xia F, Jetten AM (2007) The ubiquitin-interacting motif containing protein RAP80 interacts with BRCA1 and functions in DNA damage repair response. Cancer Res 67 (14):6647–6656

136. Sobhian B, Shao G, Lilli DR, Culhane AC, Moreau LA, Xia B, Livingston DM, Greenberg RA (2007) RAP80 targets BRCA1 to specific ubiquitin structures at DNA damage sites. Science 316(5828):1198–1202

137. Nikkila J, Coleman KA, Morrissey D, Pylkas K, Erkko H, Messick TE, Karppinen SM, Amelina A, Winqvist R, Greenberg RA (2009) Familial breast cancer screening reveals an alteration in the RAP80 UIM domain that impairs DNA damage response function. Oncogene 28(16):1843–1852

138. Brzovic PS, Meza JE, King MC, Klevit RE (2001) BRCA1 RING domain cancer-predisposing mutations. Structural consequences and effects on protein-protein interactions. J Biol Chem 276(44):41399–41406

139. Polanowska J, Martin JS, Garcia-Muse T, Petalcorin MI, Boulton SJ (2006) A conserved pathway to activate BRCA1-dependent ubiquitylation at DNA damage sites. EMBO J 25(10):2178–2188

140. Morris JR, Solomon E (2004) BRCA1: BARD1 induces the formation of conjugated ubiquitin structures, dependent on K6 of ubiquitin, in cells during DNA replication and repair. Hum Mol Genet 13(8):807–817

141. Gatti M, Pinato S, Maiolica A, Rocchio F, Prato MG, Aebersold R, Penengo L (2015) RNF168 promotes noncanonical K27 ubiquitination to signal DNA damage. Cell Rep 10 (2):226–238

142. Michel MA, Swatek KN, Hospenthal MK, Komander D (2017) Ubiquitin linkage-specific affimers reveal insights into K6-linked ubiquitin signaling. Mol Cell 68 (1):233–246.e235
143. Bennett BT, Knight KL (2005) Cellular localization of human Rad51C and regulation of ubiquitin-mediated proteolysis of Rad51. J Cell Biochem 96(6):1095–1109
144. Yard BD, Reilly NM, Bedenbaugh MK, Pittman DL (2016) RNF138 interacts with RAD51D and is required for DNA interstrand crosslink repair and maintaining chromosome integrity. DNA Repair (Amst) 42:82–93
145. Ismail IH, Gagne JP, Genois MM, Strickfaden H, McDonald D, Xu Z, Poirier GG, Masson JY, Hendzel MJ (2015) The RNF138 E3 ligase displaces Ku to promote DNA end resection and regulate DNA repair pathway choice. Nat Cell Biol 17 (11):1446–1457
146. Schmidt CK, Galanty Y, Sczaniecka-Clift M, Coates J, Jhujh S, Demir M, Cornwell M, Beli P, Jackson SP (2015) Systematic E2 screening reveals a UBE2D-RNF138-CtIP axis promoting DNA repair. Nat Cell Biol 17 (11):1458–1470
147. Chen J, Feng W, Jiang J, Deng Y, Huen MS (2012) Ring finger protein RNF169 antagonizes the ubiquitin-dependent signaling cascade at sites of DNA damage. J Biol Chem 287(33):27715–27722
148. Poulsen M, Lukas C, Lukas J, Bekker-Jensen-S, Mailand N (2012) Human RNF169 is a negative regulator of the ubiquitin-dependent response to DNA double-strand breaks. J Cell Biol 197(2):189–199
149. An L, Jiang Y, Ng HH, Man EP, Chen J, Khoo US, Gong Q, Huen MS (2017) Dual-utility NLS drives RNF169-dependent DNA damage responses. Proc Natl Acad Sci U S A 114(14):E2872–e2881
150. Liu J, Kruswick A, Dang H, Tran AD, Kwon SM, Wang XW, Oberdoerffer P (2017) Ubiquitin-specific protease 21 stabilizes BRCA2 to control DNA repair and tumor growth. Nat Commun 8(1):137
151. Ogrunc M, Martinez-Zamudio RI, Sadoun PB, Dore G, Schwerer H, Pasero P, Lemaitre JM, Dejean A, Bischof O (2016) USP1 regulates cellular senescence by controlling genomic integrity. Cell Rep 15 (7):1401–1411
152. Mertins P, Qiao JW, Patel J, Udeshi ND, Clauser KR, Mani DR, Burgess MW, Gillette MA, Jaffe JD, Carr SA (2013) Integrated proteomic analysis of post-translational modifications by serial enrichment. Nat Methods 10(7):634–637
153. Mertins P, Mani DR, Ruggles KV, Gillette MA, Clauser KR, Wang P, Wang X, Qiao JW, Cao S, Petralia F et al (2016) Proteogenomics connects somatic mutations to signalling in breast cancer. Nature 534 (7605):55–62
154. Kim W, Bennett EJ, Huttlin EL, Guo A, Li J, Possemato A, Sowa ME, Rad R, Rush J, Comb MJ et al (2011) Systematic and quantitative assessment of the ubiquitin-modified proteome. Mol Cell 44(2):325–340
155. Zhou H, Di Palma S, Preisinger C, Peng M, Polat AN, Heck AJ, Mohammed S (2013) Toward a comprehensive characterization of a human cancer cell phosphoproteome. J Proteome Res 12(1):260–271
156. Somyajit K, Basavaraju S, Scully R, Nagaraju G (2013) ATM- and ATR-mediated phosphorylation of XRCC3 regulates DNA double-strand break-induced checkpoint activation and repair. Mol Cell Biol 33 (9):1830–1844
157. Slupianek A, Jozwiakowski SK, Gurdek E, Skorski T (2009) BCR/ABL kinase interacts with and phosphorylates the RAD51 paralog, RAD51B. Leukemia 23:2308–2310
158. Takizawa Y, Kinebuchi T, Kagawa W, Yokoyama S, Shibata T, Kurumizaka H (2004) Mutational analyses of the human Rad51-Tyr315 residue, a site for phosphorylation in leukaemia cells. Genes Cells 9 (9):781–790
159. Inano S, Sato K, Katsuki Y, Kobayashi W, Tanaka H, Nakajima K, Nakada S, Miyoshi H, Knies K, Takaori-Kondo A et al (2017) RFWD3-mediated ubiquitination promotes timely removal of both RPA and RAD51 from DNA damage sites to facilitate homologous recombination. Mol Cell 66 (5):622–634.e628
160. Knies K, Inano S, Ramirez MJ, Ishiai M, Surralles J, Takata M, Schindler D (2017) Biallelic mutations in the ubiquitin ligase RFWD3 cause Fanconi anemia. J Clin Invest 127(8):3013–3027
161. Knijnenburg TA, Wang L, Zimmermann MT, Chambwe N, Gao GF, Cherniack AD, Fan H, Shen H, Way GP, Greene CS et al (2018) Genomic and molecular landscape of DNA damage repair deficiency across the cancer genome atlas. Cell Rep 23(1):239–254.e236
162. Watkins JA, Irshad S, Grigoriadis A, Tutt AN (2014) Genomic scars as biomarkers of homologous recombination deficiency and

drug response in breast and ovarian cancers. Breast Cancer Res 16(3):211

163. Pennington KP, Walsh T, Harrell MI, Lee MK, Pennil CC, Rendi MH, Thornton A, Norquist BM, Casadei S, Nord AS et al (2014) Germline and somatic mutations in homologous recombination genes predict platinum response and survival in ovarian, fallopian tube, and peritoneal carcinomas. Clin Cancer Res 20(3):764–775
164. Yard BD, Adams DJ, Chie EK, Tamayo P, Battaglia JS, Gopal P, Rogacki K, Pearson BE, Phillips J, Raymond DP et al (2016) A genetic basis for the variation in the vulnerability of cancer to DNA damage. Nat Commun 7:11428
165. Vanderstichele A, Busschaert P, Olbrecht S, Lambrechts D, Vergote I (2017) Genomic signatures as predictive biomarkers of homologous recombination deficiency in ovarian cancer. Eur J Cancer 86:5–14
166. Hillman RT, Chisholm GB, Lu KH, Futreal PA (2018) Genomic rearrangement signatures and clinical outcomes in high-grade serous ovarian cancer. J Natl Cancer Inst 110(3):265–272
167. Telli ML, Timms KM, Reid J, Hennessy B, Mills GB, Jensen KC, Szallasi Z, Barry WT, Winer EP, Tung NM et al (2016) Homologous recombination deficiency (HRD) score predicts response to platinum-containing neoadjuvant chemotherapy in patients with triple-negative breast cancer. Clin Cancer Res 22(15):3764–3773
168. Lheureux S, Lai Z, Dougherty BA, Runswick S, Hodgson DR, Timms KM, Lanchbury JS, Kaye S, Gourley C, Bowtell D et al (2017) Long-term responders on olaparib maintenance in high-grade serous ovarian cancer: clinical and molecular characterization. Clin Cancer Res 23 (15):4086–4094
169. Zou X, Owusu M, Harris R, Jackson SP, Loizou JI, Nik-Zainal S (2018) Validating the concept of mutational signatures with isogenic cell models. Nat Commun 9(1):1744
170. Davies H, Glodzik D, Morganella S, Yates LR, Staaf J, Zou X, Ramakrishna M, Martin S, Boyault S, Sieuwerts AM et al (2017) HRDetect is a predictor of BRCA1 and BRCA2 deficiency based on mutational signatures. Nat Med 23(4):517–525
171. Nik-Zainal S, Morganella S (2017) Mutational signatures in breast cancer: the problem at the DNA level. Clin Cancer Res 23:2617–2629
172. Chien J, Sicotte H, Fan JB, Humphray S, Cunningham JM, Kalli KR, Oberg AL, Hart SN, Li Y, Davila JI et al (2015) TP53 mutations, tetraploidy and homologous recombination repair defects in early stage high-grade serous ovarian cancer. Nucleic Acids Res 43 (14):6945–6958
173. Lim DS, Hasty P (1996) A mutation in mouse rad51 results in an early embryonic lethal that is suppressed by a mutation in p53. Mol Cell Biol 16(12):7133–7143
174. Shu Z, Smith S, Wang L, Rice MC, Kmiec EB (1999) Disruption of muREC2/RAD51L1 in mice results in early embryonic lethality which can Be partially rescued in a p53(−/−) background. Mol Cell Biol 19 (12):8686–8693
175. Kuznetsov SG, Haines DC, Martin BK, Sharan SK (2009) Loss of Rad51c leads to embryonic lethality and modulation of Trp53-dependent tumorigenesis in mice. Cancer Res 69(3):863–872
176. Pittman DL, Schimenti JC (2000) Midgestation lethality in mice deficient for the RecA-related gene, Rad51d/Rad51l3. Genesis 26 (3):167–173
177. Adam J, Deans B, Thacker J (2007) A role for Xrcc2 in the early stages of mouse development. DNA Repair 6(2):224–234
178. Peng G, Chun-Jen Lin C, Mo W, Dai H, Park YY, Kim SM, Peng Y, Mo Q, Siwko S, Hu R et al (2014) Genome-wide transcriptome profiling of homologous recombination DNA repair. Nat Commun 5:3361
179. Lord CJ, Ashworth A (2013) Mechanisms of resistance to therapies targeting BRCA-mutant cancers. Nat Med 19(11):1381–1388
180. Swisher EM, Lin KK, Oza AM, Scott CL, Giordano H, Sun J, Konecny GE, Coleman RL, Tinker AV, O'Malley DM et al (2017) Rucaparib in relapsed, platinum-sensitive high-grade ovarian carcinoma (ARIEL2 Part 1): an international, multicentre, open-label, phase 2 trial. Lancet Oncol 18(1):75–87
181. Norquist B, Wurz KA, Pennil CC, Garcia R, Gross J, Sakai W, Karlan BY, Taniguchi T, Swisher EM (2011) Secondary somatic mutations restoring BRCA1/2 predict chemotherapy resistance in hereditary ovarian carcinomas. J Clin Oncol 29(22):3008–3015
182. Kondrashova O, Nguyen M, Shield-Artin K, Tinker AV, Teng NNH, Harrell MI, Kuiper MJ, Ho GY, Barker H, Jasin M et al (2017) Secondary somatic mutations restoring RAD51C and RAD51D associated with acquired resistance to the PARP inhibitor

rucaparib in high-grade ovarian carcinoma. Cancer Discov 7(9):984–998

183. Bouwman P, Aly A, Escandell JM, Pieterse M, Bartkova J, van der Gulden H, Hiddingh S, Thanasoula M, Kulkarni A, Yang Q et al (2010) 53BP1 loss rescues BRCA1 deficiency and is associated with triple-negative and BRCA-mutated breast cancers. Nat Struct Mol Biol 17(6):688–695
184. Hu HM, Zhao X, Kaushik S, Robillard L, Barthelet A, Lin KK, Shah KN, Simmons AD, Raponi M, Harding TC et al (2018) A quantitative chemotherapy genetic interaction map reveals factors associated with PARP inhibitor resistance. Cell Rep 23 (3):918–929
185. Wang X, Nagl NG Jr, Flowers S, Zweitzig D, Dallas PB, Moran E (2004) Expression of p270 (ARID1A), a component of human SWI/SNF complexes, in human tumors. Int J Cancer 112(4):636
186. Shen J, Peng Y, Wei L, Zhang W, Yang L, Lan L, Kapoor P, Ju Z, Mo Q, Shih Ie M et al (2015) ARID1A deficiency impairs the DNA damage checkpoint and sensitizes cells to PARP inhibitors. Cancer Discov 5 (7):752–767
187. Shen J, Ju Z, Zhao W, Wang L, Peng Y, Ge Z, Nagel ZD, Zou J, Wang C, Kapoor P et al (2018) ARID1A deficiency promotes mutability and potentiates therapeutic antitumor immunity unleashed by immune checkpoint blockade. Nat Med 24(5):556–562
188. McGranahan N, Furness AJ, Rosenthal R, Ramskov S, Lyngaa R, Saini SK, Jamal-Hanjani M, Wilson GA, Birkbak NJ, Hiley CT et al (2016) Clonal neoantigens elicit T cell immunoreactivity and sensitivity to immune checkpoint blockade. Science 351 (6280):1463–1469
189. Germano G, Lamba S, Rospo G, Barault L, Magri A, Maione F, Russo M, Crisafulli G, Bartolini A, Lerda G et al (2017) Inactivation of DNA repair triggers neoantigen generation and impairs tumour growth. Nature 552 (7683):116–120
190. Luksza M, Riaz N, Makarov V, Balachandran VP, Hellmann MD, Solovyov A, Rizvi NA, Merghoub T, Levine AJ, Chan TA et al (2017) A neoantigen fitness model predicts tumour response to checkpoint blockade immunotherapy. Nature 551(7681):517–520
191. Snyder A, Makarov V, Merghoub T, Yuan J, Zaretsky JM, Desrichard A, Walsh LA, Postow MA, Wong P, Ho TS et al (2014) Genetic basis for clinical response to CTLA-4 blockade in melanoma. N Engl J Med 371 (23):2189–2199
192. Rizvi NA, Hellmann MD, Snyder A, Kvistborg P, Makarov V, Havel JJ, Lee W, Yuan J, Wong P, Ho TS et al (2015) Cancer immunology. Mutational landscape determines sensitivity to PD-1 blockade in non-small cell lung cancer. Science 348 (6230):124–128
193. Strickland KC, Howitt BE, Shukla SA, Rodig S, Ritterhouse LL, Liu JF, Garber JE, Chowdhury D, Wu CJ, D'Andrea AD et al (2016) Association and prognostic significance of BRCA1/2-mutation status with neoantigen load, number of tumor-infiltrating lymphocytes and expression of PD-1/PD-L1 in high grade serous ovarian cancer. Oncotarget 7(12):13587–13598
194. Hugo W, Zaretsky JM, Sun L, Song C, Moreno BH, Hu-Lieskovan S, Berent-Maoz B, Pang J, Chmielowski B, Cherry G et al (2016) Genomic and transcriptomic features of response to anti-PD-1 therapy in metastatic melanoma. Cell 165(1):35–44
195. Davoli T, Uno H, Wooten EC, Elledge SJ (2017) Tumor aneuploidy correlates with markers of immune evasion and with reduced response to immunotherapy. Science 355 (6322):eaaf8399
196. Kelderman S, Schumacher TN, Kvistborg P (2015) Mismatch repair-deficient cancers are targets for anti-PD-1 therapy. Cancer Cell 28 (1):11–13
197. Le DT, Uram JN, Wang H, Bartlett BR, Kemberling H, Eyring AD, Skora AD, Luber BS, Azad NS, Laheru D et al (2015) PD-1 blockade in tumors with mismatch-repair deficiency. N Engl J Med 372 (26):2509–2520

Chapter 2

Resolving Roadblocks to Telomere Replication

Emily Mason-Osann, Himabindu Gali, and Rachel Litman Flynn

Abstract

The maintenance of genome stability in eukaryotic cells relies on accurate and efficient replication along each chromosome following every cell division. The terminal position, repetitive sequence, and structural complexities of the telomeric DNA make the telomere an inherently difficult region to replicate within the genome. Thus, despite functioning to protect genome stability mammalian telomeres are also a source of replication stress and have been recognized as common fragile sites within the genome. Telomere fragility is exacerbated at telomeres that rely on the Alternative Lengthening of Telomeres (ALT) pathway. Like common fragile sites, ALT telomeres are prone to chromosome breaks and are frequent sites of recombination suggesting that ALT telomeres are subjected to chronic replication stress. Here, we will review the features of telomeric DNA that challenge the replication machinery and also how the cell overcomes these challenges to maintain telomere stability and ensure the faithful duplication of the human genome.

Key words Replication stress, Telomere, Alternative lengthening of telomeres (ALT), Genome maintenance, G-quadruplex, R-loop

1 Telomeric DNA

Telomeres are repetitive DNA elements located at the ends of chromosomes that function to protect genome integrity. In mammalian cells, telomeres are composed of sequential TTAGGG hexanucleotide repeats. Although predominately double-stranded, telomeres contain a single stranded 3′ overhang that loops back to invade the more centromere proximal telomeric DNA sequence, forming a T-loop structure [1]. The structure of T-loops is supported by a six-protein complex called shelterin, which binds and coats telomeric DNA [2]. Shelterin binding facilitates formation of the T-loop and functions to protect telomeric DNA from degradation, recombination, and chromosome end-to-end fusions [2]. Due to the semiconservative nature of lagging strand DNA synthesis, following each cellular division chromosomes undergo progressive shortening, losing approximately 100 nucleotides of telomeric sequence. The persistent erosion of telomeric DNA compromises the function of the telomere in end protection and

Lata Balakrishnan and Jason A. Stewart (eds.), *DNA Repair: Methods and Protocols*, Methods in Molecular Biology, vol. 1999, https://doi.org/10.1007/978-1-4939-9500-4_2,

ultimately threatens genome stability. When the telomeric DNA shortens to a critical length, the telomere end initiates a DNA damage response that triggers the cell to exit the cell cycle and enter a senescent state [3].

Given that telomere attrition eventually induces cellular senescence, faithful replication through the end of the telomere is essential to maintain telomere length and therefore the proliferative capacity of the cell. While telomeres are essential to the maintenance of genome stability, the structure and repetitive nature of mammalian telomeres pose a challenge to DNA replication, making telomeres prone to replication stress. Here, we will review the literature on mammalian telomere replication, the challenges associated with this process, and the connection between telomere replication stress and the alternative lengthening of telomeres (ALT) pathway.

2 Mammalian DNA Replication

The maintenance of genome stability in eukaryotic cells relies on accurate and efficient replication of DNA during S-phase of each cell cycle. The initiation of DNA replication is a highly coordinated and tightly regulated process that consists of two main phases, the licensing of replication origins and the subsequent firing of those origins. During licensing, a multiprotein complex referred to as the prereplicative complex (pre-RC) is assembled at origins of replication. While replication origins are well-defined consensus sequences in the yeast genome, mammalian origins of replication are not as well defined [4]. During origin licensing, the pre-RC is assembled at replication origins during late M phase into early G1 phase, poising these origins for the initiation of DNA synthesis [5–7]. The mammalian pre-RC is made up of the DNA binding origin recognition complex (ORC), Cdc6, Cdt1, and two inactive helicase complexes made of MCM2-7, which are sequentially recruited to origins [5, 8–12]. As cells enter into S-phase, components of the pre-RC are dissociated and/or degraded, preventing origins from licensing more than once per cell cycle and ultimately preventing rereplication [13]. In human cells, this is thought to be mediated primarily through inactivation/degradation of Cdt1 [14–17].

Origin firing and the initiation of DNA synthesis requires the assembly of additional regulatory proteins and the DNA replication machinery specifically at licensed origins. This process is regulated by two key kinases, S-phase CDK and Cdc7-Dbf4 (DDK) [6, 18]. These kinases phosphorylate MCM2-7 to promote recruitment of GINS and Cdc45, thereby forming the highly active CMG complex (Cdc45—MCM2-7—GINS) that travels with the replication fork and functions as the replicative helicase [19]. Full assembly of the replisome requires the recruitment of other factors, including the single stranded DNA (ssDNA) binding protein replication protein A (RPA), DNA polymerase α-primase (Pol α), leading strand polymerase (Pol ε), and lagging strand polymerase (Pol δ) [20, 21]. Origins are fired and the complete replisome is assembled following the onset of S-phase. The order of origin firing in mammalian cells has been shown to correlate with the three-dimensional structure of the chromatin [22], suggesting that origin firing may not be stochastic. The MCM2-7 complex is loaded onto origins in excess. A subset of these origins will become active and fire during S-phase [5, 23], while the remainder of these origins will remain dormant, only firing if the integrity of a neighboring replication fork is compromised [24, 25].

Upon origin firing the double stranded DNA (dsDNA) at the replication origin is melted to create a replication bubble and two bidirectional replication forks. Although the active replication fork is essential for DNA synthesis, it also promotes the formation of ssDNA, which, if left unprotected threatens genome stability [26]. Thus, the exposed ssDNA is bound by the replication protein RPA. The formation of RPA-coated ssDNA plays an important role not only in binding and protecting the ssDNA but also in stabilizing the replication fork itself [27–29]. The replication fork and active replisome, containing the CMG complex, proliferating cell nuclear antigen (PCNA), and DNA polymerases, travels bidirectionally away from the origin of replication. The CMG complex unwinds dsDNA ahead of Pol ε and Pol δ to facilitate the synthesis of new copies of each individual DNA strand [30]. Given the semiconservative nature of DNA replication, Pol ε synthesizes the leading strand of DNA continuously in the 5′–3′ direction, while Pol δ DNA synthesis is discontinuous extending Okazaki fragments generated by Pol α/Primase on the lagging strand [30]. The RNA primers generated by Pol α/Primase are later removed and filled in with DNA nucleotides, and the Okazaki fragments are ligated together to form a continuous DNA strand.

Much like the rest of the genomic DNA during S-phase, telomeric DNA is replicated in a similar fashion, with a few unique features. Previously, telomere replication was thought to only initiate from origins of replication in the subtelomeric region, creating replication forks that would travel to the terminal end of the chromosome. However, more recent data have shown that pre-RCs can assemble and origins can be activated within mammalian telomeric repeats [31, 32]. While this data demonstrates that it is possible for replication to initiate within telomeric repeats, it also shows that replication of mammalian telomeres is most often accomplished by replisomes that originate in the subtelomeric region [31]. Given that DNA synthesis is discontinuous, DNA replication on the lagging strand cannot be extended through the terminal end of the telomere. When the terminal RNA primer is degraded there is a gap at the end of the C-strand that cannot be filled in, ultimately leading to telomere shortening referred to as the "end replication problem" [3, 33]. In contrast, leading strand synthesis is processive and continues to the terminal end of the telomere, creating a blunt end at the chromosome termini. Therefore, telomeric DNA generated by leading strand synthesis must then be processed by nucleases, including Apollo and ExoI, to create the 3′ overhang that facilitates T-loop formation. Likewise, the newly synthesized lagging strand is also processed to create the mature 3′ G-rich overhangs characteristic of telomeres. Following the completion of lagging strand synthesis, the C-strand fill in complex, composed of CTC1, STN1 and TEN1 (CST) promotes 3′ overhang maturation, potentially by recruiting and activating Pol α to fill in hyperresected C-strand DNA [34–36]. In addition to its role in C-strand fill-in at resected telomeres and G-strands that have been extended by telomerase, the CST complex facilitates replication fork restart, suggesting that an important role for this complex is alleviating telomere replication stress [37].

Over the years, extensive research has demonstrated that the telomere is replicated in much the same fashion as the rest of the genomic DNA. However, it poses a number of unique challenges to the replication machinery that must be mitigated to ensure faithful duplication of the genome.

3 Challenges to DNA Replication

There are a number of inherent challenges to replication that arise as by-products of normal cellular metabolism, causing replication stress. Innately, certain parts of the genome are more prone to replication stress and therefore experience more replication fork collapse in the presence of low levels of replication stress [38–41]. These sites, called common fragile sites (CFS), are stable in unperturbed cells but under low levels of replication stress show evidence of chromosomal breaks and are frequently sites of recombination (reviewed in [38]) [42–44]. In addition to CFS loci, there are a number of other naturally occurring impediments to the replication machinery, including bulky DNA adducts, G-quadruplexes (G4s), DNA hairpins, DNA mismatches, DNA nicks and/or ssDNA gaps, transcription bubbles and nucleotide depletion [45–49]. These sources of replications stress cause the slowing, or stalling, of the replication fork and/or DNA synthesis altogether [45]. If not alleviated, replication stress can cause genome instability or even cell death. To mitigate genome instability and complete DNA replication, cells employ a variety of mechanisms to manage and resolve replication stress.

3.1 Replication Stress Response

When the replication fork slows or stalls, the replicative helicase can dissociate from the replicative polymerases, causing DNA to be unwound well ahead of DNA synthesis [50]. This uncoupling can lead to the accumulation of long stretches of ssDNA that become bound and coated by RPA. Given the affinity of RPA for ssDNA, replication stress that leads to the accumulation of ssDNA can risk depleting cellular stores of RPA, provoking a phenomenon called RPA exhaustion [51]. RPA exhaustion exposes unprotected ssDNA, which is quickly converted into DNA double strand breaks (DSBs) leading to genomic instability and ultimately, cell death in a process referred to as "replication catastrophe" [27, 51]. Even in the absence of RPA exhaustion, RPA-coated ssDNA at stalled replication forks regulates a replication stress response by recruiting and activating the DNA damage checkpoint kinase, ataxia-telangiectasia mutated and Rad3 related (ATR) [50, 52–55]. The activation of ATR induces both a global and local response to help alleviate replication stress [45]. Globally, ATR halts the cell cycle and inhibits origin firing by phosphorylating, and activating the checkpoint protein CHK1 [51]. Thus, ATR activation allows the cell time to resolve replication stress, restart stalled forks, and avoid replication catastrophe, or if the DNA damage is irreparable, induce apoptosis. Locally, ATR promotes the protection of stalled replication forks and supports the restart of stalled forks, helping the cell to faithfully complete DNA replication [56, 57].

Obstacles to the replication machinery can lead to both reversible and irreversible stalling of the replication fork. Reversibly stalled replication forks can be restarted by removing, resolving, or bypassing the source of replication stress, allowing the replication fork to proceed. The machinery required to facilitate replication restart depends on the source of the replication stress, and often includes the recruitment of specific DNA damage repair machinery to remove the lesion [45]. If the source of the replication stress cannot be removed, replication can continue across the lesions by switching out the canonical replicative polymerase for one of the translesion synthesis polymerases that can accommodate that particular lesion, promoting a DNA damage tolerance pathway. In addition, reversibly stalled forks can bypass certain DNA lesions by repriming DNA synthesis on the 5′ side of the lesion, allowing the replication fork to restart and continue past the lesion, leaving a section of ssDNA that is later filled in [58]. Finally, a reversibly stalled replication fork can be physically reversed to form a "chicken foot" structure. The chicken foot structure has been suggested to protect the stalled replication fork from nucleolytic degradation by minimizing the exposed ssDNA at the replication fork and facilitating replication fork restart [59]. After stalled forks are reversed, and the lesion is removed, resolved, or bypassed, these forks can be restarted. However, if the damage has compromised the ability of the fork to restart, the chicken foot can be processed into a DNA DSB and repaired by recombination [59].

When barriers to replication lead to an irreversibly stalled replication fork, the replisome dissociates from the DNA, collapsing the replication fork and leaving the DNA vulnerable to DNA damage. If these collapsed forks are not repaired, these regions of DNA could persist unreplicated; thus, the cell has evolved several mechanisms to rescue collapsed forks and ensure faithful duplication of the genome. One mechanism for the rescue of collapsed forks is through firing of adjacent dormant origins [24, 25, 60–62]. Dormant origins are already licensed and poised to initiate replication, but have not yet been fired. Firing of these dormant origins creates a new bidirectional replication fork which can converge on the adjacent stalled replication fork restoring replication through that region [24, 25]. While it was previously thought that there were no origins of replication in telomeres, it has recently been shown that DNA replication can initiate within telomeric repeats [63], indicating that dormant origin firing may be a mechanism to rescue collapsed replication forks in the telomeres. However, to date, dormant origins fired in response to telomere replication stress have only been observed in the subtelomeric region [63]. If collapsed replication forks are not rescued by a converging replication fork, the fork can be nucleolytically cleaved into a single-ended DNA DSBs [64, 65]. Single-ended DSBs are common substrates for break-induced replication (BIR), a specific form of homology directed repair [64, 66, 67]. By breaking, collapsed replication forks can engage homology directed repair following unresolved replication stress [65], providing cells with a backup mechanism to complete DNA replication in the presence of irreversibly stalled replication forks.

4 Challenges to Telomeric DNA Replication

Since the identification of the sequence of the telomeric DNA, telomeres have been hypothesized to pose a challenge to DNA replication, given their repetitive nature. This hypothesis was later supported by studies in yeast demonstrating that replication forks paused on telomeric DNA, both at the terminal ends of the chromosome and when telomeric repeats are artificially embedded in other regions of the genome [68–70]. Moreover, under low levels of replication stress, telomeric DNA showed signs of fragility and broken DNA suggesting that telomeres resemble CFS [70]. These studies have highlighted the complexities of replicating telomeric DNA and have provided the framework for investigating how the repetitive nature of the telomeric sequence, along with the proclivity for secondary structures and propensity for collisions with the transcriptional program affect faithful telomere replication (Fig. 1).

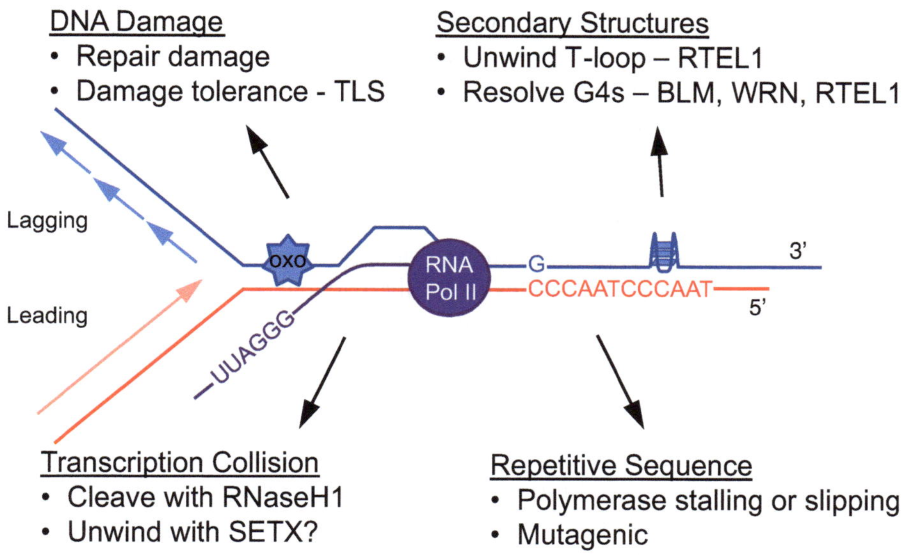

Fig. 1 Challenges associated with telomere replication. Several features of the telomere make this region of the genome prone to replication stress. Telomeres are repetitive sequences, which inherently cause polymerase stalling and slipping, and as a result can cause replication stress and lead to mutagenesis. Telomere can also accrue DNA damage following oxidative stress, which leads to the formation of oxidized guanine residues (8-oxo-G). The 8-oxo-G residue decreases shelterin binding and promotes telomere instability and can be resolved either by repairing the DNA lesion or utilizing damage tolerance pathways, such as translesion synthesis. Secondary structures including both G4s and the T-loops (not shown) block the replicative machinery, and are resolved by either helicases (RTEL1, BLM, and WRN) or nucleases (SLX4, TZAP). Finally, the transcription of TERRA at telomeres generates R-loops that can also block the replication machinery and are removed by nucleolytic cleavage (RNaseH1), or potentially by RNA–DNA helicases (i.e., SETX)

4.1 Structure

4.1.1 Shelterin

Structurally, mammalian telomeres are coated by a six-protein complex, called shelterin (TRF1, TRF2 POT1, TPP1, RAP1, and TIN2), that functions to protect mammalian telomeres from degradation and from eliciting a DNA damage response [2]. TRF1, TRF2, and POT1 bind directly to the telomeric DNA with sequence specificity. While TRF1 and TRF2 bind the double stranded regions of telomeric DNA, POT1 binds single stranded telomeric DNA. TIN2, TPP1, and RAP1 act as stabilizing proteins, and function to recruit other proteins crucial for telomere homeostasis [2]. Shelterin binding was initially thought to be a source of replication stress, acting as a roadblock to the replisome [71]. However, TRF1, TRF2, and POT1 have been shown to promote telomere replication, prevent fork stalling, and protect against telomere fragility by facilitating the resolution of common structures found at the telomeres that cause replication stress, including G-quadruplexes (G4) and the T-loop [70, 72–74].

4.1.2 Repetitive G-Rich DNA

Repetitive regions of the genome, including but not limited to telomeric DNA, make up to half of the human genome and cause inherent challenges to DNA replication. In addition to DNA polymerase slipping, which can cause expansion of repetitive regions of DNA [75] and/or introduce DNA mismatches [76], the replisome is also prone to stalling [77], highlighting an inherent challenge to replicating repetitive regions of DNA like telomeres. In addition to being repetitive, telomeric DNA is also rich in guanine nucleotides. Guanine residues are easily oxidized, making telomeres intrinsically susceptible to oxidative stress. Chronic oxidative stress can lead to telomere dysfunction, attrition, and ultimately cellular senescence [78, 79]. Mechanistically, oxidative stress causes the formation of an oxidized guanine nucleotide, 7,8-dihydro-8-oxo-guanine (8-oxoG) [80]. Modified nucleotides, such as 8-oxoG can limit shelterin binding directly by decreasing recognition of telomeric repeats by TRF1 and TRF2, [81] or indirectly by generating inappropriate base pairing that leads to GC → TA transversion mutations [82] and consequently removing shelterin binding sites within the telomere. Given that TRF1 is necessary for telomere replication and loss of TRF1 can increase telomere fragility [70], oxidative damage at guanine residues that limits TRF1 binding may contribute to replication stress at telomeres.

The G-rich sequences are not only susceptible to oxidative damage, but they can also form secondary structures that are stabilized by Hoogsteen pairing, called G-quadruplexes (G4s). The G4 structures are generated by four guanine residues that assemble into a planar quartet, which can then stack on top of one another to create an extremely stable nucleic acid structure [83, 84]. Given the G-rich nature of telomeric DNA, telomeres are prone to forming G4 structures [85, 86]. Pharmacological stabilization of G4 structures impairs replication fork progression in the telomere, inducing replication stress, and increasing telomere fragility [87–89]. These data highlight the need to resolve telomeric G4s to ensure faithful replication of the telomeric DNA [89]. In vitro G4s can be unwound by RecQ family helicases BLM and WRN [90]. Human cells lacking WRN require extended time to complete S-phase [91], and have specific defects in telomere replication leading to telomere loss and instability [92, 93]. TRF2 interacts with WRN helicase and stimulates helicase activity [94], suggesting that WRN may be recruited to the telomere to unwind structures like G4s. Human cells lacking BLM helicase show evidence of telomere fragility that is epistatic with loss of TRF1 [70], suggesting that BLM and TRF1 functions are coordinated to unwind secondary structures during lagging strand DNA synthesis at the telomeres [95]. In addition to BLM and WRN, the regulator of telomere length (RTEL1) helicase has been implicated in resolution of secondary structures at telomeres, including G4s [96]. RTEL1 associates with the replisome by interacting with

PCNA through a PCNA interacting protein (PIP) box domain [97] and RTEL1 depletion causes additive telomere fragility when combined with BLM depletion, suggesting multiple, nonoverlapping mechanisms for resolution of G4s at telomeres by RTEL1 and BLM [96, 97]. There are several other helicases and enzymes that have been identified to possess G4 unwinding activity and are also known associate with mammalian telomeres, including PIF1, FANCJ and the CST complex, however the exact functions and mechanisms of these helicases at mammalian telomeres is still emerging [98–106].

4.1.3 T-Loop

Another structure found at the telomere that can be an impediment to completing replication is the T-loop. The T-loop serves an important function in protecting telomere ends, but can also act as a source of replication stress, preventing the replisome from gaining access to the terminal end of the chromosome. In addition to its role in G4 resolution, RTEL1 supports telomere replication by disassembling the T-loop structure [96, 97]. TRF2 recruits RTEL1 specifically to the telomeres to promote T-loop resolution during S-phase [107], separate from the PCNA mediated recruitment of RTEL1 to resolve G4 structures. In addition, both BLM and WRN helicases have been suggested to unwind the T-loop structure in vitro [94]. In the absence of RTEL1, T-loops are cleaved by the SLX4 nuclease, which creates extrachromosomal telomeric repeats (ECTRs) and causes telomere shortening, thereby expediting length dependent telomere dysfunction [96]. More recently, the protein TZAP was also shown to cleave T-loops structures from excessively long telomeres [108, 109]. Long telomeres are susceptible to replication stress [110], indicating that cells maintain mechanisms to trim these telomeres in order to promote successful telomere replication. However, whether TZAP functions to mitigate replication stress at telomeric DNA has not been investigated.

4.2 TERRA R-Loops

The telomere is transcribed into a long noncoding RNA called TERRA [111, 112]. In mammalian cells, TERRA is cell cycle regulated and plays a role in reloading the shelterin component POT1 following replication, ensuring proper end protection of telomeres [113–116]. TERRA is transcribed by RNA Polymerase II, beginning in the subtelomeric region and continuing distally toward the telomeric repeats. Once transcribed, a fraction of TERRA is chromatin bound [115], and can hybridize with DNA [111, 117–119] forming a triple stranded DNA–RNA hybrid structure called an R-loop [120, 121]. R-loops have many functions within the genome, both productive and deleterious, and deregulation of R-loops has been tied to many diseases, including cancer [120, 122]. R-loops can pose an obstacle to the DNA replication machinery and induce replication fork stalling, which, if unresolved leads to replication fork collapse, DNA DSBs and homology directed repair through BIR [120, 123–125]. Induction of TERRA transcription leads to an increase in telomere fragility in mammalian cells [117], indicating that TERRA R-loops may contribute to replication stress at telomeres. Therefore, cells must maintain mechanisms to remove TERRA R-loops. The endonuclease RNaseH1 cleaves the RNA moiety of R-loops, and has known activity at telomeric DNA [117, 125, 126]. In addition to RNA-seH1, several helicases that localize to mammalian telomeres have been implicated in the resolution of R-loops (i.e., SETX) and consequently, may prevent replication fork stalling at telomeres [105, 125, 127–131]. In mammalian cells, TERRA is transcribed from the C-strand and thus, TERRA hybridizes specifically with the C-strand of telomeric DNA generating a unique challenge for leading strand DNA synthesis. In addition, formation of TERRA R-loops on the C-strand causes the displacement of the G-rich strand of DNA from the duplex. The displaced G-rich ssDNA may be prone to the formation of G4 structures, potentially inducing replication stress on the lagging strand during DNA synthesis. This suggest that the combination of TERRA R-loops and G4s in parallel presents challenges to both leading and lagging strand synthesis and indicate that both R-loop and G4 helicases may be required to resolve replication stress caused by TERRA R-loops.

4.3 Telomeric Damage and Repair Following Replication Stress

Given the telomere's inherent fragility, replication stress has the potential to cause DNA damage within the telomeres. While replication stress does not refer to a specific damaged DNA structure, DNA can be damaged as a result of stalled or collapsed replication forks. However, prior to the cleavage of a stalled or collapsed replication fork into a DNA break, replication stress alone can activate DNA damage repair pathways. During replication stress, ssDNA is exposed and coated with RPA. Persistent RPA-coated ssDNA at stalled replication forks recruits ATR and the ATR-interacting protein (ATRIP) complex ultimately leading to activation of checkpoint signaling [29, 52, 57, 132]. Cells with persistent replication stress at the telomeres, such as cells that utilize the Alternative Lengthening of Telomere pathway, show a persistence of RPA foci at the telomeres, and a dependence on ATR signaling [113, 133, 134].

If replication forks stall irreversibly, they collapse into single-ended DNA double-strand breaks [135, 136] which are then susceptible to homology directed repair processes including break-induced replication that function to reestablish an active replication fork and ensure genome stability. The exact mechanism by which DNA double-strand breaks are initially formed at irreversibly stalled replication forks has not been fully elucidated. However, the prevalence of ssDNA at stalled replication forks has suggested that both passive breakage and/or endonucleolytic cleavage can contribute to the formation of these single ended double strand breaks [45, 64, 137].

In addition to the direct formation of DNA double strand breaks at collapsed replication forks, slowing or stalling of replication forks can lead to incomplete DNA replication during S-phase. Cells that then enter mitosis with underreplicated DNA are susceptible to the formation of anaphase bridges during chromosomal segregation [138–141]. If left unresolved, anaphase bridges can lead to chromosomal breakage and genomic instability in the following cell cycle [138, 141–143]. Given the propensity of the telomeres to undergo replication stress, and therefore be underreplicated as cells enter into mitosis, specific enzymes, such as BLM, WRN and SLX4, are recruited to telomeres to prevent the formation of ultra-fine anaphase bridges at the telomere [141, 144]. Both BLM localization and ultra-fine bridges at the telomeres increase upon induction of replication stress or by increasing telomere fragility [141, 144]. Thus, unresolved replication stress at fragile sites, including the telomere, have the ability to produce both isolated DNA breaks at collapsed replication forks and/or underreplicated regions, which can then induce mitotic arrest or lead to the accumulation of micronuclei [45, 139, 142, 143, 145].

5 Alternative Lengthening of Telomeres

In mammalian cells telomeres shorten with each cell division, eventually inducing cellular senescence, a metabolically active but non-dividing state. Cancer cells evade cellular senescence by activating mechanisms to continually elongate telomeric DNA, thereby acquiring replicative immortality [146]. It is estimated that 90% of cancers achieve this by reactivating the enzyme telomerase, that is minimally composed of a reverse transcriptase (TERT) and internal RNA component (TERC), which progressively adds TTAGGG repeats onto the ends of telomeres. The remaining 10% of cancers achieve replicative immortality by activating the alternative lengthening of telomeres (ALT) pathway [147].

The identification of cancer cells that maintain telomere length in the absence of telomerase activity led to the identification of the ALT pathway [148]. Cells with an active ALT mechanism are characterized by a number of unique cellular phenotypes. These phenotypes have not only served as surrogates for ALT activity, but have also been instrumental in developing our understanding of the ALT mechanism. One of the first observations made in ALT cells was that the telomeric DNA was incredibly dynamic, constantly undergoing rapid attrition and subsequent elongation. This extreme heterogeneity in telomere length was one of the first indications that telomeres in ALT cells might be elongated by a recombination-based mechanism [148]. Later experiments confirmed the role of recombination in maintaining ALT telomeres by demonstrating that a DNA tag incorporated into a single telomere was copied into other chromosome ends [149], highlighting the role of homology directed repair in promoting telomere elongation. These findings were later supported by studies demonstrating that the telomeres in ALT cells are recruited into nuclear foci, that in addition to the promyelocytic leukemia (PML) protein, contain a number of recombination and repair factors, including, but not limited to, RAD51, RAD52, RPA, BLM, and WRN [150–152]. These ALT-associated PML bodies (APBs) are also believed to be associated with the extrachromosomal telomeric repeat (ECTR) DNA generated in ALT cells. ECTR DNA exists in both linear and circular forms. A subset of these ECTR DNA are likely generated as by-products of recombination, and they may also perpetuate the ALT phenotype by functioning as templates for elongation [153, 154]. Together, these cellular phenotypes highlight the importance of recombination for telomere elongation in the ALT pathway.

Recent studies have demonstrated that the ALT mechanism promotes telomere elongation through a process resembling BIR [88, 144, 155, 156]. BIR, also called recombination-dependent replication, is a specific form of homology directed repair that is well known to be important for the repair of collapsed replication forks in bacteria and yeast [67, 157]. Using oncogene induced replication stress models, more recent evidence demonstrate that BIR, mediated by RAD52 dependent strand invasion and POLD3 dependent DNA synthesis, facilitates repair and restart of collapsed replication forks in mammalian cells [158, 159]. Likewise, inducing breaks at ALT telomeres leads to increased POLD3 mediated telomere elongation [155], indicating that ALT telomeres, in part, engage BIR as a mechanism of elongation.

Although it is unclear exactly how the recombinogenic state of ALT telomeres is established, telomere fragility is exacerbated in ALT cells by defects in nucleosome assembly, challenges to the replication machinery, and increased telomere dysfunction suggesting that chronic replication stress contributes to ALT activity. Here, we will review further evidence supporting the connection between telomere replication stress and the ALT mechanism.

5.1 Replication Stress Associated with the Alternative Lengthening of Telomeres

ALT telomeres maintain high levels of spontaneous DNA damage and harbor more fragile telomeres than syngeneic telomerase positive cell lines [88], suggesting that ALT telomeres may experience chronic replication stress. In support of this, ALT cells also show an accumulation of RPA at telomeric DNA [113, 133, 134]. RPA-coated ssDNA is not only an intermediate of replication and recombination, but also facilitates the recruitment of the DNA damage response kinase, ATR, suggesting that ATR may function to mitigate replication stress at ALT telomeres. Recently, studies have demonstrated that ALT cells require ATR to promote telomere elongation and subsequently maintain ALT activity [88, 113, 134]. The persistence of RPA foci in APBs and the reliance on ATR suggest that ALT telomeres are prone to replication stress.

The DNA annealing helicase SMARCAL1 (SWI/SNF Related, Matrix Associated, Actin Dependent Regulator of Chromatin, Subfamily A-like 1) is one of the most enriched proteins bound to sites of replication stress [135] and functions to promote replication restart and/or repair by mediating replication fork reversal [160–163]. More recently it was demonstrated that SMARCAL1 not only functions to remodel stalled replication forks at telomeric DNA, but is significantly enriched at ALT telomeres [164, 165]. In the absence of SMARCAL1 replication fork restart is compromised and stalled replication forks collapse into DSBs, promoting recombination (Fig. 2). In support of this, SMARCAL1 depletion was shown to lead to an accumulation of phosphorylated RPA at ALT telomeres, suggesting ALT telomeres are experiencing an increase in replication fork stalling. It was recently shown that a subset of glioblastoma cases that are wild type for both the TERT promoter and isocitrate dehydrogenase (IDH) gene contain somatic loss-of-function mutations in the SMARCAL1 gene, suggesting that cells lacking this replication stress response protein may be prone to breakage and subsequent recombination events at telomeric DNA [166]. The characterization of SMARCAL1 function at telomeric DNA coupled with the identification of SMARCAL1 mutations in ALT positive cancers further support the hypothesis that replication stress contributes to the ALT mechanism (Fig. 2).

Irreversibly stalled replication forks at telomeric DNA are cleaved into DNA DSBs and must be repaired to ensure telomere length, and consequently maintain genome stability. BIR is a common homology directed repair mechanism to repair single-ended DNA double strand breaks, the structure created by the collapsed and broken replication fork [66, 67, 157]. Recent studies have demonstrated that DNA breaks at ALT telomeres engaged homology directed repair via the BIR mechanism [88, 155, 156]. BIR at ALT telomeres is mediated by the recombination proteins RAD51 and RAD52 that facilitate the search and capture, or telomere clustering, of a homologous telomeric DNA sequence for repair [53, 88, 155, 167]. Forcing replication fork collapse by depleting SMARCAL1 in ALT cells promotes telomere clustering [164], highlighting the requirement for SMARCAL1 in the restart of stalled replication forks. Following search and capture, ALT telomeres are bound by PCNA, DNA polymerase η [105], and the POLD3 subunit of DNA polymerase δ, the putative BIR polymerase to facilitate DNA damage repair and ultimately, the elongation of telomeric DNA [155]. (Fig. 2) Given that BIR is frequently engaged following the collapse of a stalled replication fork into a single-ended DNA DSB along with the role of BIR in ALT

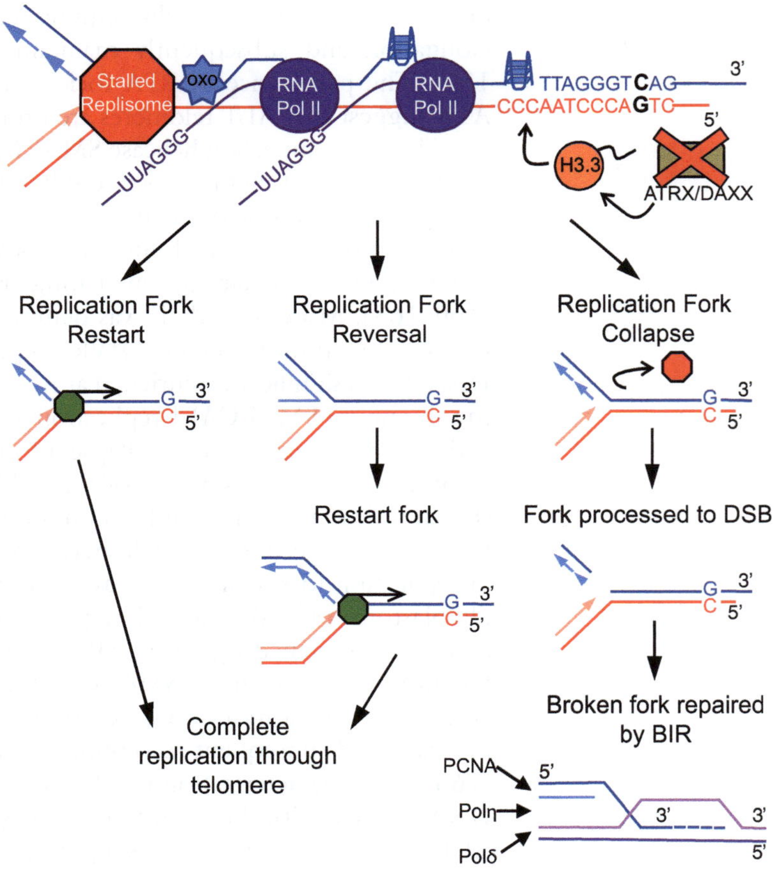

Fig. 2 Chronic replication stress at ALT telomeres. ALT telomeres have deregulated TERRA expression, incorporation of variant repeats, altered chromatin environments due to frequent loss of ATRX/DAXX/H3.3 and can be excessively long (not shown), compounding the replication stress normally present at telomeres. Stalled replication forks (stalled replisome indicated by red octagon) at ALT telomeres can be processed in several ways. The replication fork can be restarted (restarted replisome indicated by green octagon) by removing the source of replication stress, such as cleaving an R-loop using RNaseH1, or unwinding a G4 using BLM/WRN/RTEL1. Replication forks can be reversed with the help of reannealing helicase SMARCAL1, thus protecting the replication fork and aiding in fork restart. However, if the fork stalls irreversibly, the replisome dissociates and the replication fork collapses. A collapsed fork can then be processed into a DNA double-strand break. The broken fork can engage BIR, utilizing PCNA, Polη, and Polδ to synthesize new DNA, using homologous DNA, such as another telomere or ECTR, as a template. This repair process results in completion of replication and also the elongation of the broken telomere

telomere elongation, it is reasonable to propose chronic replication stress, and collapsed replication forks as the source of DNA damage that promotes telomere recombination in ALT cells. Supporting this hypothesis, inducing replication stress at ALT telomeres ultimately leads to telomere elongation [88, 155, 167].

Induced damage or exacerbated replication stress at ALT telomeres increases the abundance of a subset of ECTR that are C-rich and partially single stranded (C-circles) [164, 167]. C-circles are unique to ALT cells and are arguably the most specific marker of ALT activity [168]. It was recently demonstrated that elongation of ALT telomeres happens preferentially at lagging strand overhangs, suggesting that C-circles are produced as a replication dependent by-product of recombination [169], further supporting a connection between replication stress and the ALT mechanism.

5.2 Specific Sources of Replication Stress at ALT Telomeres

5.2.1 Distinct Telomere Structure and Sequence

In addition to the general challenges associated with telomere replication, ALT telomeres are susceptible to several additional sources of replication stress (Fig. 2). Although telomere length is variable across cells and/or tissue, ALT telomeres can reach up to 50 kilobases in length. Long telomeres are inherently more susceptible to replication stress [110, 170, 171], making ALT telomeres predisposed to frequent replication fork stalling. In addition, ALT telomeres contain both canonical and degenerate, or variant, repeats. Variant repeats are noncanonical telomeric sequences, including TCAGGG (C-type) or TGAGGG (G-type), interspersed within the canonical TTAGGG telomeric repeats [172–174]. These variant repeats spread stochastically throughout telomeres in ALT cells through homology directed repair. Variant repeats bind shelterin component TRF2 with five- to sixfold less affinity than canonical repeats [173], contributing to telomere deprotection and ultimately, inducing telomeric DNA damage. This was supported by studies demonstrating that the introduction of variant repeats leads to an elevation in telomere dysfunction-induced foci (TIFs), and decreased TRF2 binding [173]. This moderate deprotection increases replication stress, as measured by an increase in telomere fragility, and contributes to DNA damage at ALT telomeres [70, 95, 172, 173]. Variant repeats bind nuclear receptors COUP-TF2 and TF4 with high affinity [173, 175], which can recruit the nucleosome remodeling and histone deacetylation (NuRD) complex, decompacting chromatin at ALT telomeres [175]. The combination of decreased shelterin binding and decompacted chromatin at ALT telomeres creates an environment that perpetuates chronic replication stress and recombination. Although decreased shelterin binding is a consequence of the incorporation of variant repeats, it does not appear sufficient to activate ALT activity, as incorporation of variant repeats in a telomerase positive cell line causes some ALT-associated characteristics, but does not cause detectable recombination events between telomeres [173].

5.2.2 Changes in Chromatin

The chromatin environment at ALT telomeres has also been hypothesized to contribute to the replication stress and telomere fragility in ALT cells. Specifically, defects and genetic mutations in the chromatin remodeling complex ATRX (α-thalassemia/mental retardation syndrome X-linked)/DAXX (death-domain associated protein) and histone variant H3.3 are highly correlated with ALT positive cancers [176, 177]. ATRX is a chromatin-remodeling enzyme that is part of the SWI/SNF2 family. In conjunction with DAXX, ATRX functions to deposit histone variant H3.3 into heterochromatic regions, including the telomere. Loss of ATRX in non-ALT cells renders cells susceptible to replication stress, especially G4 stabilizing agents, leading to replication fork stalling and collapse [178–180]. Ectopic expression of ATRX in ALT cells causes a decrease in ALT activity and ALT associated phenotypes [181, 182]. Moreover, the suppression of ALT activity by expression of ATRX relies on functional DAXX, suggesting that ATRX and DAXX together suppress recombination at ALT telomeres [181]. However, to date, loss of ATRX or DAXX alone has not been demonstrated to fully activate the ALT mechanism [182]. Finally, loss of the histone chaperone ASF1 (anti-silencing factor 1) in mammalian cells, which promotes histone transfer during replication, leads to the induction of ALT phenotypes [134], highlighting changes in the regulation of telomeric chromatin as a source of replication perturbations and subsequent replication stress in ALT cells.

5.2.3 Telomere Repeat Containing RNA (TERRA)

Changes in the chromatin environment and ATRX inactivation have also been linked with the deregulation of the telomere transcript, TERRA [113, 183]. TERRA is cell cycle regulated, with levels decreasing during late S phase before reaccumulating in the next cell cycle [114–116]. The cell cycle regulation of TERRA promotes a switch between telomere replication and end-protection by facilitating the exchange between RPA and POT1 [114]. The decrease, and subsequent reaccumulation of TERRA promotes the displacement of RPA following S-phase and reloading of POT1 onto telomeres to suppress the DNA damage response and facilitate telomere end-protection. When cell cycle regulation of TERRA is lost, RPA remains bound to single stranded telomeric DNA [113]. RPA coated ssDNA recruits and activates ATR [52], and the RAD51 recombinase that initiates the search and capture of a homologous template [167], thus promoting a DNA damage response and recombinogenic state at ALT telomeres.

The levels of TERRA are generally increased in ALT cells compared to telomerase positive cells, and at least in a subset of ALT cells, cell cycle regulation of TERRA is lost [113, 184, 185]. This loss of TERRA cell cycle regulation has the potential to induce replication stress at ALT telomeres by increasing telomeric R-loops and altering telomere end protection. In support of this hypothesis, TERRA R-loops have been demonstrated at mammalian and yeast telomeres [117–119, 184]. Moreover, deregulation of TERRA has been shown to increase R-loops at telomeric DNA [117]. In contrast, overexpression of the enzyme RNAseH1, which degrades R-loops, causes a decrease in TERRA R-loop formation and a subsequent decrease in recombination at ALT telomeres [117]. Given that TERRA deregulation in ALT cells is likely caused by an increase in transcription, not a change in stability [185], it can be reasoned that the increased level of TERRA transcription in ALT cells causes an increase in R-loops. Unresolved TERRA R-loops lead to an increase in replication stress and an increase in recombination at ALT telomeres [117–119]. While RNaseH1 is known to degrade TERRA R-loops, it has yet to be investigated whether other R-loop resolving enzymes are involved in the resolution of TERRA R-loops and/or whether there are any defects in R-loop resolution specific to ALT cells.

6 Methods to Measure Replication Stress at Telomeres

There are several techniques that are commonly used to measure replication stress at mammalian telomeres. Measuring the phosphorylation of unique protein targets can be used as a surrogate for ATR activity and ultimately, an indicator of potential replication stress. First, phosphorylation of the histone variant H2AX (γH2AX) can be measured either by Western blot or by immunofluorescence [132]. Using immunofluorescence for γH2AX combined with immunofluorescence for shelterin components, or fluorescence in situ hybridization (FISH) for telomeric repeats, DNA damage, including replication stress, can be analyzed specifically at the telomeres. While H2AX is phosphorylated by ATR in response to replication stress, it can also be phosphorylated by the DNA damage response kinases ATM or DNA-PK in response to other types of DNA damage, making it a less specific marker of replication stress and more indicative of general DNA damage [132]. Second, RPA [54] and CHK1 [132] are more specific targets of ATR, and RPA phosphorylated at serine 33 and CHK1 phosphorylated at serine 345 and serine 317 can both be measured as markers of ATR activation. Third, the accumulation of ssDNA in response to polymerase-helicase uncoupling can be directly measured by the incorporation of the thymidine analog BrdU. When added directly to cell culture media and incorporated into DNA, ssDNA containing BrdU can be visualized by native immunofluorescence using a BrdU specific antibody. This method can be combined with immunofluorescence for shelterin components to specifically detect ssDNA at the telomeres. These three methods can be used as indirect readouts of replication stress. However, Single-Molecule Analysis of Replicating DNA (SMARD) has emerged as a more direct measure of the stalling or slowing of the replication fork during DNA synthesis [55, 186]. Similar to BrdU incorporation assays, SMARD relies on the incorporation of halogenated nucleotide analogs into actively replicating DNA. However, in SMARD, the genomic DNA is then extracted from cells and stretched across coverslips, or microscope slides, before detecting the incorporated nucleotide analogs by immunofluorescence. Following immunodetection, the stretched fibers can then be analyzed using standard microscopy techniques to determine the kinetics of DNA replication. More recently, this technique has been combined with telomere FISH to measure replication stress directly at the telomeres [31, 63]. This provides a powerful direct technique to measure replication fork slowing or stalling at distinct genomic loci, including telomeric DNA.

7 Concluding Remarks

DNA replication is a remarkably complex and organized process that cells must fully and faithfully complete to ensure continued cellular proliferation. While cells have evolved to accomplish this function routinely, DNA replication does not always proceed unchallenged, and cells must overcome various obstacles to replication to completely duplicate the genome. Telomeres present unique challenges to DNA replication, due to their repetitive sequence, unique secondary structures, and the propensity for collision between the transcription program and replication machinery. Understanding how highly repetitive and structurally complex sequences, like telomeric DNA, are faithfully replicated has been the focus of much research. Defects in DNA replication, or in the replication stress response, have the potential to cause DNA damage at the telomeres and consequently threaten genome stability. Additionally, in some contexts chronic replication stress at the telomeres may promote tumorigenesis by facilitating telomere elongation through the ALT pathway. Together, these studies highlight the need for additional research investigating the molecular, genetic, and epigenetic facets of telomere replication stress, and how it is managed to protect genome stability.

References

1. Griffith JD, Comeau L, Rosenfield S et al (1999) Mammalian telomeres end in a large duplex loop. Cell 97:503–514. https://doi.org/10.1016/S0092-8674(00)80760-6
2. Palm W, de Lange T (2008) How shelterin protects mammalian telomeres. Annu Rev Genet 42:301–334. https://doi.org/10.1146/annurev.genet.41.110306.130350
3. Levy MZ, Allsopp RC, Futcher AB et al (1992) Telomere end-replication problem and cell aging. J Mol Biol 225:951–960
4. Prioleau M-N, MacAlpine DM (2016) DNA replication origins-where do we begin? Genes Dev 30:1683–1697. https://doi.org/10.1101/gad.285114.116

5. Donovan S, Harwood J, Drury LS, Diffley JF (1997) Cdc6p-dependent loading of Mcm proteins onto pre-replicative chromatin in budding yeast. Proc Natl Acad Sci U S A 94:5611–5616
6. Bell SP, Dutta A (2002) DNA replication in eukaryotic cells. Annu Rev Biochem 71:333–374. https://doi.org/10.1146/annurev.biochem.71.110601.135425
7. Diffley JF, Cocker JH, Dowell SJ, Rowley A (1994) Two steps in the assembly of complexes at yeast replication origins in vivo. Cell 78:303–316
8. Cocker JH, Piatti S, Santocanale C et al (1996) An essential role for the Cdc6 protein in forming the pre-replicative complexes of budding yeast. Nature 379:180–182. https://doi.org/10.1038/379180a0
9. Speck C, Chen Z, Li H, Stillman B (2005) ATPase-dependent cooperative binding of ORC and Cdc6 to origin DNA. Nat Struct Mol Biol 12:965–971. https://doi.org/10.1038/nsmb1002
10. Chen S, de Vries MA, Bell SP (2007) Orc6 is required for dynamic recruitment of Cdt1 during repeated Mcm2-7 loading. Genes Dev 21:2897–2907. https://doi.org/10.1101/gad.1596807
11. Bell SP, Stillman B (1992) ATP-dependent recognition of eukaryotic origins of DNA replication by a multiprotein complex. Nature 357:128–134. https://doi.org/10.1038/357128a0
12. Evrin C, Clarke P, Zech J et al (2009) A double-hexameric MCM2-7 complex is loaded onto origin DNA during licensing of eukaryotic DNA replication. Proc Natl Acad Sci 106:20240–20245. https://doi.org/10.1073/pnas.0911500106
13. Arias EE, Walter JC (2007) Strength in numbers: preventing rereplication via multiple mechanisms in eukaryotic cells. Genes Dev 21:497–518. https://doi.org/10.1101/gad.1508907
14. Tada S, Li A, Maiorano D et al (2001) Repression of origin assembly in metaphase depends on inhibition of RLF-B/Cdt1 by geminin. Nat Cell Biol 3:107–113. https://doi.org/10.1038/35055000
15. Wohlschlegel JA, Dwyer BT, Dhar SK et al (2000) Inhibition of eukaryotic DNA replication by geminin binding to Cdt1. Science 290:2309–2312. https://doi.org/10.1126/science.290.5500.2309
16. Nishitani H, Sugimoto N, Roukos V et al (2006) Two E3 ubiquitin ligases, SCF-Skp2 and DDB1-Cul4, target human Cdt1 for proteolysis. EMBO J 25:1126–1136. https://doi.org/10.1038/sj.emboj.7601002
17. Li X, Zhao Q, Liao R et al (2003) The SCF (Skp2) ubiquitin ligase complex interacts with the human replication licensing factor Cdt1 and regulates Cdt1 degradation. J Biol Chem 278:30854–30858. https://doi.org/10.1074/jbc.C300251200
18. Stillman B (2005) Origin recognition and the chromosome cycle. FEBS Lett 579:877–884. https://doi.org/10.1016/j.febslet.2004.12.011
19. Ilves I, Petojevic T, Pesavento JJ, Botchan MR (2010) Activation of the MCM2-7 helicase by association with Cdc45 and GINS proteins. Mol Cell 37:247–258. https://doi.org/10.1016/j.molcel.2009.12.030
20. Parker MW, Botchan MR, Berger JM (2017) Mechanisms and regulation of DNA replication initiation in eukaryotes. Crit Rev Biochem Mol Biol 52:107–144. https://doi.org/10.1080/10409238.2016.1274717
21. Yeeles JTP, Deegan TD, Janska A et al (2015) Regulated eukaryotic DNA replication origin firing with purified proteins. Nature 519:431–435. https://doi.org/10.1038/nature14285
22. Rhind N, Gilbert DM (2013) DNA replication timing. Cold Spring Harb Perspect Biol 5:a010132. https://doi.org/10.1101/cshperspect.a010132
23. Edwards MC, Tutter AV, Cvetic C et al (2002) MCM2-7 complexes bind chromatin in a distributed pattern surrounding the origin recognition complex in Xenopus egg extracts. J Biol Chem 277:33049–33057. https://doi.org/10.1074/jbc.M204438200
24. Ge XQ, Jackson DA, Blow JJ (2007) Dormant origins licensed by excess Mcm2-7 are required for human cells to survive replicative stress. Genes Dev 21:3331–3341. https://doi.org/10.1101/gad.457807
25. Woodward AM, Göhler T, Luciani MG et al (2006) Excess Mcm2-7 license dormant origins of replication that can be used under conditions of replicative stress. J Cell Biol 173:673–683. https://doi.org/10.1083/jcb.200602108
26. Masai H, Matsumoto S, You Z et al (2010) Eukaryotic chromosome DNA replication: where, when, and how? Annu Rev Biochem 79:89–130. https://doi.org/10.1146/annurev.biochem.052308.103205
27. Toledo L, Neelsen KJ, Lukas J (2017) Replication catastrophe: when a checkpoint fails because of exhaustion. Mol Cell

66:735–749. https://doi.org/10.1016/j.molcel.2017.05.001

28. Deng SK, Chen H, Symington LS (2015) Replication protein A prevents promiscuous annealing between short sequence homologies: implications for genome integrity. BioEssays 37:305–313. https://doi.org/10.1002/bies.201400161
29. Zou Y, Liu Y, Wu X, Shell SM (2006) Functions of human replication protein A (RPA): from DNA replication to DNA damage and stress responses. J Cell Physiol 208:267–273. https://doi.org/10.1002/jcp.20622
30. Burgers PMJ, Kunkel TA (2017) Eukaryotic DNA replication fork. Annu Rev Biochem 86:417–438. https://doi.org/10.1146/annurev-biochem-061516-044709
31. Drosopoulos WC, Kosiyatrakul ST, Yan Z et al (2012) Human telomeres replicate using chromosome-specific, rather than universal, replication programs. J Cell Biol 197:253–266. https://doi.org/10.1083/jcb.201112083
32. Kurth I, Gautier J (2010) Origin-dependent initiation of DNA replication within telomeric sequences. Nucleic Acids Res 38:467–476. https://doi.org/10.1093/nar/gkp929
33. Martínez P, Blasco MA (2015) Replicating through telomeres: a means to an end. Trends Biochem Sci 40:504–515
34. Huang C, Dai X, Chai W (2012) Human Stn1 protects telomere integrity by promoting efficient lagging-strand synthesis at telomeres and mediating C-strand fill-in. Cell Res 22:1681–1695. https://doi.org/10.1038/cr.2012.132
35. Wu P, Takai H, de Lange T (2012) Telomeric 3′ overhangs derive from resection by Exo1 and Apollo and fill-in by POT1b-associated CST. Cell 150:39–52. https://doi.org/10.1016/j.cell.2012.05.026
36. Wang F, Stewart JA, Kasbek C et al (2012) Human CST has independent functions during telomere duplex replication and C-strand fill-in. Cell Rep 2:1096–1103. https://doi.org/10.1016/j.celrep.2012.10.007
37. Stewart JA, Wang F, Chaiken MF et al (2012) Human CST promotes telomere duplex replication and general replication restart after fork stalling. EMBO J 31:3537–3549. https://doi.org/10.1038/emboj.2012.215
38. Durkin SG, Glover TW (2007) Chromosome fragile sites. Annu Rev Genet 41:169–192. https://doi.org/10.1146/annurev.genet.41.042007.165900
39. Sutherland GR (1979) Heritable fragile sites on human chromosomes I. Factors affecting expression in lymphocyte culture. Am J Hum Genet 31:125–135
40. Sutherland GR (1977) Fragile sites on human chromosomes: demonstration of their dependence on the type of tissue culture medium. Science 197:265–266
41. Glover TW, Berger C, Coyle J, Echo B (1984) DNA polymerase alpha inhibition by aphidicolin induces gaps and breaks at common fragile sites in human chromosomes. Hum Genet 67:136–142
42. Glover TW, Stein CK (1987) Induction of sister chromatid exchanges at common fragile sites. Am J Hum Genet 41:882–890
43. Feichtinger W, Schmid M (1989) Increased frequencies of sister chromatid exchanges at common fragile sites (1)(q42) and (19)(q13). Hum Genet 83:145–147
44. Hecht F, Sutherland GR (1984) Fragile sites and cancer breakpoints. Cancer Genet Cytogenet 12:179–181. https://doi.org/10.1016/0165-4608(84)90132-8
45. Zeman MK, Cimprich KA (2014) Causes and consequences of replication stress. Nat Cell Biol 16:2–9. https://doi.org/10.1038/ncb2897
46. Bermejo R, Lai MS, Foiani M (2012) Preventing replication stress to maintain genome stability: resolving conflicts between replication and transcription. Mol Cell 45:710–718. https://doi.org/10.1016/j.molcel.2012.03.001
47. Helmrich A, Ballarino M, Nudler E, Tora L (2013) Transcription-replication encounters, consequences and genomic instability. Nat Struct Mol Biol 20:412–418. https://doi.org/10.1038/nsmb.2543
48. Aguilera A, García-Muse T (2012) R loops: from transcription byproducts to threats to genome stability. Mol Cell 46:115–124. https://doi.org/10.1016/j.molcel.2012.04.009
49. Bester AC, Roniger M, Oren YS et al (2011) Nucleotide deficiency promotes genomic instability in early stages of cancer development. Cell 145:435–446. https://doi.org/10.1016/j.cell.2011.03.044
50. Byun TS, Pacek M, Yee M et al (2005) Functional uncoupling of MCM helicase and DNA polymerase activities activates the ATR-dependent checkpoint. Genes Dev 19:1040–1052. https://doi.org/10.1101/gad.1301205
51. Toledo LI, Altmeyer M, Rask MB et al (2014) ATR prohibits replication catastrophe by preventing global exhaustion of RPA. Cell 156:374

52. Zou L, Elledge SJ (2003) Sensing DNA damage through ATRIP recognition of RPA-ssDNA complexes. Science 300:1542–1548. https://doi.org/10.1126/science.1083430
53. Ward IM, Minn K, Chen J (2004) UV-induced ataxia-telangiectasia-mutated and Rad3-related (ATR) activation requires replication stress. J Biol Chem 279:9677–9680. https://doi.org/10.1074/jbc.C300554200
54. Vassin VM, Anantha RW, Sokolova E et al (2009) Human RPA phosphorylation by ATR stimulates DNA synthesis and prevents ssDNA accumulation during DNA-replication stress. J Cell Sci 122:4070–4080. https://doi.org/10.1242/jcs.053702
55. Bianco JN, Poli J, Saksouk J et al (2012) Analysis of DNA replication profiles in budding yeast and mammalian cells using DNA combing. Methods 57:149–157. https://doi.org/10.1016/j.ymeth.2012.04.007
56. Nam EA, Cortez D (2011) ATR signalling: more than meeting at the fork. Biochem J 436:527–536. https://doi.org/10.1042/BJ20102162
57. Cimprich KA, Cortez D (2008) ATR: an essential regulator of genome integrity. Nat Rev Mol Cell Biol 9:616–627. https://doi.org/10.1038/nrm2450
58. Elvers I, Johansson F, Groth P et al (2011) UV stalled replication forks restart by re-priming in human fibroblasts. Nucleic Acids Res 39:7049–7057. https://doi.org/10.1093/nar/gkr420
59. Neelsen KJ, Lopes M (2015) Replication fork reversal in eukaryotes: from dead end to dynamic response. Nat Rev Mol Cell Biol 16:207–220
60. McIntosh D, Blow JJ (2012) Dormant origins, the licensing checkpoint, and the response to replicative stresses. Cold Spring Harb Perspect Biol 4:a012955–a012955. https://doi.org/10.1101/cshperspect.a012955
61. Rybak P, Waligórska A, Bujnowicz Ł et al (2015) Activation of new replication foci under conditions of replication stress. Cell Cycle 14:2634–2647. https://doi.org/10.1080/15384101.2015.1064566
62. Ibarra A, Schwob E, Méndez J (2008) Excess MCM proteins protect human cells from replicative stress by licensing backup origins of replication. Proc Natl Acad Sci 105:8956–8961. https://doi.org/10.1073/pnas.0803978105
63. Drosopoulos WC, Kosiyatrakul ST, Schildkraut CL (2015) BLM helicase facilitates telomere replication during leading strand synthesis of telomeres. J Cell Biol 210:191–208. https://doi.org/10.1083/jcb.201410061
64. Hanada K, Budzowska M, Davies SL et al (2007) The structure-specific endonuclease Mus81 contributes to replication restart by generating double-strand DNA breaks. Nat Struct Mol Biol 14:1096–1104. https://doi.org/10.1038/nsmb1313
65. Petermann E, Orta ML, Issaeva N et al (2010) Hydroxyurea-stalled replication forks become progressively inactivated and require two different RAD51-mediated pathways for restart and repair. Mol Cell 37:492–502. https://doi.org/10.1016/j.molcel.2010.01.021
66. Mehta A, Haber JE (2014) Sources of DNA double-strand breaks and models of recombinational DNA repair. Cold Spring Harb Perspect Biol 6:a016428. https://doi.org/10.1101/cshperspect.a016428
67. Llorente B, Smith CE, Symington LS (2008) Break-induced replication: what is it and what is it for? Cell Cycle 7:859–864. https://doi.org/10.4161/cc.7.7.5613
68. Miller KM, Rog O, Cooper JP (2006) Semi-conservative DNA replication through telomeres requires Taz1. Nature 440:824–828. https://doi.org/10.1038/nature04638
69. Makovets S, Herskowitz I, Blackburn EH (2004) Anatomy and dynamics of DNA replication fork movement in yeast telomeric regions. Mol Cell Biol 24:4019–4031. https://doi.org/10.1128/MCB.24.9.4019-4031.2004
70. Sfeir A, Kosiyatrakul ST, Hockemeyer D et al (2009) Mammalian telomeres resemble fragile sites and require TRF1 for efficient replication. Cell 138:90–103. https://doi.org/10.1016/j.cell.2009.06.021
71. Ohki R, Ishikawa F (2004) Telomere-bound TRF1 and TRF2 stall the replication fork at telomeric repeats. Nucleic Acids Res 32:1627–1637. https://doi.org/10.1093/nar/gkh309
72. Pinzaru AM, Hom RA, Beal A et al (2016) Telomere replication stress induced by POT1 inactivation accelerates tumorigenesis. Cell Rep 15:2170–2184. https://doi.org/10.1016/j.celrep.2016.05.008
73. Ye J, Lenain C, Bauwens S et al (2010) TRF2 and Apollo cooperate with topoisomerase 2α to protect human telomeres from

replicative damage. Cell 142:230–242. https://doi.org/10.1016/j.cell.2010.05.032

74. Martínez P, Thanasoula M, Muñoz P et al (2009) Increased telomere fragility and fusions resulting from TRF1 deficiency lead to degenerative pathologies and increased cancer in mice. Genes Dev 23:2060–2075. https://doi.org/10.1101/gad.543509
75. Hancock JM (1996) Simple sequences and the expanding genome. BioEssays 18:421–425. https://doi.org/10.1002/bies.950180512
76. Li Y-C, Korol AB, Fahima T et al (2002) Microsatellites: genomic distribution, putative functions and mutational mechanisms: a review. Mol Ecol 11:2453–2465
77. Fouché N, Ozgür S, Roy D, Griffith JD (2006) Replication fork regression in repetitive DNAs. Nucleic Acids Res 34:6044–6050. https://doi.org/10.1093/nar/gkl757
78. Jurk D, Wilson C, Passos JF et al (2014) Chronic inflammation induces telomere dysfunction and accelerates ageing in mice. Nat Commun 2:4172. https://doi.org/10.1038/ncomms5172
79. von Zglinicki T (2002) Oxidative stress shortens telomeres. Trends Biochem Sci 27:339–344. https://doi.org/10.1016/S0968-0004(02)02110-2
80. van Loon B, Markkanen E, Hübscher U (2010) Oxygen as a friend and enemy: how to combat the mutational potential of 8-oxo-guanine. DNA Repair (Amst) 9:604–616. https://doi.org/10.1016/j.dnarep.2010.03.004
81. Opresko PL, Fan J, Danzy S et al (2005) Oxidative damage in telomeric DNA disrupts recognition by TRF1 and TRF2. Nucleic Acids Res 33:1230–1239. https://doi.org/10.1093/nar/gki273
82. Bruner SD, Norman DPG, Verdine GL (2000) Structural basis for recognition and repair of the endogenous mutagen 8-oxoguanine in DNA. Nature 403:859–866. https://doi.org/10.1038/35002510
83. Sen D, Gilbert W (1988) Formation of parallel four-stranded complexes by guanine-rich motifs in DNA and its implications for meiosis. Nature 334:364–366. https://doi.org/10.1038/334364a0
84. Tarsounas M, Tijsterman M (2013) Genomes and G-quadruplexes: for better or for worse. J Mol Biol 425:4782–4789. https://doi.org/10.1016/J.JMB.2013.09.026
85. Sundquist WI, Klug A (1989) Telomeric DNA dimerizes by formation of guanine tetrads between hairpin loops. Nature 342:825–829. https://doi.org/10.1038/342825a0
86. Rodriguez R, Miller KM, Forment JV et al (2012) Small-molecule-induced DNA damage identifies alternative DNA structures in human genes. Nat Chem Biol 8:301–310. https://doi.org/10.1038/nchembio.780
87. Rizzo A, Salvati E, Porru M et al (2009) Stabilization of quadruplex DNA perturbs telomere replication leading to the activation of an ATR-dependent ATM signaling pathway. Nucleic Acids Res 37:5353–5364. https://doi.org/10.1093/nar/gkp582
88. Min J, Wright WE, Shay JW (2017) Alternative lengthening of telomeres mediated by mitotic DNA synthesis engages break-induced replication processes. Mol Cell Biol 37:e00226–e00217. https://doi.org/10.1128/MCB.00226-17
89. Salvati E, Scarsella M, Porru M et al (2010) PARP1 is activated at telomeres upon G4 stabilization: possible target for telomere-based therapy. Oncogene 29:6280–6293. https://doi.org/10.1038/onc.2010.344
90. Mohaghegh P, Karow JK, Brosh RM et al (2001) The Bloom's and Werner's syndrome proteins are DNA structure-specific helicases. Nucleic Acids Res 29:2843–2849
91. Sidorova JM, Li N, Folch A, Monnat RJ Jr (2008) The RecQ helicase WRN is required for normal replication fork progression after DNA damage or replication fork arrest. Cell Cycle 7:796–807. https://doi.org/10.4161/cc.7.6.5566
92. Crabbe L, Verdun RE, Haggblom CI, Karlseder J (2004) Defective telomere lagging strand synthesis in cells lacking WRN helicase activity. Science 306:1951–1953. https://doi.org/10.1126/science.1103619
93. Damerla RR, Knickelbein KE, Strutt S et al (2012) Werner syndrome protein suppresses the formation of large deletions during the replication of human telomeric sequences. Cell Cycle 11:3036–3044. https://doi.org/10.4161/cc.21399
94. Opresko PL, von Kobbe C, Laine J-P et al (2002) Telomere-binding protein TRF2 binds to and stimulates the Werner and Bloom syndrome helicases. J Biol Chem 277:41110–41119. https://doi.org/10.1074/jbc.M205396200
95. Zimmermann M, Kibe T, Kabir S, de Lange T (2014) TRF1 negotiates TTAGGG repeat-associated replication problems by recruiting

the BLM helicase and the TPP1/POT1 repressor of ATR signaling. Genes Dev 28:2477–2491. https://doi.org/10.1101/gad.251611.114

96. Vannier J-B, Pavicic-Kaltenbrunner V, Petalcorin MIR et al (2012) RTEL1 dismantles T loops and counteracts telomeric G4-DNA to maintain telomere integrity. Cell 149:795–806. https://doi.org/10.1016/J.CELL.2012.03.030
97. Vannier J-B, Sandhu S, Petalcorin MI et al (2013). RTEL1 is a replisome-associated helicase that promotes telomere and genome-wide replication) Science 342:239–242. https://doi.org/10.1126/science.1241779
98. Sanders CM (2010) Human Pif1 helicase is a G-quadruplex DNA-binding protein with G-quadruplex DNA-unwinding activity. Biochem J 430:119–128. https://doi.org/10.1042/BJ20100612
99. Paeschke K, Bochman ML, Garcia PD et al (2013) Pif1 family helicases suppress genome instability at G-quadruplex motifs. Nature 497:458–462. https://doi.org/10.1038/nature12149
100. Snow BE, Mateyak M, Paderova J et al (2007) Murine Pif1 interacts with telomerase and is dispensable for telomere function in vivo. Mol Cell Biol 27:1017–1026. https://doi.org/10.1128/MCB.01866-06
101. Ribeyre C, Lopes J, Boulé J-B et al (2009) The yeast Pif1 helicase prevents genomic instability caused by G-quadruplex-forming CEB1 sequences in vivo. PLoS Genet 5: e1000475. https://doi.org/10.1371/journal.pgen.1000475
102. Wu Y, Shin-ya K, Brosh RM (2008) FANCJ helicase defective in Fanconia anemia and breast cancer unwinds G-quadruplex DNA to defend genomic stability. Mol Cell Biol 28:4116–4128. https://doi.org/10.1128/MCB.02210-07
103. Castillo Bosch P, Segura-Bayona S, Koole W et al (2014) FANCJ promotes DNA synthesis through G-quadruplex structures. EMBO J 33:2521–2533. https://doi.org/10.15252/embj.201488663
104. Déjardin J, Kingston RE (2009) Purification of proteins associated with specific genomic Loci. Cell 136:175–186. https://doi.org/10.1016/j.cell.2008.11.045
105. Garcia-Exposito L, Bournique E, Bergoglio V et al (2016) Proteomic profiling reveals a specific role for translesion DNA polymerase η in the alternative lengthening of telomeres. Cell Rep 17:1858–1871. https://doi.org/10.1016/j.celrep.2016.10.048
106. Bhattacharjee A, Wang Y, Diao J, Price CM (2017) Dynamic DNA binding, junction recognition and G4 melting activity underlie the telomeric and genome-wide roles of human CST. Nucleic Acids Res 45:12311–12324. https://doi.org/10.1093/nar/gkx878
107. Sarek G, Vannier J-B, Panier S et al (2015) TRF2 recruits RTEL1 to telomeres in S phase to promote T-loop unwinding. Mol Cell 57:622–635. https://doi.org/10.1016/j.molcel.2014.12.024
108. Jahn A, Rane G, Paszkowski-Rogacz M et al (2017) ZBTB48 is both a vertebrate telomere-binding protein and a transcriptional activator. EMBO Rep 18:929–946. https://doi.org/10.15252/embr.201744095
109. Li JSZ, Miralles Fusté J, Simavorian T et al (2017) TZAP: a telomere-associated protein involved in telomere length control. Science 355:638–641. https://doi.org/10.1126/science.aah6752
110. Rivera T, Haggblom C, Cosconati S, Karlseder J (2017) A balance between elongation and trimming regulates telomere stability in stem cells. Nat Struct Mol Biol 24:30–39. https://doi.org/10.1038/nsmb.3335
111. Azzalin CM, Reichenbach P, Khoriauli L et al (2007) Telomeric repeat containing RNA and RNA surveillance factors at mammalian chromosome ends. Science 318:798–801. https://doi.org/10.1126/science.1147182
112. Schoeftner S, Blasco MA (2008) Developmentally regulated transcription of mammalian telomeres by DNA-dependent RNA polymerase II. Nat Cell Biol 10:228–236. https://doi.org/10.1038/ncb1685
113. Flynn RL, Cox KE, Jeitany M et al (2015) Alternative lengthening of telomeres renders cancer cells hypersensitive to ATR inhibitors. Science 347:273–277. https://doi.org/10.1126/science.1257216
114. Flynn RL, Centore RC, O'Sullivan RJ et al (2011) TERRA and hnRNPA1 orchestrate an RPA-to-POT1 switch on telomeric single-stranded DNA. Nature 471:532–536. https://doi.org/10.1038/nature09772
115. Porro A, Feuerhahn S, Reichenbach P, Lingner J (2010) Molecular dissection of telomeric repeat-containing RNA biogenesis unveils the presence of distinct and multiple regulatory pathways. Mol Cell Biol 30:4808–4817. https://doi.org/10.1128/MCB.00460-10
116. Arnoult N, Van Beneden A, Decottignies A (2012) Telomere length regulates TERRA levels through increased trimethylation of

telomeric H3K9 and HP1α. Nat Struct Mol Biol 19:948–956. https://doi.org/10.1038/nsmb.2364

117. Arora R, Lee Y, Wischnewski H et al (2014) RNaseH1 regulates TERRA-telomeric DNA hybrids and telomere maintenance in ALT tumour cells. Nat Commun 5:5220. https://doi.org/10.1038/ncomms6220
118. Balk B, Maicher A, Dees M et al (2013) Telomeric RNA-DNA hybrids affect telomere-length dynamics and senescence. Nat Struct Mol Biol 20:1199–1205. https://doi.org/10.1038/nsmb.2662
119. Balk B, Dees M, Bender K, Luke B (2014) The differential processing of telomeres in response to increased telomeric transcription and RNA-DNA hybrid accumulation. RNA Biol 11:95–100. https://doi.org/10.4161/rna.27798
120. Costantino L, Koshland D (2015) The Yin and Yang of R-loop biology. Curr Opin Cell Biol 34:39–45
121. Santos-Pereira JM, Aguilera A (2015) R loops: new modulators of genome dynamics and function. Nat Rev Genet 16:583–597. https://doi.org/10.1038/nrg3961
122. Richard P, Manley JL (2017) R loops and links to human disease. J Mol Biol 429:3168–3180
123. Houlard M, Artus J, Léguillier T et al (2011) DNA-RNA hybrids contribute to the replication dependent genomic instability induced by Omcg1 deficiency. Cell Cycle 10:108–117. https://doi.org/10.4161/cc.10.1.14379
124. Neil AJ, Liang MU, Khristich AN et al (2018) RNA–DNA hybrids promote the expansion of Friedreich's ataxia (GAA)n repeats via break-induced replication. Nucleic Acids Res 46:3487–3497. https://doi.org/10.1093/nar/gky099
125. Sollier J, Cimprich KA (2015) Breaking bad: R-loops and genome integrity. Trends Cell Biol 25:514–522
126. Graf M, Bonetti D, Lockhart A et al (2017) Telomere length determines TERRA and R-loop regulation through the cell cycle. Cell 170:72–85.e14. https://doi.org/10.1016/j.cell.2017.06.006
127. De Amicis A, Piane M, Ferrari F et al (2011) Role of senataxin in DNA damage and telomeric stability. DNA Repair (Amst) 10:199–209. https://doi.org/10.1016/j.dnarep.2010.10.012
128. Skourti-Stathaki K, Proudfoot NJ, Gromak N (2011) Human senataxin resolves RNA/DNA hybrids formed at transcriptional pause sites to promote Xrn2-dependent termination. Mol Cell 42:794–805. https://doi.org/10.1016/j.molcel.2011.04.026
129. Hatchi E, Skourti-Stathaki K, Ventz S et al (2015) BRCA1 recruitment to transcriptional pause sites is required for R-loop-driven DNA damage repair. Mol Cell 57:636–647. https://doi.org/10.1016/j.molcel.2015.01.011
130. Hodroj D, Recolin B, Serhal K et al (2017) An ATR-dependent function for the Ddx19 RNA helicase in nuclear R-loop metabolism. EMBO J 36:1182–1198. https://doi.org/10.15252/embj.201695131
131. Mischo HE, Gómez-González B, Grzechnik P et al (2011) Yeast Sen1 helicase protects the genome from transcription-associated instability. Mol Cell 41:21–32. https://doi.org/10.1016/j.molcel.2010.12.007
132. Maréchal A, Zou L (2013) DNA damage sensing by the ATM and ATR kinases. Cold Spring Harb Perspect Biol 5:a012716. https://doi.org/10.1101/cshperspect.a012716
133. Grudic A, Jul-Larsen Å, Haring SJ et al (2007) Replication protein A prevents accumulation of single-stranded telomeric DNA in cells that use alternative lengthening of telomeres. Nucleic Acids Res 35:7267–7278. https://doi.org/10.1093/nar/gkm738
134. O'Sullivan RJ, Arnoult N, Lackner DH et al (2014) Rapid induction of alternative lengthening of telomeres by depletion of the histone chaperone ASF1. Nat Struct Mol Biol 21:167–174. https://doi.org/10.1038/nsmb.2754
135. Sirbu BM, McDonald WH, Dungrawala H et al (2013) Identification of proteins at active, stalled, and collapsed replication forks using isolation of proteins on nascent DNA (iPOND) coupled with mass spectrometry. J Biol Chem 288:31458–31467. https://doi.org/10.1074/jbc.M113.511337
136. Sirbu BM, Couch FB, Feigerle JT et al (2011) Analysis of protein dynamics at active, stalled, and collapsed replication forks. Genes Dev 25:1320–1327. https://doi.org/10.1101/gad.2053211
137. Lopes M, Foiani M, Sogo JM (2006) Multiple mechanisms control chromosome integrity after replication fork uncoupling and restart at irreparable UV lesions. Mol Cell 21:15–27. https://doi.org/10.1016/j.molcel.2005.11.015
138. Burrell RA, McClelland SE, Endesfelder D et al (2013) Replication stress links structural

and numerical cancer chromosomal instability. Nature 494:492–496. https://doi.org/10.1038/nature11935
139. Chan KL, Palmai-Pallag T, Ying S, Hickson ID (2009) Replication stress induces sister-chromatid bridging at fragile site loci in mitosis. Nat Cell Biol 11:753–760. https://doi.org/10.1038/ncb1882
140. Macheret M, Halazonetis TD (2015) DNA replication stress as a hallmark of cancer. Annu Rev Pathol 10:425–448. https://doi.org/10.1146/annurev-pathol-012414-040424
141. Barefield C, Karlseder J (2012) The BLM helicase contributes to telomere maintenance through processing of late-replicating intermediate structures. Nucleic Acids Res 40:7358–7367. https://doi.org/10.1093/nar/gks407
142. Gelot C, Magdalou I, Lopez B (2015) Replication stress in mammalian cells and its consequences for mitosis. Genes (Basel) 6:267–298. https://doi.org/10.3390/genes6020267
143. Naim V, Rosselli F (2009) The FANC pathway and BLM collaborate during mitosis to prevent micro-nucleation and chromosome abnormalities. Nat Cell Biol 11:761–768. https://doi.org/10.1038/ncb1883
144. Sobinoff AP, Allen JA, Neumann AA et al (2017) BLM and SLX4 play opposing roles in recombination-dependent replication at human telomeres. EMBO J 36:2907–2919. https://doi.org/10.15252/embj.201796889
145. Mankouri HW, Huttner D, Hickson ID (2013) How unfinished business from S-phase affects mitosis and beyond. EMBO J 32:2661–2671
146. Hanahan D, Weinberg RA (2011) Hallmarks of cancer: the next generation. Cell 144:646–674. https://doi.org/10.1016/j.cell.2011.02.013
147. Heaphy CM, Subhawong AP, Hong SM et al (2011) Prevalence of the alternative lengthening of telomeres telomere maintenance mechanism in human cancer subtypes. Am J Pathol 179:1608–1615. https://doi.org/10.1016/j.ajpath.2011.06.018
148. Bryan TM, Englezou A, Gupta J et al (1995) Telomere elongation in immortal human cells without detectable telomerase activity. EMBO J 14:4240–4248
149. Dunham MA, Neumann AA, Fasching CL, Reddel RR (2000) Telomere maintenance by recombination in human cells. Nat Genet 26:447–450. https://doi.org/10.1038/82586
150. Nabetani A, Yokoyama O, Ishikawa F (2004) Localization of hRad9, hHus1, hRad1, and hRad17 and caffeine-sensitive DNA replication at the alternative lengthening of telomeres-associated promyelocytic leukemia body. J Biol Chem 279:25849–25857. https://doi.org/10.1074/jbc.M312652200
151. Jiang W-Q, Zhong Z-H, Henson JD et al (2005) Suppression of alternative lengthening of telomeres by Sp100-mediated sequestration of the MRE11/RAD50/NBS1 complex. Mol Cell Biol 25:2708–2721. https://doi.org/10.1128/MCB.25.7.2708-2721.2005
152. Yeager TR, Neumann AA, Englezou A et al (1999) Telomerase-negative immortalized human cells contain a novel type of promyelocytic leukemia (PML) body. Cancer Res 59:4175–4179
153. Tokutake Y, Matsumoto T, Watanabe T et al (1998) Extra-chromosomal telomere repeat DNA in telomerase-negative immortalized cell lines. Biochem Biophys Res Commun 247:765–772. https://doi.org/10.1006/bbrc.1998.8876
154. Nabetani A, Ishikawa F (2009) Unusual telomeric DNAs in human telomerase-negative immortalized cells. Mol Cell Biol 29:703–713. https://doi.org/10.1128/MCB.00603-08
155. Dilley RL, Verma P, Cho NW et al (2016) Break-induced telomere synthesis underlies alternative telomere maintenance. Nature 539:54–58. https://doi.org/10.1038/nature20099
156. Roumelioti F, Sotiriou SK, Katsini V et al (2016) Alternative lengthening of human telomeres is a conservative DNA replication process with features of break-induced replication. EMBO Rep 17:1731–1737. https://doi.org/10.15252/embr.201643169
157. Anand RP, Lovett ST, Haber JE (2013) Break-induced DNA replication. Cold Spring Harb Perspect Biol 5:a010397. https://doi.org/10.1101/cshperspect.a010397
158. Sotiriou SK, Kamileri I, Lugli N et al (2016) Mammalian RAD52 functions in break-induced replication repair of collapsed DNA replication forks. Mol Cell 64:1127–1134. https://doi.org/10.1016/j.molcel.2016.10.038
159. Costantino L, Sotiriou SK, Rantala JK et al (2014) Break-induced replication repair of damaged forks induces genomic duplications in human cells. Science 343:88–91. https://doi.org/10.1126/science.1243211

160. Ciccia A, Bredemeyer AL, Sowa ME et al (2009) The SIOD disorder protein SMARCAL1 is an RPA-interacting protein involved in replication fork restart. Genes Dev 23:2415–2425. https://doi.org/10.1101/gad.1832309
161. Bansbach CE, Bétous R, Lovejoy CA et al (2009) The annealing helicase SMARCAL1 maintains genome integrity at stalled replication forks. Genes Dev 23:2405–2414. https://doi.org/10.1101/gad.1839909
162. Jingsong Y, Ghosal G, Junjie C (2009) The annealing helicase HARP protects stalled replication forks. Genes Dev 23:2394–2399. https://doi.org/10.1101/gad.1836409
163. Couch FB, Bansbach CE, Driscoll R et al (2013) ATR phosphorylates SMARCAL1 to prevent replication fork collapse. Genes Dev 27:1610–1623. https://doi.org/10.1101/gad.214080.113
164. Cox KE, Maréchal A, Flynn RL (2016) SMARCAL1 resolves replication stress at ALT telomeres. Cell Rep 14:1032–1040. https://doi.org/10.1016/j.celrep.2016.01.011
165. Poole LA, Zhao R, Glick GG et al (2015) SMARCAL1 maintains telomere integrity during DNA replication. Proc Natl Acad Sci 112:14864–14869. https://doi.org/10.1073/pnas.1510750112
166. Diplas BH, He X, Brosnan-Cashman JA et al (2018) The genomic landscape of TERT promoter wildtype-IDH wildtype glioblastoma. Nat Commun 9:2087. https://doi.org/10.1038/s41467-018-04448-6
167. Cho NW, Dilley RL, Lampson MA, Greenberg RA (2014) Interchromosomal homology searches drive directional ALT telomere movement and synapsis. Cell 159:108–121. https://doi.org/10.1016/j.cell.2014.08.030
168. Henson JD, Cao Y, Huschtscha LI et al (2009) DNA C-circles are specific and quantifiable markers of alternative-lengthening-of-telomeres activity. Nat Biotechnol 27:1181–1185. https://doi.org/10.1038/nbt.1587
169. Min J, Wright WE, Shay JW (2017) Alternative lengthening of telomeres can be maintained by preferential elongation of lagging strands. Nucleic Acids Res 45:2615–2628. https://doi.org/10.1093/nar/gkw1295
170. Pickett HA, Reddel RR (2012) The role of telomere trimming in normal telomere length dynamics. Cell Cycle 11:1309–1315. https://doi.org/10.4161/cc.19632
171. Pickett HA, Cesare AJ, Johnston RL et al (2009) Control of telomere length by a trimming mechanism that involves generation of t-circles. EMBO J 28:799–809. https://doi.org/10.1038/emboj.2009.42
172. Lee M, Hills M, Conomos D et al (2014) Telomere extension by telomerase and ALT generates variant repeats by mechanistically distinct processes. Nucleic Acids Res 42:1733–1746. https://doi.org/10.1093/nar/gkt1117
173. Conomos D, Stutz MD, Hills M et al (2012) Variant repeats are interspersed throughout the telomeres and recruit nuclear receptors in ALT cells. J Cell Biol 199:893–906. https://doi.org/10.1083/jcb.201207189
174. Varley H, Pickett HA, Foxon JL et al (2002) Molecular characterization of inter-telomere and intra-telomere mutations in human ALT cells. Nat Genet 30:301–305. https://doi.org/10.1038/ng834
175. Conomos D, Reddel RR, Pickett HA (2014) NuRD–ZNF827 recruitment to telomeres creates a molecular scaffold for homologous recombination. Nat Struct Mol Biol 21:760–770. https://doi.org/10.1038/nsmb.2877
176. Lovejoy CA, Li W, Reisenweber S et al (2012) Loss of ATRX, genome instability, and an altered DNA damage response are hallmarks of the alternative lengthening of telomeres pathway. PLoS Genet 8:e1002772. https://doi.org/10.1371/journal.pgen.1002772
177. Heaphy CM, De Wilde RF, Jiao Y et al (2011) Altered telomeres in tumors with ATRX and DAXX mutations. Science 333:425
178. Leung JW-C, Ghosal G, Wang W et al (2013) Alpha thalassemia/mental retardation syndrome X-linked gene product ATRX is required for proper replication restart and cellular resistance to replication stress. J Biol Chem 288:6342–6350. https://doi.org/10.1074/jbc.M112.411603
179. Huh MS, Ivanochko D, Hashem LE et al (2016) Stalled replication forks within heterochromatin require ATRX for protection. Cell Death Dis 7:e2220. https://doi.org/10.1038/cddis.2016.121
180. Watson LA, Solomon LA, Li JR et al (2013) Atrx deficiency induces telomere dysfunction, endocrine defects, and reduced life span. J Clin Invest 123:2049–2063. https://doi.org/10.1172/JCI65634
181. Clynes D, Jelinska C, Xella B et al (2015) Suppression of the alternative lengthening of telomere pathway by the chromatin

remodelling factor ATRX. Nat Commun 6:7538. https://doi.org/10.1038/ncomms8538
182. Napier CE, Huschtscha LI, Harvey A et al (2015) ATRX represses alternative lengthening of telomeres. Oncotarget 6:16543–16558. https://doi.org/10.18632/oncotarget.3846
183. Goldberg AD, Banaszynski LA, Noh K-M et al (2010) Distinct factors control histone variant H3.3 localization at specific genomic regions. Cell 140:678–691. https://doi.org/10.1016/j.cell.2010.01.003
184. Arora R, Azzalin CM (2015) Telomere elongation chooses TERRA ALTernatives. RNA Biol 12:938–941. https://doi.org/10.1080/15476286.2015.1065374
185. Episkopou H, Draskovic I, Van Beneden A et al (2014) Alternative lengthening of telomeres is characterized by reduced compaction of telomeric chromatin. Nucleic Acids Res 42:4391–4405. https://doi.org/10.1093/nar/gku114
186. Herrick J, Bensimon A (1999) Single molecule analysis of DNA replication. Biochimie 81:859–871

Part II

Cellular Assays to Detect and Measure DNA Damage and Damage Response

Chapter 3

In Time and Space: Laser Microirradiation and the DNA Damage Response

Jae Jin Kim, Ramhari Kumbhar, Fade Gong, and Kyle M. Miller

Abstract

Maintenance of genomic integrity depends on the spatiotemporal recruitment and regulation of DNA damage response and repair proteins at DNA damage sites. These highly dynamic processes have been widely studied using laser microirradiation coupled with fluorescence microscopy. Laser microirradiation has provided a powerful methodology to identify and determine mechanisms of DNA damage response pathways. Here we describe methods used to analyze protein recruitment dynamics of fluorescently tagged or endogenous proteins to laser-induced DNA damage sites using laser scanning and fluorescence microscopy. We further describe multiple applications employing these techniques to study additional processes at DNA damage sites including transcription. Together, we aim to provide robust visualization methods employing laser-microirradiation to detect and determine protein behavior, functions and dynamics in response to DNA damage in mammalian cells.

Key words Laser microirradiation, DNA repair, DNA damage, Transcription, Fluorescence microscopy, Live-cell imaging

1 Introduction

Genome integrity is constantly challenged by endogenous, as well as exogenous sources of DNA damage including replication stress, reactive cellular metabolites, UV, ionizing radiation, and other mutagens including various chemicals. These threats generate a multitude of different DNA lesions including various base damages and single as well as double-strand breaks in the DNA. These lesions must be identified and faithfully repaired in order to avoid mutations that can result in genomic instability [1]. Defects in the ability to maintain genome stability are known to contribute to several human diseases and conditions including neurodegeneration, aging, and cancer.

To protect the genome, organisms employ highly orchestrated surveillance and repair mechanisms that are collectively known as DNA damage response (DDR) pathways [2, 3]. These pathways

Lata Balakrishnan and Jason A. Stewart (eds.), *DNA Repair: Methods and Protocols*, Methods in Molecular Biology, vol. 1999,
https://doi.org/10.1007/978-1-4939-9500-4_3,

consist of DNA damage sensor proteins that detect damage, which then activate signaling pathways that promote checkpoint responses and DNA repair reactions. The spatiotemporal regulation of these processes is vital for coordinating these activities within chromatin to ensure genome integrity and cell survival [4, 5]. Following DNA damage, a multitude of proteins are localized to DNA damage sites in an ordered manner [4, 6]. For example, the MRE11-RAD50-NBS1 (MRN) complex and PARP detect DNA double-strand break (DSBs) and DNA single-strand breaks respectively [7]. For DSBs, these sensor proteins promote the phosphorylation of histone H2AX on serine 139 residue (γH2AX) via activation and recruitment of DDR kinases including ATM [8, 9]. The mediator of DNA damage checkpoint 1 (MDC1) binds to phosphorylated histone H2AX amplifying downstream DNA damage signaling. The ATM kinase also phosphorylates MDC1 and this modification is recognized by the RNF8 E3 ubiquitin ligase. RNF8 initiates an ubiquitin signaling cascade at the DNA damage site for recruitment of the DNA repair factors 53BP1 and BRCA1 [10–12].

Significant progress has been made to identify and characterize key proteins involved in these DNA damage response pathways. A key method that has been integral to our identification of these proteins has been the use of laser microirradiation. This technique allows localized tracks of DNA damage to be generated, including DSBs, in cells using lasers coupled to fluorescent microscopes. Upon DNA damage instigation, proteins can be analyzed in live or fixed cells to study the spatial, temporal and coordinated dynamics of DDR proteins and the pathways that they regulate at DNA damage sites.

In addition to studying DNA damage response factor recruitment using laser microirradiation, this technique can also be used to study other nuclear processes at DNA damage sites and protein dynamics within cells [13, 14]. For example, various types of DNA lesions result in local inhibition of transcription, which is thought to be critical for limiting the production of anomalous transcripts and to avoid conflicts between transcription and DDR machinery [15]. Several groups have shown that cells coordinate transcriptional responses with the DDR. For example, the ATM kinase promotes transcriptional silencing through DNA damage responsive histone ubiquitylations by RNF8 and RNF168 [16]. In another example, KDM5A, a H3K4 lysine demethylase, is recruited to laser-induced DNA damage sites where it acts to demethylate chromatin to facilitate recruitment of the ZMYND8-NuRD chromatin-remodeling complex [17–19]. Together, these factors enforce transcriptional repression at DNA damage sites, which promotes homologous recombination repair of DSBs, potentially by the removal of transcriptional conflicts. In addition to recruitment dynamics of DDR proteins, laser microirradiation has provided key experimental evidence for understanding transcriptional responses that occur within damaged regions of chromatin, an important process that is as yet not well defined.

In this chapter, we describe detailed methods for employing laser-induced DNA damage in living cells to study recruitment dynamics to DNA lesions of DDR factors. We also provide an assay to study nascent transcription at laser-induced DNA damage sites. These methods can be utilized for detecting, quantifying, and analyzing protein recruitment and transcriptional dynamics at DNA damage sites. These methods also provide insights into the spatiotemporal dynamics of DDR and transcription proteins that occur at DNA damage sites, processes that require coordinated responses of these pathways to ensure genome and epigenome stability.

2 Materials

2.1 Equipment

1. Microscope equipped with a 405 nm 50 mW laser and CO_2/humidity chamber for live imaging (*see* **Note 1**).
2. A 37 °C incubator with 5% CO_2 for growing human cells.

2.2 Cell Culture

1. Glass bottomed dishes with glass well size 12 mm (WillCo Wells, GWST-3512) or 35 mm high grid-500 (ibidi, 81168).
2. 5-Bromo-2′-deoxyuridine (BrdU).
3. Hoechst 33342.
4. Adherent human cell line (e.g., U2OS, HeLa, RPE).
5. Cells are cultured in DMEM medium supplemented with 10% FBS and penicillin–streptomycin–glutamine.
6. Opti-MEM for transfection.
7. 0.25% trypsin.
8. Phosphate-buffered saline.
9. Transfection reagent (e.g., Fugene HD [Promega] or other lipid-based reagent).
10. Expression vector for fluorescence-tagged gene of interest.

2.3 Immunofluorescence

1. Fixation buffer: 4% formaldehyde in PBS.
2. 2% Formaldehyde: dilute 4% formaldehyde to 2% with 1× PBS.
3. 0.5% Triton X-100 in PBS.
4. Microscope coverslips.
5. CSK (cytoskeleton) buffer: 10 mM PIPES, pH 6.8; 100 mM NaCl, 300 mM sucrose, 3 mM $MgCl_2$, 1 mM EGTA, 0.5% (v/v) Triton X-100.

6. Blocking and antibody solution (1× PBS + 3% BSA).
7. Anti-fade mounting medium.
8. 4′,6-Diamidino-2-phenylindole (DAPI).
9. Diamond tipped pen.

2.4 Antibodies

1. Anti-γH2AX (Millipore, 05-636).
2. Anti-ZMYND8 (Bethyl, A302-089A).
3. Alexa Fluor 488 goat anti-rabbit IgG (Invitrogen, A11034).
4. Alexa Fluor 594 goat anti-mouse IgG (Invitrogen, A11032).

2.5 Nascent Transcription Analysis

1. Click-iT RNA Imaging Kit (Invitrogen, Cat No. C10330)*
   C10329 labels EU with Alexa Fluor 488.
   C10330 labels EU with Alexa Fluor 594.

   *Prepare stock solutions following the instructions provided in the kit.

3 Methods

3.1 Recruitment of Fluorescently Tagged Proteins to DNA Damage Sites Using Laser Microirradiation and Live-Cell Imaging

1. Day 1: Seeding cells. Seed 5–7 × 10^4 cells/1.5 mL in a 12 mm glass bottomed dishes. These experiments were optimized for human U2OS cancer cells.
2. Incubate at 37 °C in 5% CO_2 for 24 h.
3. Day 2: Transfection of fluorescently tagged gene of interest. Transfect vector containing fluorescence tagged target gene for protein expression (as explained below).
4. Remove medium from target cells.
5. Add 1.5 mL regular medium without Penicillin/Streptomycin.
6. Prepare transfection mixture: add 1.5 μg of expression construct to 150 μL of Opti-MEM and add 4.5 μL of Fugene HD transfection reagent. Mix gently by pipetting and incubate tubes for 15 min at room temperature (*see* **Note 2**).
7. Add the transfection mixture dropwise onto the target cells, mix by gently moving the plate forward and backward, side to side. Once mixed, the plate is placed back into the incubator.
8. Day 3: Sensitization of transfected cells for laser microirradiation (*see* **Note 3**). Cells are incubated with photo sensitizer containing media before laser microirradiation. This treatment allows for the generation of DNA damage using lower laser powers, which helps to reduce phototoxicity to the cell. For example, it has been calculated that six to ten times higher laser power is required to induce DNA damage using various lasers

without presensitization [14]. This can be done using BrdU or Hoechst 33342.

(a) BrdU sensitization: BrdU is a synthetic nucleoside that is an analog of thymidine and is incorporate into newly synthesized DNA. For BrdU, the CH_3 group of thymidine is substituted with a bromine atom, which makes BrdU incorporated cells are more sensitive to DNA damage. Incubate cells with 10 μM BrdU containing media 24 h prior to laser microirradiation to allow all of the cells to complete one S-phase and therefore all cells will have incorporated BrdU and be presensitized to laser damage.

(b) Hoechst 33342 sensitization: This compound binds to the minor groove of DNA, which results in sensitizing to laser damage. Incubate cells with 10 μg/mL Hoechst 33342 containing media 10 min prior to laser microirradiation.

9. Day 4: Laser Microirradiation: Laser induced DNA damage. Turn on the CO_2/humidity/temperature-controlled chamber and the confocal laser scanning microscope. Once the chamber has reached the desired parameters (e.g., 5% CO_2 and 37 °C), place the dish containing the cells into the chamber (*see* **Note 4**).

10. Visualize the fluorescent protein using the appropriate filters/ lasers to locate and focus on the cells that will be damaged (*see* **Notes 5** and **6**).

11. Adjust laser power, HV (current voltage to PMT), gain (amplification of the signal from the PMT), and offset (cut off the higher or lower signal) control. Normally, 0.1% laser power and 500 HV setting are a good starting point for scanning. Adjustment of these parameters can be used to get a better quality image after starting the image scan.

12. Once the image settings are set, laser DNA damage can be induced using the stimulation settings for the 405 nm laser. A defined region of interest (ROI) is drawn within the cell of interest (Fig. 1a). ROIs can include a line, circle, or box. For the laser, set the scan speed to 20.0 μs/pixel first. Activate the 405 laser using 150 frames at 60% laser power (*see* **Note 7**). This is a general setting that will change depending on the application and set up of the microscope including objectives. These settings will need to be empirically optimized by the user based on experimental design and goals.

13. After putting in the settings and defining the ROI, the laser damage is delivered and imaging and quantification of the fluorescence intensity of the sample is performed, including an identical ROI in an undamaged region of the same cell (Fig. 1b–e, *see* **Notes 8** and **9**).

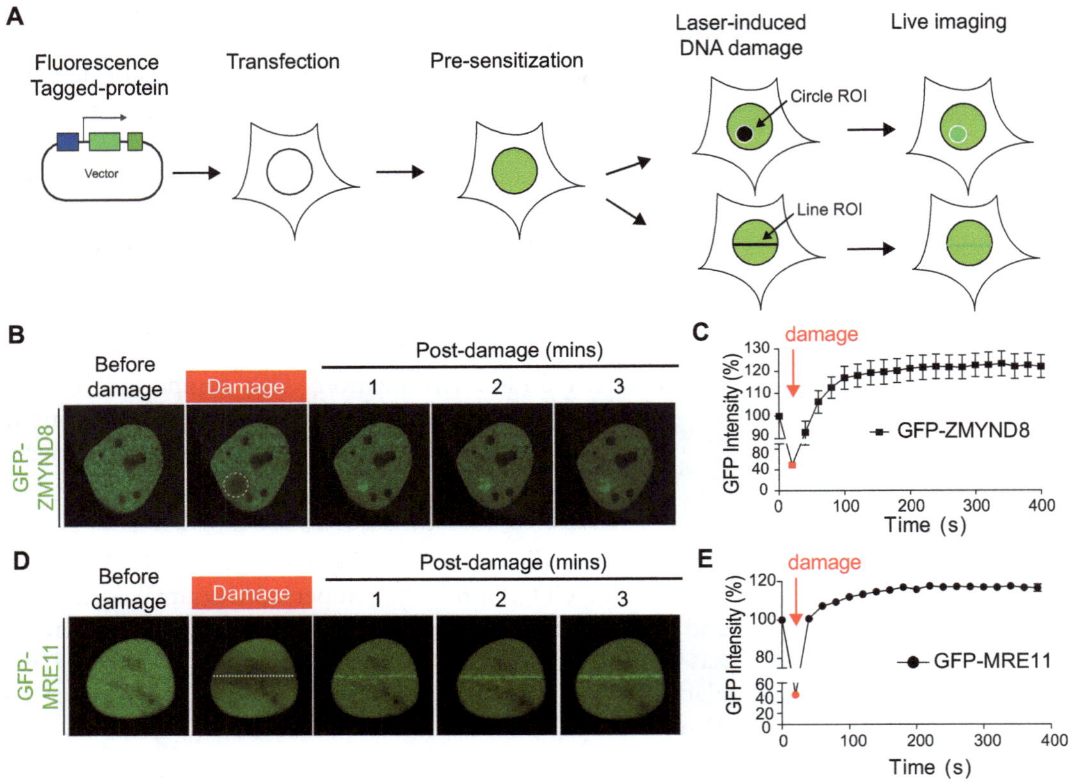

Fig. 1 Laser microirradiation and live imaging of fluorescently tagged repair protein translocation to DNA damage sites. (**a**) Schematic illustration of live imaging for protein recruitment to DNA damage sites. (**b**) Laser damage-induced recruitment of GFP-tagged DDR protein ZMYND8. The white circle indicates the laser path defined by the ROI. (**c**) Quantification of the GFP intensity using laser microirradiation. The fluorescence intensity of GFP-ZMYND8 in **b** of damaged versus undamaged regions are plotted as the relative GFP intensity at each time point. Error bars indicate SEM; $n > 10$. (**d**) Laser damage-induced recruitment of GFP-tagged DDR protein MRE11. The white line indicates the laser path defined by the ROI. (**e**) Quantification of **d** as in **c**

3.2 Data Analysis of Subheading 3.1

1. At least ten cells are needs for quantification of kinetics for repair protein recruitment to DNA damage site.
2. Fluorescence intensities values that are measured by FluoView FV3000 software at DNA damage site and an undamaged region are compared. Intensity of damaged region (ROI_D)/intensity of undamaged region (ROI_U). These are normalized to fluorescence intensity at time zero of the experiment predamage.
3. Relative fluorescence quantified for each time points are normalized with relative fluorescence predamage (Fig. 1c, e). After DNA damage (ROI_D/ROI_u)/before DNA damage (ROI_D/ROI_u) $\times$ 100.
4. Fluorescence intensities values measured by FluoView FV3000 software and relative fluorescence are calculated using

Microsoft Excel and Prism software to create graphs showing the fluorescence intensity as a measure of time post-damage (Fig. 1c, e).

3.3 Endogenous Protein Recruitment to DNA Damage Sites Using Laser Microirradiation

1. Day 1: Seed Cells. Cells are counted and $5–7 \times 10^4$ cells are seeded/dish in glass bottom dishes. For endogenous protein recruitment, it is useful to laser damage cells that are in a defined region. For this a diamond pen is used to etch a cross in the glass (Fig. 2a) or identifiable grids can be used. This makes it easier to locate the damaged cells for imaging once processing of the samples has been performed (*see* **Note 10**).
2. Day 2: Presensitize cells as in Subheading 3.1, **step 8**. Cells are then microirradiated using the same procedures as in Subheading 3.1 (Fig. 2b, *see* **Note 6**).
3. Day 3: Perform laser damage followed by immunofluorescence. After laser damage, cells are treated with CSK buffer or Triton X-100 to permeabilize the cells followed by fixation for preparation for antibody staining (*see* **Note 11**).

   CSK extraction:

   (a) Wash cells three times with 1 mL of 1× PBS, 5 min each wash.

   (b) Place dish with cells on ice. Add 500 μL cold CSK buffer and incubate for 5 min. (This is the only step that is required to be performed on ice; other steps are done at room temperature [RT].)

   (c) Remove the CSK buffer following three times wash with 1 mL of 1× PBS, 5 min each wash.

   (d) Add 500 μL of fixation buffer at RT for 15 min to fix the cells.

   (e) Remove the fixation buffer followed by washing in 1 mL of PBS three times, 5 min each wash.

   (f) Add 500 μL of 3% BSA blocking solution for 15 min at RT.

   Triton X-100 permeabilization:

   (a) Wash cells with 1 mL of PBS three times, 5 min each wash.

   (b) Add 500 μL of fixation buffer for 15 min to fix cells at RT.

   (c) Remove fixation buffer and wash in 1 mL of PBS three times, 5 min each wash.

   (d) Add 0.5% Triton X-100 for 15 min at RT.

   (e) Remove Triton X-100 buffer and wash in 1 mL of PBS three times, 5 min each wash.

   (f) Add 500 μL of 3% BSA blocking solution for 15 min at RT.

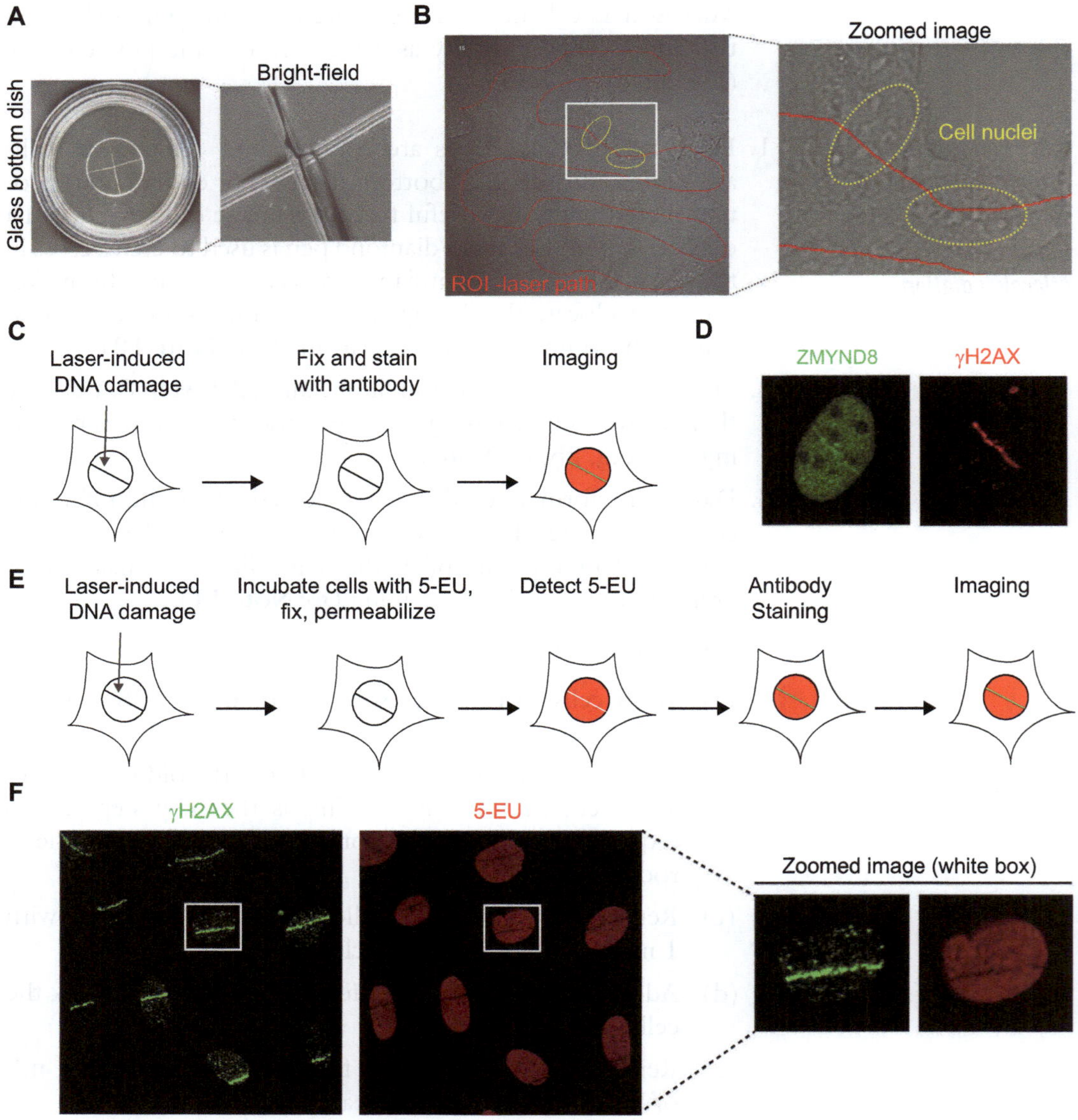

Fig. 2 Immunofluorescence analysis of endogenous protein recruitment and transcription in the DNA damage site. (**a**) Experimental setup for laser microirradiation of fixed samples. Left: etched crossed line on glass bottom dish. Right: zoomed image using 60× oil immersion objective and bright-field microscopy. (**b**) Free drawn ROI line on U2OS human cancer cells within glass bottomed dishes viewed by bright-field microscopy. Red line indicates the ROI and yellow circle indicates the cell nuclei. (**c**) Schematic illustration for analysis of endogenous protein recruitment and transcription in DNA damage site. (**d**) Endogenous ZMYND8 translocation to DNA damage site. DNA damage was induced by laser microirradiation and stain with ZMYND8 and γH2AX antibodies. (**e**) Schematic of nascent transcription and laser microirradiation technique. (**f**) Transcription analysis in the DNA damage site using 5-EU staining. γH2AX is marker for the DNA damage region

4. During the blocking step, dilute primary antibodies with 3% BSA (in 1× PBS) solution. If dual staining with two antibodies is required, ensure the two antibodies are derived from different hosts to avoid cross-contamination. For example, to stain

with ZMYND8 and γH2AX antibodies at the same time, use mouse anti-γH2AX and rabbit anti-ZMYND8.

5. After blocking, add 100 μL of antibody-BSA mix into the small well of the glass bottom dish.
6. Incubate with primary antibodies for 1 h at RT. Certain antibodies may need incubation at 4 °C overnight (*see* **Notes 12** and **13**).
7. Wash cells with 1 mL of PBS three times, 5 min each wash.
8. Incubate with secondary antibodies for 1 h at RT. Since secondary antibodies are light sensitive, this step should be done in reduced light conditions, which may be accomplished by putting the reaction container in the dark or by covering the reaction with aluminum foil.
9. Wash cells with 1 mL of PBS three times, 5 min each wash.
10. Incubate with 0.1–1 μg/mL DAPI for 5 min.
11. Wash cells 1 mL of PBS three times, 5 min each wash.
12. Add 10 μL of mounting solution to the small well of the glass bottom dish and cover with a 10 mm coverslip.
13. Nail polish can be used to seal the coverslip at the bottom of the dish. If the samples will be analyzed immediately, this step is not necessary and dishes can be stored in a humidified box at 4 °C for up to a week.
14. Analyze the samples by fluorescence microscopy to detect the desired fluorophores used for the experiment (Fig. 2d).

3.4 Data Analysis of Subheading 3.3

1. Immunofluorescence intensity in the DNA damage site can be measured by ImageJ software.
2. The number of cells displaying the labeled protein of interest that is colocalized with a DNA damage marker, for example γH2AX, is determined and compared to the total number of damaged cells. At least 100 damaged cells should be analyzed.

3.5 Analysis of Transcription at Laser-Induced DNA Damage Sites (Fig. 2e, f)

1. Day 1: Seed cells. Cells are split, counted, and seeded as in Subheading 3.1, **step 1**.
2. Day 2: Laser damage and transcriptional analysis. After 5 min following DNA damage induction, remove medium and add 100 μL of medium with EU (1 mM) to the glass well. 100 μL is enough to cover the 12 mm glass well of the dish; use more volume if a large dish is used to ensure that the cells are covered by the medium.
3. Incubate the cells for 1 h, remove the medium and wash cells three times with 1 mL of PBS, 5 min each wash. Fix cells with 500 μL of 2% formaldehyde in PBS at room temperature for 10 min.

4. After fixation, remove the fixative and wash three times with 1 mL of PBS, 5 min each wash. Add 500 μL of 0.5% Triton X-100 in PBS to each well to permeabilize the cells. Incubate 15 min at room temperature.
5. During the permeabilization step, prepare the Click-iT reaction cocktail mix following the manufacturer's instructions.
6. After permeabilization, wash each well three times with 1 mL of PBS, 5 min each wash. Add 100 μL of Click-iT cocktail mix to the 12 mm glass dish well and incubate 30 min. After this step, all procedures should be performed to reduce light exposure.
7. After incubation, remove the Click-iT cocktail mix. Rinse each well with 500 μL of Click-iT reaction rinse buffer.
8. Perform immunofluorescence staining procedures with desired antibodies as described in Subheading 3.3 (*see* **Notes 11–13**).
9. Image and analyze samples by fluorescence microscopy (Fig. 2e, f) as described in Subheading 3.3. For example, γH2AX staining can be used to verify and detect damaged DNA.

4 Notes

1. These protocols and settings for laser damage have been optimized and conducted using a FLUOVIEW FV1000 and FV3000 Olympus confocal laser scanning microscope system with a 405 nm laser. LSM systems with additional lasers can also be used to generate laser microirradiation. We refer users to these protocols, resources and additional information [20–24].
2. This protocol is optimized for U2OS cells. Transfection efficiency will be different depending on the cell type and transfection reagent. The transfection method should be modified or used with different transfection reagents that are determined empirically for each cell type. For laser microirradiation in difficult to transfect cell types, nucleofector™ kit (Lonza) or other electroporation techniques can be used [25, 26].
3. Experiments can also be performed without presensitization to ensure that the laser damage is dependent on this procedure, which reduces the laser power required and therefore the potential cellular toxicity. Cell cycle analysis of damage recruitment, including RPA and BRCA1, which should only occur in S/G2 phases of the cell cycle, can also be used to monitor damage strength and biological effects of laser damage [27]. For recruitment of endogenous proteins, DNA damage markers including γH2AX or 53BP1 are used to control for

DSB formation. Other markers can be used for other types of DNA damage [6].

4. Allow heat chamber to stabilize for 5–10 min after setting required temperature and CO_2. Maintain humidity of chamber by adding and replenishing water as needed. The laser source should be turned on for 10–15 min before use, which should be consistent for experiments to ensure that the laser is stable when used.
5. Choosing which cells to analyze by microirradiation is critical for obtaining consistent results for both live and fixed cell imaging and analysis. It is important to select cells with good expression of the protein of interest as well as cells that have normal morphology. Cells with very high expression of ectopically expressed tagged protein or abnormal cell shape or size should be avoided to reduce potential artifacts due to overexpression. However, it is always a good idea to check recruitment between different expression levels as well as compare with tagged and endogenously expressed proteins.
6. A very important detail for this experiment is to ensure that cells are in focus before damaging the cells. For fluorescence experiments, this can be done while imaging the tagged protein of interest. For recruitment studies of endogenous proteins, a focused cell should have the nucleus distinguished clearly using bright-field microscopy; while the nucleoli will appear bright and white in color (Fig. 2b). Having properly focused cells ensures that the laser path is concentrated within the center of the cell, which will provide better laser stripes of DNA damage for analysis.
7. Intensity of 405 nm lasers can vary due to laser power and microscope design. It is important to calibrate the optimum laser intensity required to induce DNA damage stripes or marks. To maintain the same laser power between experiments, a laser power meter can be used to measure 405 nm laser intensity.
8. Phenol red in culture medium is a chromophore and can increase the level of background fluorescence. This can be avoided by using phenol red free cell culture medium such as FluoroBrite DMEM (Thermo Scientific, #A1896701).
9. Many proteins have been shown to be recruited to the sites of DNA damage. These can act as positive controls to test the settings and conditions for these experiments. Good positive controls for DNA double-strand breaks include 53BP1, RNF168, KAP1, BRCA1, RAD51 and others, which can be found in the literature.
10. Before seeding cells, use a diamond tipped pen to carefully etch a cross on the glass bottom, which can help to track cells that

have been damaged for downstream analyses including immunofluorescence and imaging. Alternatively, use a grid glass bottom dish, which contains grids with 500 μm repeat distance on the bottom glass. It can help to locate damaged cells without the need to use a diamond pen but these dishes are larger which requires more antibodies than the 12 mm dishes. Make a cross gently, as the glass bottom is very fragile. Applying excessive pressure while etching the glass could damage glass bottom resulting in breakage and leakage.

11. Proper fixation and permeabilization are important for successful immunofluorescence staining. Choice of fixative and permeabilization reagent and incubation time should be optimized based on cell type, antibody and protein of interest. For example, CSK removes all soluble proteins so this condition works well for chromatin associated proteins. Antibody datasheets may provide information about suitable reagents and incubation times.
12. Concentration of primary antibody, blocking solution and incubation time should be optimized based on the antibody. For anti-γH2AX, a dilution of 1:500 in 3% BSA/1× PBS is used for an incubation time of 1 h at room temperature.
13. If the imaging cannot be performed immediately, the dishes can be stored at 4 °C in humidified chamber in the dark for at least a week. Waiting for longer times could result in dried out samples and/or a loss of fluorescence signal.

Acknowledgments

The K.M.M. laboratory is supported by the NIH National Cancer Institute (R01CA198279 and RO1CA201268) and the American Cancer Society (RSG-16-042-01-DMC).

References

1. Hoeijmakers JH (2001) Genome maintenance mechanisms for preventing cancer. Nature 411 (6835):366–374. https://doi.org/10.1038/35077232
2. Ciccia A, Elledge SJ (2010) The DNA damage response: making it safe to play with knives. Mol Cell 40(2):179–204. https://doi.org/10.1016/j.molcel.2010.09.019
3. Jackson SP, Bartek J (2009) The DNA-damage response in human biology and disease. Nature 461(7267):1071–1078. https://doi.org/10.1038/nature08467
4. Polo SE, Jackson SP (2011) Dynamics of DNA damage response proteins at DNA breaks: a focus on protein modifications. Genes Dev 25 (5):409–433. https://doi.org/10.1101/gad.2021311
5. Sulli G, Di Micco R, d'Adda di Fagagna F (2012) Crosstalk between chromatin state and DNA damage response in cellular senescence and cancer. Nat Rev Cancer 12(10):709–720. https://doi.org/10.1038/nrc3344
6. Aleksandrov R, Dotchev A, Poser I, Krastev D, Georgiev G, Panova G, Babukov Y, Danovski G, Dyankova T, Hubatsch L, Ivanova A, Atemin A, Nedelcheva-Veleva MN, Hasse S, Sarov M, Buchholz F, Hyman AA, Grill SW, Stoynov SS (2018) Protein

dynamics in complex DNA lesions. Mol Cell 69 (6):1046–1061.e1045. https://doi.org/10.1016/j.molcel.2018.02.016
7. Petrini JH, Stracker TH (2003) The cellular response to DNA double-strand breaks: defining the sensors and mediators. Trends Cell Biol 13(9):458–462
8. Lee JH, Paull TT (2005) ATM activation by DNA double-strand breaks through the Mre11-Rad50-Nbs1 complex. Science 308 (5721):551–554. https://doi.org/10.1126/science.1108297
9. Rogakou EP, Pilch DR, Orr AH, Ivanova VS, Bonner WM (1998) DNA double-stranded breaks induce histone H2AX phosphorylation on serine 139. J Biol Chem 273 (10):5858–5868
10. Jackson SP, Durocher D (2013) Regulation of DNA damage responses by ubiquitin and SUMO. Mol Cell 49(5):795–807. https://doi.org/10.1016/j.molcel.2013.01.017
11. Kolas NK, Chapman JR, Nakada S, Ylanko J, Chahwan R, Sweeney FD, Panier S, Mendez M, Wildenhain J, Thomson TM, Pelletier L, Jackson SP, Durocher D (2007) Orchestration of the DNA-damage response by the RNF8 ubiquitin ligase. Science 318 (5856):1637–1640. https://doi.org/10.1126/science.1150034
12. Wang B, Elledge SJ (2007) Ubc13/Rnf8 ubiquitin ligases control foci formation of the Rap80/Abraxas/Brca1/Brcc36 complex in response to DNA damage. Proc Natl Acad Sci U S A 104(52):20759–20763. https://doi.org/10.1073/pnas.0710061104
13. Gassman NR, Wilson SH (2015) Microirradiation tools to visualize base excision repair and single-strand break repair. DNA Repair (Amst) 31:52–63. https://doi.org/10.1016/j.dnarep.2015.05.001
14. Mistrik M, Vesela E, Furst T, Hanzlikova H, Frydrych I, Gursky J, Majera D, Bartek J (2016) Cells and stripes: a novel quantitative photo-manipulation technique. Sci Rep 6:19567. https://doi.org/10.1038/srep19567
15. Svejstrup JQ (2010) The interface between transcription and mechanisms maintaining genome integrity. Trends Biochem Sci 35 (6):333–338. https://doi.org/10.1016/j.tibs.2010.02.001
16. Shanbhag NM, Rafalska-Metcalf IU, Balane-Bolivar C, Janicki SM, Greenberg RA (2010) ATM-dependent chromatin changes silence transcription in cis to DNA double-strand breaks. Cell 141(6):970–981. https://doi.org/10.1016/j.cell.2010.04.038
17. Gong F, Chiu LY, Cox B, Aymard F, Clouaire T, Leung JW, Cammarata M, Perez M, Agarwal P, Brodbelt JS, Legube G, Miller KM (2015) Screen identifies bromodomain protein ZMYND8 in chromatin recognition of transcription-associated DNA damage that promotes homologous recombination. Genes Dev 29(2):197–211. https://doi.org/10.1101/gad.252189.114
18. Gong F, Clouaire T, Aguirrebengoa M, Legube G, Miller KM (2017) Histone demethylase KDM5A regulates the ZMYND8-NuRD chromatin remodeler to promote DNA repair. J Cell Biol 216 (7):1959–1974. https://doi.org/10.1083/jcb.201611135
19. Gong F, Miller KM (2018) Double duty: ZMYND8 in the DNA damage response and cancer. Cell Cycle 17(4):414–420. https://doi.org/10.1080/15384101.2017.1376150
20. Adam S, Dabin J, Chevallier O, Leroy O, Baldeyron C, Corpet A, Lomonte P, Renaud O, Almouzni G, Polo SE (2016) Real-time tracking of parental histones reveals their contribution to chromatin integrity following DNA damage. Mol Cell 64(1):65–78. https://doi.org/10.1016/j.molcel.2016.08.019
21. Britton S, Coates J, Jackson SP (2013) A new method for high-resolution imaging of Ku foci to decipher mechanisms of DNA double-strand break repair. J Cell Biol 202 (3):579–595. https://doi.org/10.1083/jcb.201303073
22. Kong X, Cruz GMS, Silva BA, Wakida NM, Khatibzadeh N, Berns MW, Yokomori K (2018) Laser microirradiation to study in vivo cellular responses to simple and complex DNA damage. J Vis Exp (131). https://doi.org/10.3791/56213
23. Lukas C, Bartek J, Lukas J (2005) Imaging of protein movement induced by chromosomal breakage: tiny 'local' lesions pose great 'global' challenges. Chromosoma 114(3):146–154. https://doi.org/10.1007/s00412-005-0011-y
24. Xie S, Mortusewicz O, Ma HT, Herr P, Poon RY, Helleday T, Qian C (2015) Timeless interacts with PARP-1 to promote homologous recombination repair. Mol Cell 60 (1):163–176. https://doi.org/10.1016/j.molcel.2015.07.031
25. Dobbin MM, Madabhushi R, Pan L, Chen Y, Kim D, Gao J, Ahanonu B, Pao PC, Qiu Y, Zhao Y, Tsai LH (2013) SIRT1 collaborates with ATM and HDAC1 to maintain genomic stability in neurons. Nat Neurosci 16

(8):1008–1015. https://doi.org/10.1038/nn.3460

26. Wang WY, Pan L, Su SC, Quinn EJ, Sasaki M, Jimenez JC, Mackenzie IR, Huang EJ, Tsai LH (2013) Interaction of FUS and HDAC1 regulates DNA damage response and repair in neurons. Nat Neurosci 16(10):1383–1391. https://doi.org/10.1038/nn.3514

27. Hustedt N, Durocher D (2016) The control of DNA repair by the cell cycle. Nat Cell Biol 19 (1):1–9. https://doi.org/10.1038/ncb3452

Chapter 4

Quantification of Double-Strand Breaks in Mammalian Cells Using Pulsed-Field Gel Electrophoresis

Kelvin W. Pond and Nathan A. Ellis

Abstract

The double-strand break (DSB) is the most cytotoxic type of DNA damage and measurement of DSBs in cells is essential to understand their induction and repair. Pulsed-field gel electrophoresis (PFGE) allows for quantitative measurement of DSBs in a cell population generated by DNA damaging agents. PFGE has the capacity to separate DNA molecules from several hundred base pairs to over six million base pairs. In the method described here, molecules from five hundred thousand to three million base pairs are consolidated into a single band on the gel that is readily analyzed.

Key words Double-strand breaks, Pulsed-field gel electrophoresis, DNA damage, Agarose gel, Intercalating agent, Radiation

1 Introduction

Double-strand breaks (DSBs) are a particularly cytotoxic type of DNA damage [1]. Generation of this toxic DNA damage in humans has important clinical applications, especially in cancer treatments [2]. DSBs can be generated using DNA damaging agents, such as γ-irradiation and topoisomerase inhibitors, that are exogenously administered to cells in culture or to whole animals [3]. They can also be generated endogenously by normal cellular process (e.g., during immune cell gene rearrangements, DNA repair, and meiotic recombination) [4–6]. Accurate measurement of DSBs in mammalian cells is a critical methodology, because understanding both how DSBs are generated and how they are repaired is a cornerstone in basic scientific investigations and in novel clinical applications.

Clinically relevant doses of γ-irradiation (1–2 Gray) generate approximately 40 DSBs per cell [7, 8], which is equivalent to 1–2 breaks per chromosome. Human chromosomes range from 50 to 250 million base pairs in length. Consequently, for a method to be truly useful in experimental investigations, it should be able to

Lata Balakrishnan and Jason A. Stewart (eds.), *DNA Repair: Methods and Protocols*, Methods in Molecular Biology, vol. 1999, https://doi.org/10.1007/978-1-4939-9500-4_4, © Springer Science+Business Media, LLC, part of Springer Nature 2019

achieve a sensitivity to resolve DNA fragments in the ten million base pair (Mb) range or longer. The pulsed-field gel electrophoresis (PFGE) method is the only technology that approaches this range. The PFGE method relies on the capacity to separate DNA fragments on the basis of their migration through an agarose gel. In standard agarose gel electrophoresis, which relies on application of an electric field of constant strength parallel to the direction of DNA migration, the limit of DNA length resolution is approximately 25 kb [8]. The limitation of mobility is thought to arise from the parallel alignment of DNA molecules and an associated reduction in thermal motion of the molecules such that the larger molecules slither through the agarose sieve with the same mobility. PFGE differs from standard agarose gel electrophoresis in that DNA fragments are separated by application of varying field strengths at different angles for different times. The variation in field direction along with variation of both field strength and time prevents the parallel alignment of the DNA molecules and permits reorientation of the DNA molecules relative to the pores in the agarose sieve. Once the gel system and associated materials are acquired, PFGE provides a simple and quantitative method to measure DSBs in a population of cells.

The PFGE method was developed in tandem with methods that preserve chromosome DNA intact. In standard DNA preparations, the DNA molecules are sheared to lengths shorter than 500,000 bp. The method used for PFGE relies on embedding cells in low-melt agarose then lysing cells and digesting the proteins on the DNA in the agarose. This procedure spares the DNA from shear forces, such as pipetting, that would break the DNA.

When considering methods to quantify DSBs, there are alternative assays that may be better applications depending on the situation. Constant-field gel electrophoresis (CFGE) is a more cost effective alternative to PFGE (that is, the CFGE apparatus is less expensive than the PFGE apparatus) [9], and it has been suggested that it is comparable in sensitivity [10]; however, it has not yet been fully embraced by the scientific community. The neutral comet assay [11] has the capacity to measure the levels of DSBs in mammalian culture in individual cells. The method avoids a 50 °C incubation step used in the processing of embedded cells for PFGE, which can cause excess DNA breakage after radiation [12]. The neutral comet assay does not require a large number of cells and does not require the purchase of an expensive PFGE apparatus, but it requires expertise in the image analysis, and the method does not have the capacity to define the range of lengths of broken DNA fragments. Quantification of DNA repair proteins, such as measurement of the levels of phosphorylated H2AX, which is the substrate for activated ATR, ATM, and DNA-PK protein kinases, has been widely used as a readout for the levels of DNA damage [13]; however, due to the fact that exposure of both ssDNA and dsDNA results in phosphorylation of H2AX, it is not possible to determine the type of DNA damage that has been induced when using this method. Recently, RNA transcripts have been shown to be mediate DSB repair under certain circumstances, and assays have been developed which can quantify RNA-dependent DSBs [14, 15]. Small double-stranded RNAs have also been recently implicated in DSB repair [16]. However, assays to detect these specific repair RNAs to quantify DSB levels have not yet been developed.

We describe here a PFGE protocol that consolidates fragments of DNA of lengths ranging from 500 kb to approximately 2 Mb into a single band for easier quantification [6]. DNA fragments from 200 kb to greater than 6 Mb are also visible in the lane [17]. The PFGE protocol allows for a population-based analysis of DSBs without laborious image analysis. Importantly, additional steps can be added to this protocol, such as Southern or immunoblot analysis, in order to quantify the amount of DSBs associated with the replication fork [18].

2 Materials

2.1 Tissue Culture and Tissues

1. For culturing cells in vitro, a sterile environment is critical. Investigators should use a standard tissue culture hood, incubator, media, and growth conditions optimized for mammalian cell culture in which assays are to be run. Mammalian cells should be incubated in a water-jacketed incubator at 37 °C and 5% CO_2. Manipulations of cells outside the incubator should be done in a Class II (HEPA-filtered) laminar flow biosafety cabinet.
2. Tissue samples from animals and humans can also be used for PFGE analysis. The main requirement for the method is to obtain a single-cell suspension. Hence, nucleated cells from blood or disaggregated cells from solid tissues are also a source of material.

2.2 Buffers and Solutions

1. Agarose insert buffer: 10 mM Tris–HCl pH 7.5, 20 mM NaCl, 50 mM EDTA, diluted in ultrapure water. 50 mL can be made and stored at room temperature indefinitely.
2. Lysis buffer (with proteinase K; prepared fresh for each use): 100 mM EDTA pH 8.0, 0.2% (wt/vol) sodium deoxycholate, 1% (wt/vol) sodium lauryl sarcosine, and 1 mg/mL proteinase K.
3. Tris–acetate–EDTA (TAE) running buffer for gel electrophoresis: Prepare 1800 mL of 1× running buffer from 50× stock (2 M Tris, 1 M acetate, 50 mM EDTA) in ultrapure water. This reagent can be reused once after first use.

4. Agarose that is certified for use in pulsed-field gel electrophoresis. Prepare fresh for each gel run 1% (wt/vol) agarose in 1× TAE buffer.
5. Tris-EDTA (TE) buffer: 10 mM Tris–HCl (adjusted to pH 8.0), 1 mM EDTA pH 8.0 (prepared and sterilized by filtration through 0.2 μm filter). A 500 mL stock of TE buffer can be made and kept at room temperature indefinitely.
6. Agarose blocks containing DNA size markers. We use *S. cerevisiae* chromosomes.
7. DNA staining solution optimized for agarose gel imaging. We use SYBR gold detection reagent, mixing 10 μL of detection reagent with 100 mL of double distilled water.
8. Low-melting point agarose. Add 0.8 g of low-melt agarose to 40 mL of ultrapure water. Equilibrate the 2% solution at 50 °C until the agarose is entirely dissolved and prepare 4-mL aliquots using 15-mL round bottom tubes. Store at 4 °C. Round bottom tubes should be closed tightly to prevent evaporation.
9. 302 nm UV transilluminator and gel imaging system.

2.3 Gel System and Hardware

1. PFGE gel system: Electrophoresis chamber, power module, variable speed pump, cooling module, casting stand, comb, and disposable sample plug molds. The instructions that we give in this protocol for gel set up are based on the Bio-Rad CHEF-DR III system. Other PFGE systems can be used, and the experimenter should refer to the manufacturer's instructions for gel setup in their system.

3 Methods

3.1 Cell Culture, Treatment, and Harvest

1. The overall procedure depends on the DNA damage or cell stress protocol. For chemical treatments or ultraviolet irradiation, cells are seeded into dishes or flasks, treated or not with chemical agent or irradiation, allowed to recover and repair, and then prepared for analysis by embedding the cells in low-melt agarose plugs (*see* **Note 1** and Subheading 2.2, **item 8**). The protocol that we describe here generates three plugs for each sample. A typical scheme for monitoring damage and repair in tissue culture cells is shown in Fig. 1.
2. The PFGE method can be used on both attached and suspension cell cultures. Calculate the number of cells needed at the beginning of the experiment so that one million cells are obtained at harvest, taking into account the cell proliferation expected over the course of the experiment.
3. Seed cells into dishes so that cells are subconfluent and growing at the maximal rate throughout the experiment.

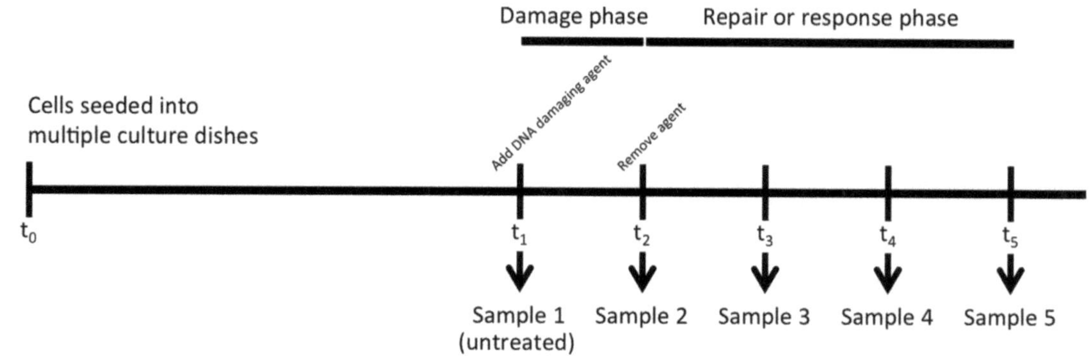

Fig. 1 A possible schema for an experiment with a treatment phase followed by a response or repair phase. Cells are seeded into dishes at the beginning of the experiment. The cell density is chosen so that the cells are growing at their maximal rate and so that there are at least one million cells at the point of harvest. A dish of cells is prepared as an untreated control to be collected at the end of the experiment. At least 12 h is allowed for the cells to recover from culturing (t_0). Cells at treated with agent at t_1. The treatment runs from t_1 to t_2. Dishes of cells are harvested immediately after treatment (t_2) and then and times after treatment (t_3, t_4, t_5, etc.) to monitor accrual of additional DSBs, repair of DSBs, or both

4. Allow at least 12 h recovery after seeding, expose cells to agent, and harvest cells at different times following the accrual of DSBs and repair at each time point (Fig. 2; *see* **Note 2**).
5. If the cells are grown in suspension, disaggregate the cells, centrifuge them to make a pellet (our standard condition is 300 × *g* for 5 min). Attached cells must be dislodged from the substrate and a single-cell suspension must be obtained for an accurate cell count. Wash cells with PBS (*see* **Note 3**), expose the cells to the agent used to detach the cells (typically trypsin-EDTA) until they are easily detached and disaggregated, and resuspend the cells in media, ensuring a thoroughly homogeneous single-cell suspension. Wash suspension culture or attached cells twice in PBS, and then count the cells using a hemocytometer. Ensuring an accurate count of cell numbers across conditions is critical. Keep cells on ice until ready to embed the cells in plugs (*see* **Note 4**).

3.2 Preparing Low-Melting Point Agarose Plugs

1. Place the 4-mL aliquot of 2% low-melt agarose in a 100-mL beaker with 40 mL of water and microwave the aliquot for 10–30 s until agarose is melted. Make sure the cap to the round bottom tube is loose and avoid boiling the agarose. Use the minimum amount of time required for melting and ensure that the agarose is thoroughly melted by checking for opacities in the solution.
2. Place the round bottom tube in a beaker of 42 °C water and allowed several minutes to equilibrate.

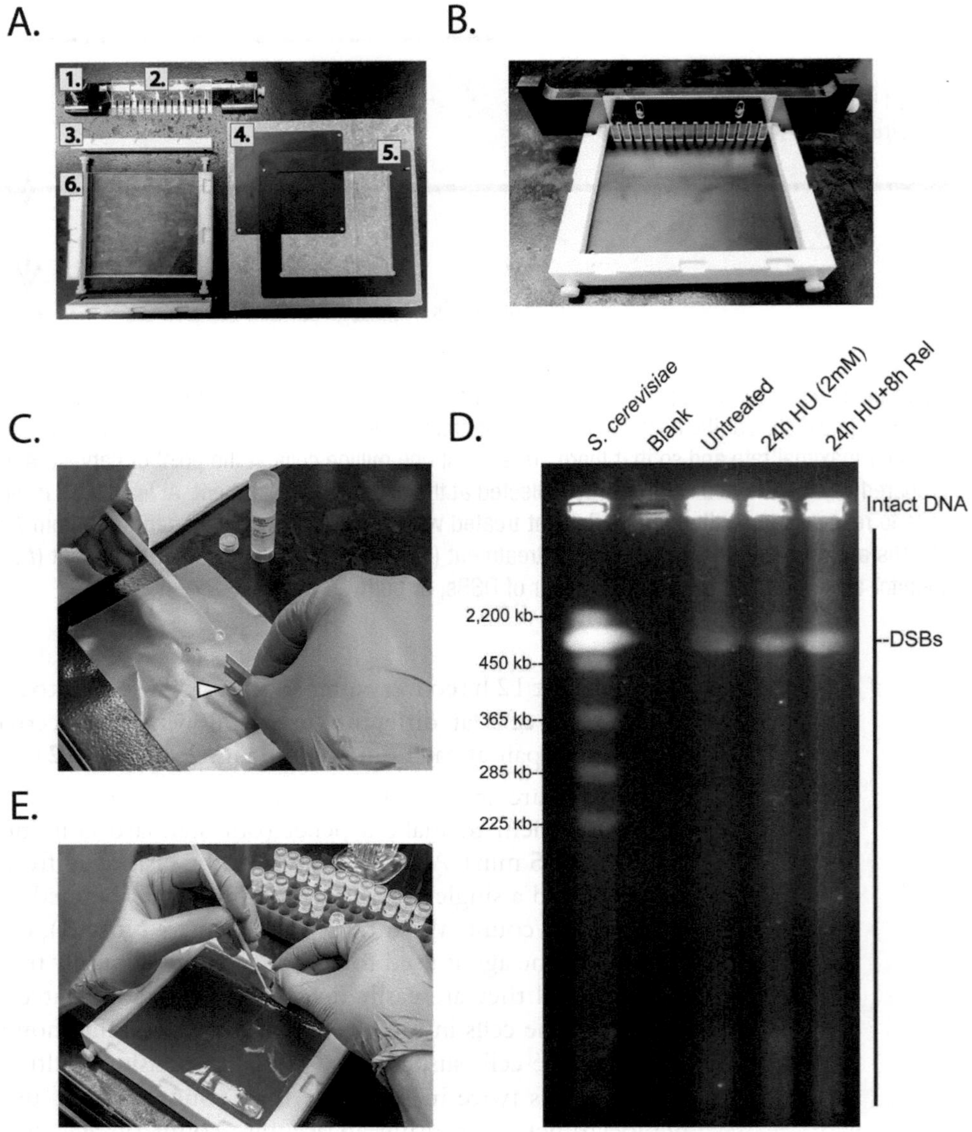

Fig. 2 (**a**) Components of the casting stand and comb holder: (*1*) comb holder, (*2*) comb, (*3*) casting stand end gate, (*4*) platform, (*5*) platform frame, (*6*) casting stand. (**b**) Assembled casting stand and comb holder prior to casting the gel. (**c**) Cutting of a marker plug prior to gel loading. (**d**) Representative gel of HeLa cells treated under the following conditions: cells were untreated, treated with 2 mM HU for 24 h, or treated for 24 h prior to HU washout and release for 8 h. (**e**) Gentle sliding of the cut plugs from the razor into the gel prior to running

3. Centrifuge 7.5×10^5 cells at $300 \times g$ for 5 min at 4 °C (*see* **Note 5**). Aspirate PBS and resuspend the cells (entire pellet) in 150 μL of agarose insert buffer.
4. Add 150 μL of the low-melt agarose for each 150 μL of cell suspension and mix the solution well using a large diameter pipette tip (*see* **Note 6**).

5. Molds are purchased with the bottom side covered, but they can be reused (after cleaning in soapy water, rinsing, and drying) by covering the bottom side with standard adhesive laboratory tape. Fill three wells of the disposable sample plug mold with 90 μL of the low-melt agarose-cell suspension (*see* **Note 7**). Avoid introducing bubbles by filling from one corner of the mold. The mold should not be over-filled.
6. After filling, place the molds at 4 °C for 10–15 min, but not longer, to allow the agarose to solidify.
7. Extrude the plugs from each sample into a 15-mL round bottom tube containing 1.5 mL of lysis (with proteinase K added) buffer and incubate in a stationary water bath at 50 °C overnight (*see* **Note 8**).

3.3 Preparing DNA Samples and Loading the Gel

1. Wash the plugs with 1.5 mL of TE buffer for at least 30 min three times (gentle rocking or nutation can be used). Use a disposable inoculating loop to prevent loss of plugs while decanting during washing steps. Decant into a small beaker to allow for recovery of lost plugs. Plugs can be kept at 4 °C in TE buffer for up to 1 year. During the lysis and protein digestion steps, the plugs should have gone from cloudy-looking to clear and are more difficult to see while in the TE buffer.
2. To prepare the agarose gel for running, assemble the casting stand (*see* Fig. 2b). The gel will sit on the black platform (*see* Fig. 2a) that holds the gel in place in the gel running tank. Place the platform in the casting stand and attach the white casting stand end gates so that the platform slides into the end gate grooves. Ensure that the screws are tight and the platform is firmly fitted into the grooves of the end gates. Adjust the casting stand so that the space between the platform and the casting stand is minimal to minimize liquid agarose leaking under the platform. Adjust the comb to at least 2 mm above the platform. Before pouring the gel, ensure the casting stand is placed on a level surface.
3. Prepare 100 mL of 1% pulsed-field certified agarose gel solution using a microwave or a hot plate. Combine 1.0 g agarose and 100 mL 1× TAE in a clean 250-mL Erlenmeyer flask, cover the flask tightly with plastic wrap to minimize evaporation, and microwave at medium intensity. Ensure all the agarose is melted and avoid excess boiling. Cool the agarose solution to <60 °C. Carefully pour the solution into the prepared casting stand. Avoid bubbles and make sure to allow the gel to cool and solidify fully before loading. Any persistent bubbles can be swept to the periphery with an inoculating loop or Pasteur pipet.

4. Gently pull the comb from the gel and ensure the wells are fully intact.
5. Place one of the *S. cerevisiae* DNA size standard plugs on a 10 cm × 10 cm square of Parafilm laid on the bench. Cut a small square of the size standard and load it, leaving an empty well between the standard and the first sample to avoid contamination of signal (*see* **Note 9**).
6. To prepare the plugs for loading on the gel, use a plastic inoculating loop to gently remove one of the plugs from the tube and place on the 10 cm × 10 cm square of Parafilm on the bench (*see* **Note 10**). To ensure a standard size of plug to run in each experiment, plugs should be cut in half (*see* Fig. 2c and **Note 11**). Next, slide the plug fragment onto a razor, and gently load each plug fragment into each well (*see* Fig. 2e). Take care not to break the plugs as they are fragile at this step.

3.4 Preparing the Gel Box

1. Add 1.8–2 L of 1× TAE buffer to the gel box and cool the buffer to 14 °C by engaging the cooler and pump. Set the pump speed to 0.8–1 L per min.
2. The black platform frame (*see* Fig. 2a) is a black metal sheet with a 20 cm × 20 cm square cut out and holds the platform in place in the gel box. Place the platform barrier in the center of the gel box and ensure the white plastic fastener pegs are engaged in the bottom of the gel box to prevent the black gel platform from moving.

3.5 Running the Gel

1. After the buffer is cooled, slide the platform from the casting stand and carefully remove any solidified gel from the bottom of the platform using a wipe. Do not remove the gel from the platform. Gently place the platform into the black platform frame (*see* Fig. 2a and **Note 12**). Ensure the gel is completely submerged in buffer (*see* **Note 13**).
2. Perform the electrophoresis for 21 h at 14 °C using the following protocol: block I: 9 h, 120° included angle, 5.5 V/cm, 30 to 18-s switch; block II: 6 h, 117° included angle, 4.5 V/cm, 18 to 9-s switch; block III: 6 h, 112° included angle, 4.0 V/cm, 9 to 5-s switch.

3.6 DNA and Double-Strand Break Detection

1. Prepare 100 mL of 1× SYBR gold detection solution in a dish slightly larger than the gel. Remove the gel from the gel box and gently slide the gel off of the platform and into the container. Cover the container with aluminium foil and gently rock for 10–15 min at room temperature.
2. Image the gel using UV illumination and acquire high resolution images adjusted for the exposure of the DSB bands. Ensure that the bands in image are captured in the linear range of exposure.

3.7 Gel Quantification

1. Quantify the gel images using Image J densitometry measurements with background correction. Use the rectangle tool to draw a region of interest around the single band that contains the consolidated DNA fragments. Measure the intensity of each band in the column (*see* **Note 14**). Subtract the background from an empty lane and represent data as either fold change compared to untreated samples or in arbitrary units.

4 Notes

1. If a baseline measurement of DSBs is to be quantified after gamma- or X-irradiation, then a different procedure for cell preparation is followed because the repair process for DSBs in response to gamma- or X-irradiation exposure is very rapid. Here, the cells are embedded in plugs prior to exposure, bathed in growth medium, and exposed to irradiation on ice. A set of plugs can be incubated at 37 °C in medium for varying times to follow the repair process. This method can also be adapted to calibrate the number of DSBs an experimenter achieves in their DNA damaging system by exposing the plugs in ice to increasing doses of irradiation (e.g., 2, 4, 8, 12 Gray).
2. The levels of DSBs are determined by the rate at which breaks are generated by the treatment and the rate at which they are repaired. In order to determine the rate at which DSBs are generated by an agent, it is necessary to inhibit the repair process. This can be accomplished by using pharmacological agents to inhibit specific repair proteins (e.g., inhibition of DNA-PK) or by testing isogenic cell lines in which mutations have been introduced into specific repair pathways.
3. DSBs can occur during apoptosis and cell death, apoptotic or dead cells will usually float or become loosely attached. Thorough washing will allow for greater accuracy of apoptosis-independent DSBs quantification. Repeating the experiment using an inhibitor of apoptosis or a parallel apoptosis assay can be used to ensure that apoptosis-independent DSBs are being observed. In general, it is best to avoid conditions in which more than 10% of the cells are apoptotic.
4. Counting the cells is the longest active step (up to an hour for 12 samples). Keeping cells on ice during sample accrual, subsequent processing, and the counting process is essential to prevent DNA repair during the preparation of the samples.
5. In order to even out the shape of the pellet, the investigator can pellet the cells as described in Subheading 3.2, **step 3**, and then rotate the tubes 180° in the rotor and centrifuge for an additional 1 min. This modification can result in more reliable cell recovery during the aspiration step.

6. A P200 tip can be cut at an angle with a razor blade for gentler pipetting to avoid exposing the cells to excess shear force.
7. Remove the samples in agarose insert buffer from ice and place at room temperature for at least 1 min before mixing with the low-melt agarose to avoid sample solidification prior to being placed into the mold. Do not add low-melt agarose to the cell samples in batches. Add the low-melt agarose to each sample one at a time and immediately fill the plug mold with the sample. This procedure will avoid sample solidification during processing.
8. DNA preparation in plugs can be processed in batches. Alternatively, plugs can be extruded into TE buffer and processed for DNA preparation at a later time. During protein digestion and for storage of plugs after processing, use of an air-tight tube will allow for less loss of buffer through evaporation. Use printed labels or avoid labeling areas of the tubes that will be in contact to the water bath.
9. The plug used for the ladder can be significantly smaller than the well because the signal intensity is usually very high.
10. Take care not to crush or split the plugs as they are removed from the tube.
11. Use a ruler to cut the plug in half. Cutting the plugs exactly in half allows for duplicate runs. To insert the plug into the well in the gel, add the plugs first at an angle then turn them upright to avoid air bubbles; you can also use a small syringe to extract a bubble that gets trapped under the plug. Top up the well with 1% agarose solution and allow time to solidify before placing the gel in the gel box. This prevents plugs from floating out of the wells.
12. The platform should be flush against the bottom of the tank and the frame should prevent the gel from moving during the run. Make sure that none of the plugs floated out of the wells during this step.
13. Ensure that the buffer lines running in to the pump are circulating properly and devoid of air bubbles. Run bubbles through the tubing and into the tank. Air bubbles can accumulate during the run, so the tank should be monitored several times after starting the run. A large air bubble can prevent proper buffer circulation.
14. Make sure to use the same region of interest for each measurement and that the region of interest is wide enough to encompass the entire width of all the bands.

References

1. Bennett CB, Lewis AL, Baldwin KK et al (1993) Lethality induced by a single site-specific double-strand break in a dispensable yeast plasmid. Proc Natl Acad Sci U S A 90:5613–5617
2. O'Connor MJ (2015) Targeting the DNA damage response in cancer. Mol Cell 60:547–560
3. van Gent DC, Hoeijmakers JH, Kanaar R (2001) Chromosomal stability and the DNA double-stranded break connection. Nat Rev Genet 2:196–206
4. Nitiss JL (2009) Targeting DNA topoisomerase II in cancer chemotherapy. Nat Rev Cancer 9:338–350
5. Ashour ME, Atteya R, El-Khamisy SF (2015) Topoisomerase-mediated chromosomal break repair: an emerging player in many games. Nat Rev Cancer 15:137–151
6. Hanada K, Budzowska M, Davies SL et al (2007) The structure-specific endonuclease Mus81 contributes to replication restart by generating double-strand DNA breaks. Nat Struct Mol Biol 14:1096–1104
7. Blocher D (1988) DNA double-strand break repair determines the RBE of alpha-particles. Int J Radiat Biol 54:761–771
8. Elia MC, DeLuca JG, Bradley MO (1991) Significance and measurement of DNA double strand breaks in mammalian cells. Pharmacol Ther 51:291–327
9. Saleh EM, El-Awady RA, Anis N et al (2012) Induction and repair of DNA double-strand breaks using constant-field gel electrophoresis and apoptosis as predictive markers for sensitivity of cancer cells to cisplatin. Biomed Pharmacother 66:554–562
10. Wlodek D, Banath J, Olive PL (1991) Comparison between pulsed-field and constant-field gel electrophoresis for measurement of DNA double-strand breaks in irradiated Chinese hamster ovary cells. Int J Radiat Biol 60:779–790
11. Olive PL, Banath JP (2006) The comet assay: a method to measure DNA damage in individual cells. Nat Protoc 1:23–29
12. Rydberg B (2000) Radiation-induced heat-labile sites that convert into DNA double-strand breaks. Radiat Res 153:805–812
13. Bonner WM, Redon CE, Dickey JS et al (2008) GammaH2AX and cancer. Nat Rev Cancer 8:957–967
14. Keskin H, Shen Y, Huang F et al (2014) Transcript-RNA-templated DNA recombination and repair. Nature 515:436–439
15. Keskin H, Storici F (2018) An approach to detect and study DNA double-strand break repair by transcript RNA using a spliced-antisense RNA template. Methods Enzymol 601:59–70
16. Michelini F, Pitchiaya S, Vitelli V et al (2017) Damage-induced lncRNAs control the DNA damage response through interaction with DDRNAs at individual double-strand breaks. Nat Cell Biol 19:1400–1411
17. Ouyang KJ, Woo LL, Zhu J et al (2009) SUMO modification regulates BLM and RAD51 interaction at damaged replication forks. PLoS Biol 7:e1000252
18. Kawashima Y, Yamaguchi N, Teshima R et al (2017) Detection of DNA double-strand breaks by pulsed-field gel electrophoresis. Genes Cells 22:84–93

Chapter 5

Methods to Study Trinucleotide Repeat Instability Induced by DNA Damage and Repair

Yanhao Lai, Ruipeng Lei, Yaou Ren, and Yuan Liu

Abstract

Trinucleotide repeat (TNR) instability (expansion and deletion) is associated with more than 42 human neurodegenerative diseases and cancer and mediated by DNA replication, repair, recombination, and gene transcription. Somatic TNR instability is involved in the progression of TNR expansion diseases and can be modulated by DNA damage repair and gene transcription. Recent studies from our group and others have shown that DNA base damage and its repair play an active role in modulating TNR instability and are responsible for somatic age-dependent CAG repeat expansion in neurons of Huntington's disease mice induced by oxidative DNA damage. However, it remains to be elucidated how DNA damage, non-B form DNA structures, and DNA repair enzymes and cofactors can coordinate to regulate somatic TNR instability. Understanding the molecular mechanisms underlying DNA damage and repair-mediated somatic TNR instability is critically important for identification of new therapeutic targets for treatment and prevention of TNR-related diseases. Here we describe the methods to study the locations and distribution of DNA base lesions and their effects on TNR instability through DNA base excision repair in in vitro reconstituted human systems.

Key words Trinucleotide repeat instability, DNA base excision repair, DNA damage in repeated sequences, DNA base damage location and distribution, Ligation-mediated PCR

1 Introduction

Trinucleotide repeat (TNR) instability associated with human neurodegenerative diseases and cancer [1, 2] can be modulated through DNA metabolism and gene transcription [3–6]. Recent studies from our group and others have shown that DNA base damage and the base excision repair (BER) pathway are actively involved in modulating somatic TNR instability [7–15] and promoting age-dependent somatic TNR expansion [16]. In our studies, we have designed synthesized oligonucleotide- and plasmid-based systems to study how DNA base lesions induce

Yanhao Lai and Ruipeng Lei contributed equally to this work.

Lata Balakrishnan and Jason A. Stewart (eds.), *DNA Repair: Methods and Protocols*, Methods in Molecular Biology, vol. 1999, https://doi.org/10.1007/978-1-4939-9500-4_5, © Springer Science+Business Media, LLC, part of Springer Nature 2019

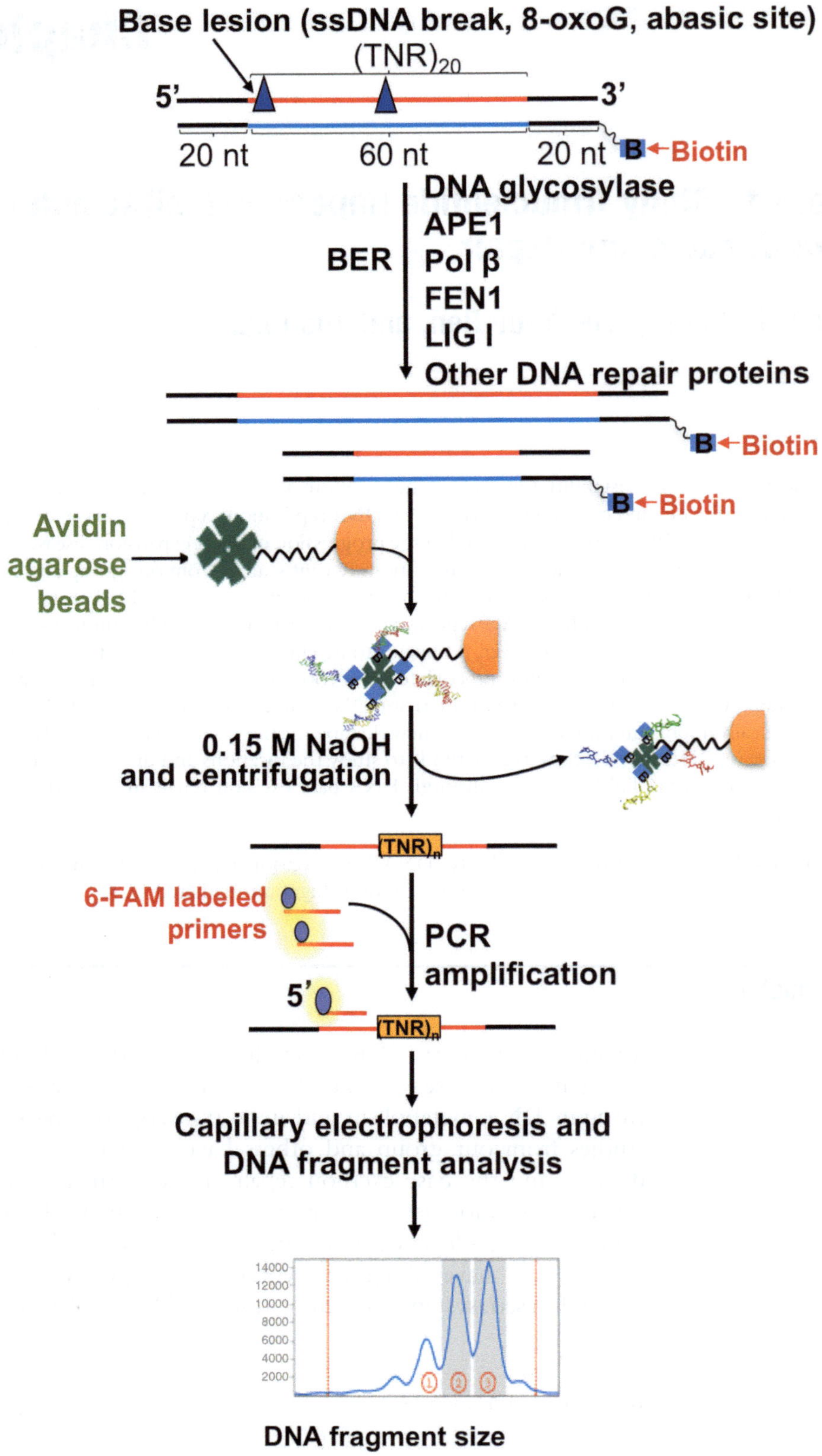

Fig. 1 A scheme of the steps for determining the position effect of a DNA base lesion on TNR instability through BER with synthesized oligonucleotide substrates. A synthesized DNA base lesion is inserted at the 5′-end or in

TNR instability through BER [17, 18]. Also, we have recently established an assay to detect DNA base lesions that occur in a long TNR tract in a site-specific manner with ligation-mediated polymerase chain reaction (LM-PCR) [19]. The methods allow for the determination of the position effects of DNA base lesions on TNR instability upon the specific damage locations [18] and distributions in TNR tracts.

For the oligonucleotide-based method, we have designed the oligonucleotide substrates that contained 20 TNRs with a base lesion (8-oxoguanine or abasic site, tetrahydrofuran, THF) located at 5′-side, middle or 3′-side of a TNR repeat tract (Figs. 1 and 2) [18]. These substrates are constructed by annealing a synthesized 100 nt-oligonucleotide containing 20 TNRs and a base lesion located at different locations of repeat tracts with its template strand. The template strand has a biotin residue attached at its 5′-end [20]. The principal of the method is to reconstitute BER by incubating TNR oligonucleotide substrates with purified BER core enzymes along with enzymes and proteins from other repair pathways to characterize the roles of BER and its interplay with other repair pathways in regulating TNR instability. Reconstituted BER removes a base lesion and results in the production of repaired products with different lengths of TNRs. The repaired products are then incubated with avidin–agarose beads allowing avidin to bind to a biotin residue attached to the 5′-end of the template strand. The repaired strand is separated from the template strand by denaturation of double-strand DNA with a low concentration of NaOH (0.15 M). The repaired products are recovered and amplified by PCR with a forward primer containing a fluorescence tag, 6-carboxyfluorescein (6-FAM) along with a reserve primer. The PCR-amplified fluorescence-labeled repair products are then subject to capillary electrophoresis and DNA fragment analysis for determination of the length change of TNRs in the repaired products (Fig. 1). The method is illustrated in a scheme shown in Fig. 1. Employing the approach, we have demonstrated that a base lesion located at the 5′-end of a TNR tract leads to repeat

Fig. 1 (continued) the middle of a TNR tract that contains 20 repeat units. The repeats are flanked by 20 nt random sequences at both the 5′- and 3′-side. Substrates are incubated with purified BER enzymes in the absence or presence of BER cofactors at 37 °C for 30 min allowing for BER of base lesions in the repeats that subsequently result in repeat instability. Unrepaired substrates and repair products with a 5′-biotin at the template strand are then incubated with avidin–agarose beads. The avidin beads are spun down and washed, and double-strand DNA is denatured by 0.15 M NaOH. The avidin–agarose beads attached to the template strand are then spun down, and the template strand is removed. The repaired strands with different lengths of TNRs, which stay in the supernatant are precipitated with ethanol and dissolved into water. Repair products are then subject to PCR amplification with a 6-FAM-tagged forward primer and a reverse primer. The 6-FAM-labeled PCR products with various lengths of TNRs are separated by capillary electrophoresis. The lengths of TNRs are determined by DNA fragment analysis with the GeneMapper v5.0 software

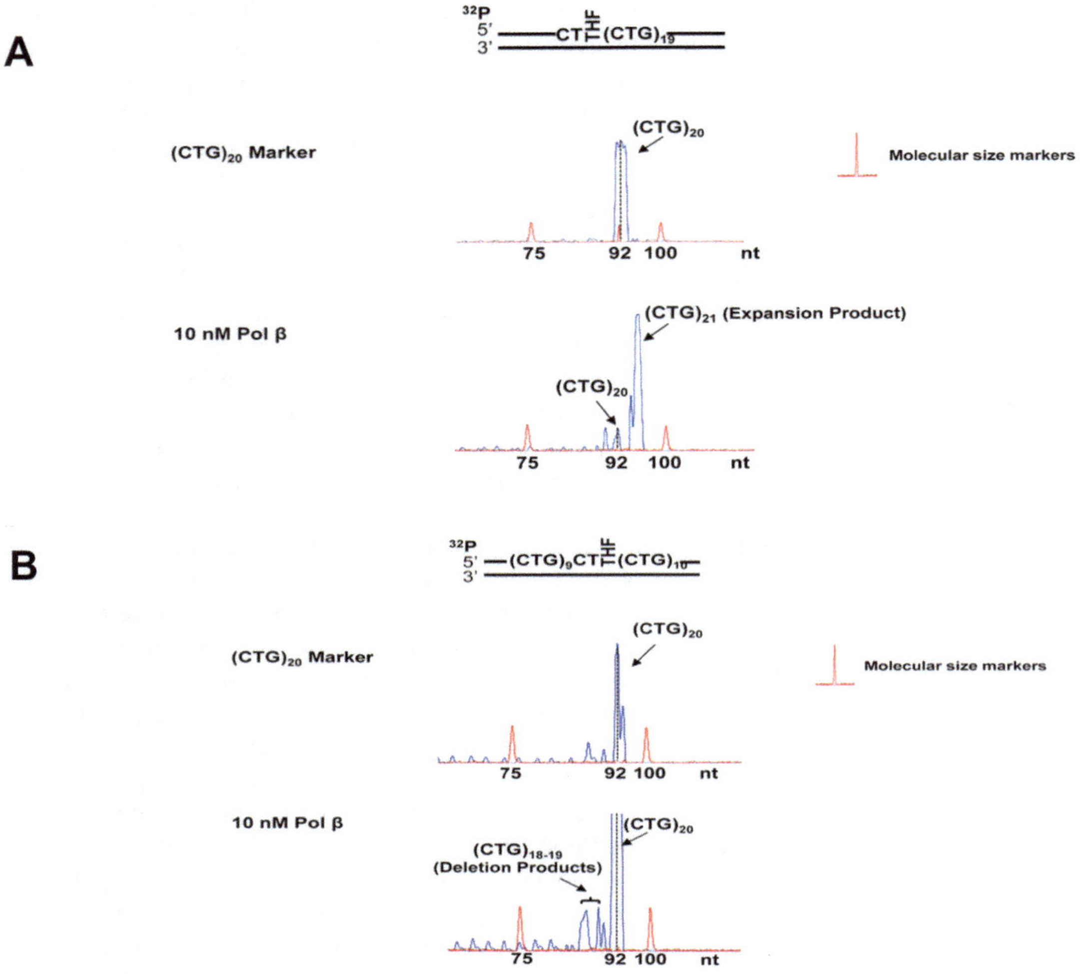

Fig. 2 Experimental data showing the effects of a DNA base lesion on TNR instability during BER when it is located at the 5′-end (**a**) or in the middle of a $(CTG)_{20}$ tract (**b**) (reprinted from ref. [18] with a permission from PLoS One)

expansion, whereas a lesion located in the middle of the repeat tract causes deletion [18] (Fig. 2).

To determine how the distribution of multiple DNA base damage and single-strand DNA (ssDNA) breaks may govern TNR expansion and deletion, we have developed a plasmid-based in vitro method. This method employs a plasmid containing a $(CAG)_{35}/(CTG)_{35}$ tract, which is exposed to reactive oxygen species (ROS), specifically hydroxyl radicals produced by Fenton reaction (Fig. 3a). Hydroxyl radicals can induce ssDNA breaks and 8-oxoGs at multiple sites in CAG repeats that are repaired by BER. The site-specific distribution of ssDNA breaks in CAG repeats resulting directly from ROS and/or removal of hydroxyl radical-induced 8-oxoGs by BER, can be determined by LM-PCR (Fig. 3b). BER of base lesions in CAG repeats is reconstituted by

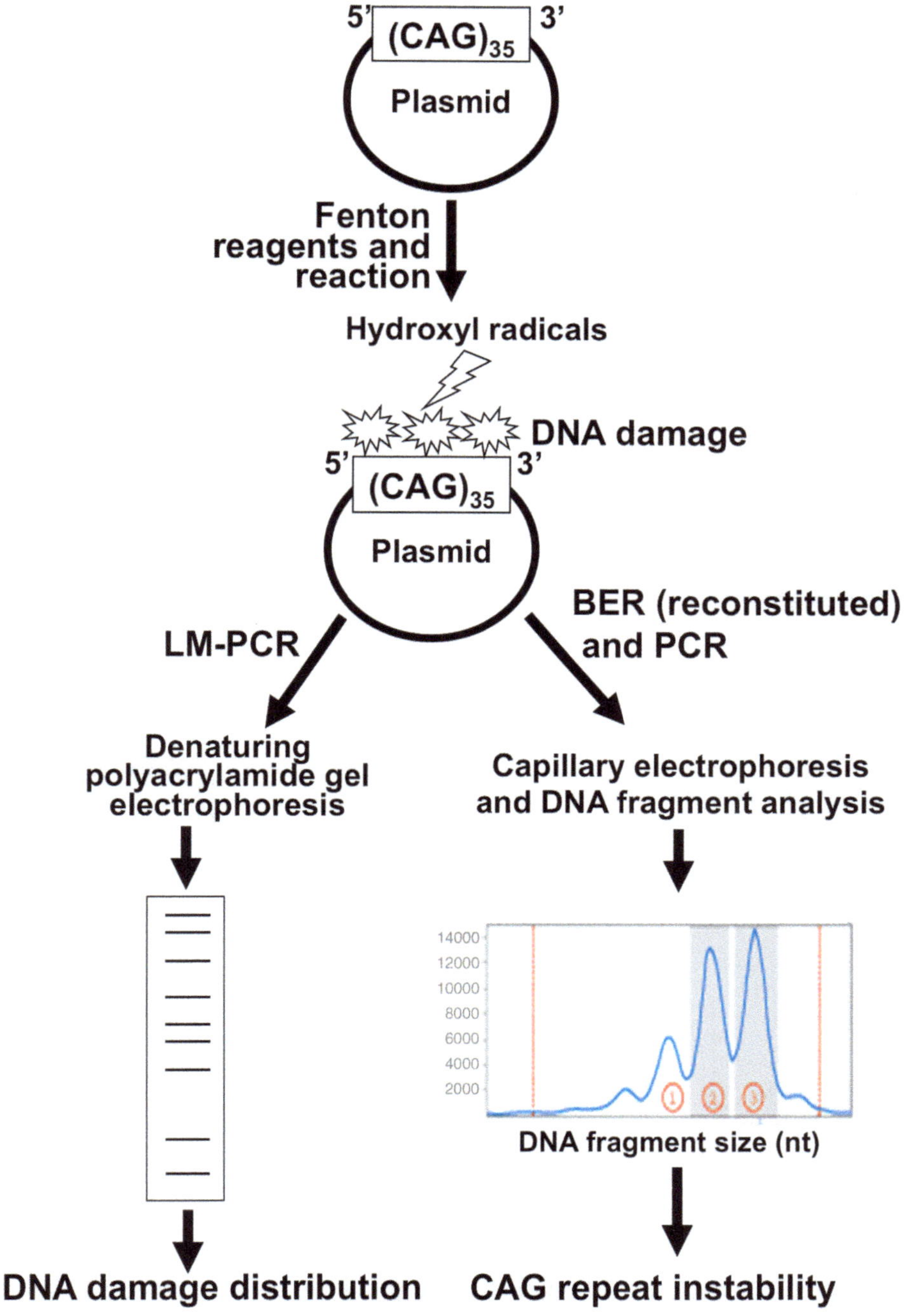

Fig. 3 A schematic representation of the steps for determining the effects of the distribution of DNA base lesions in a long CAG repeat tract on repeat instability through BER. Plasmids containing $(CAG)_{35}$ repeats are incubated with Fenton reagents allowing for generation of hydroxyl radicals that subsequently create ssDNA breaks in the $(CAG)_{35}$ repeat tract. A portion of damaged plasmids is subject to LM-PCR, and PCR products are separated by denaturing urea polyacrylamide gel electrophoresis. The distribution of ssDNA in CAG repeats is determined by the mobility of PCR products according to the mobility of the DNA size markers. Another portion of damaged plasmids is subject to reconstituted BER allowing for production of BER repair products. Repair products are then amplified by PCR with a 6-FAM-tagged forward primer and a reverse primer. PCR products are separated by capillary electrophoresis. The sizes of CAG repeats in the repair products are determined by the GeneMapper v5.0 software

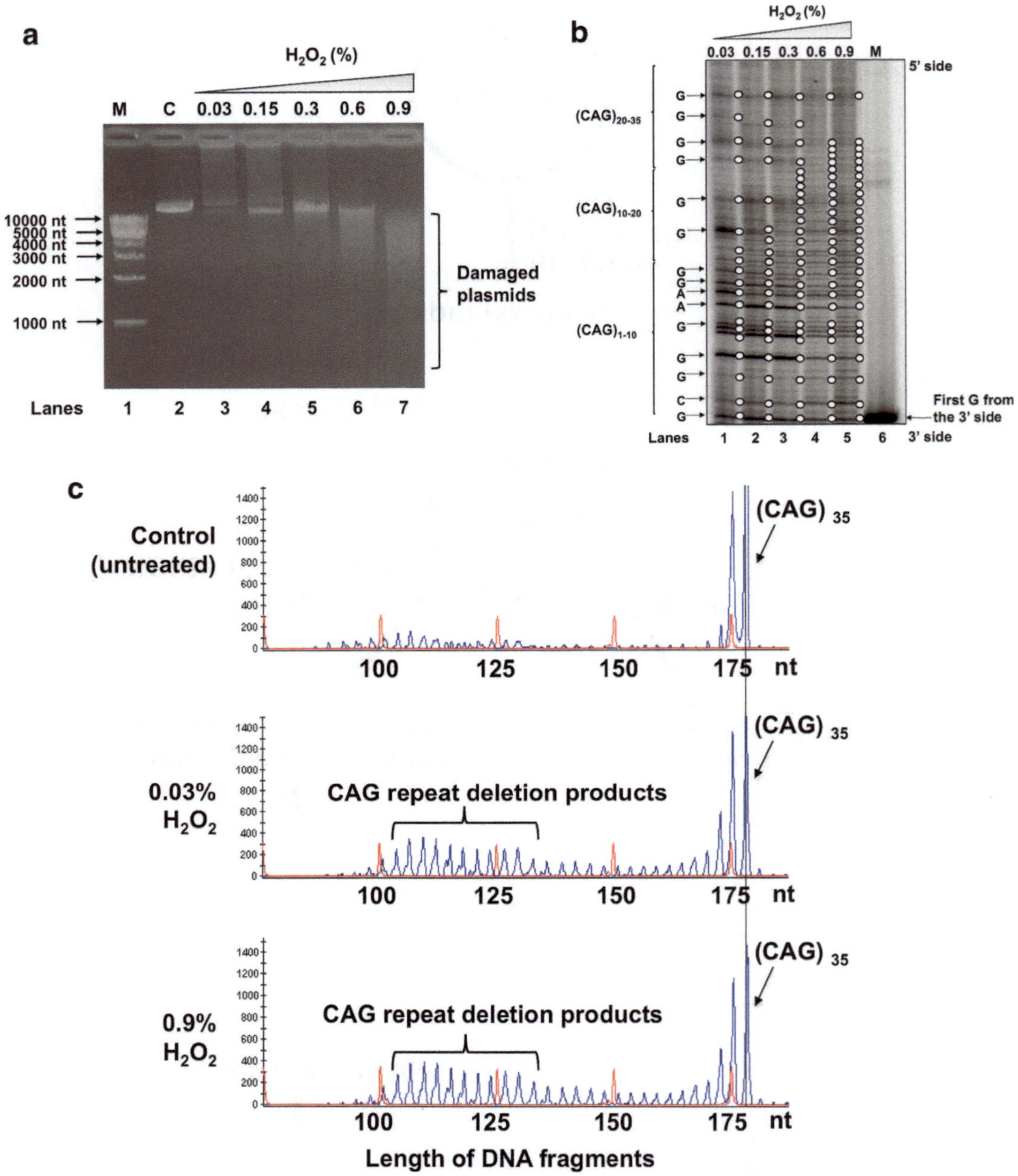

Fig. 4 DNA strand breaks results in CAG repeat instability through BER in a long repeat tract (**a**) DNA strand breaks in pcDNA3.1/CT-GFP-TOPO-CAG_{35} plasmids induced by hydroxyl radicals generated by Fenton reaction at different percentage of H_2O_2 was detected by agarose gel electrophoresis. Lane 1 represents the DNA size markers. Lane 2 represents undamaged pcDNA3.1/CT-GFP-TOPO-CAG_{35} plasmids. Lanes 3–7 indicate damaged pcDNA3.1/CT-GFP-TOP-CAG_{35} plasmids under treatment of Fenton reagents with various percentage of H_2O_2 (0.03–0.9%). (**b**) The results of urea-denaturing polyacrylamide gel electrophoresis of the products from LM-PCR, illustrate the multiple ssDNA breaks located at specific bases in 30 CAG repeats. The damaged bases are indicated by a series of white filled circles. The results showed that a low percentage of H_2O_2 at 0.03% resulted in the localized distribution of ssDNA breaks in CAG repeats (lane 1). Increasing percentage of H_2O_2 led to increasing number of DNA damage sites and even distribution of ssDNA breaks in

incubating damaged plasmids with purified BER enzymes and cofactors. Repaired plasmids are then subject to PCR amplification with a 6-FAM-tagged forward primer and a reverse primer. Repaired products with different lengths of CAG repeats are separated by capillary electrophoresis, and repeat length is determined by DNA fragment analysis with the GeneMapper v. 5.0 software. The site-specific distribution of DNA base lesions in CAG repeats is then correlated with repeat expansion and deletion. With this method, we have identified the hotspots and distribution of ssDNA breaks in a $(CAG)_{35}$ repeat tract that induced by hydroxyl radicals. We have found that hydroxyl radicals-induced ssDNA breaks predominantly distributed in the middle of the repeat tract, thereby leading to CAG repeat deletion via BER (Fig. 4).

2 Materials

2.1 TNR Oligonucleotide Substrates

Polyacrylamide gel purified synthesized oligonucleotides containing a TNR repeat tract with a base lesion at various locations in repeat tract.

1. Polyacrylamide gel purified synthesized oligonucleotide template strands that are tagged with a biotin residue at the 5′-end [20].
2. TE buffer: 10 mM Tris–HCl, 1 mM EDTA, pH 8.0, store at 25 °C.
3. 10× DNA annealing buffer: 100 mM Tris–HCl, 1 M NaCl, 10 mM EDTA, pH 7.5, store at −20 °C.

2.2 A Plasmid-Based TNR Substrate Containing a Long CAG Repeat Tract

1. 1 mM Fe^{2+} solution, freshly made and stored at 25 °C (*see* **Note 1**).
2. 1 mM EDTA, pH 8.0, freshly made and stored at 25 °C (*see* **Note 1**).
3. 100 mM sodium ascorbate, freshly made and stored at 25 °C (*see* **Note 1**).
4. 30% H_2O_2 stock solution, stored at 4 °C.
5. Thiourea-EDTA quenching buffer: 95 mM thiourea, 48 mM EDTA, pH 8.0, freshly made and stored at 25 °C.

Fig. 4 (continued) the CAG repeat tract. Lanes 1–5 indicate the LM-PCR results from damaged plasmids treated with Fenton reagents along with 0.03–0.9% H_2O_2. Lanes 6 and 7 are the markers for indicating the locations of specific DNA bases. (**c**) The instability of CAG repeats induced by hydroxyl radicals generated by Fenton reaction through BER. The results showed that little CAG repeat deletion and expansion products were detected from the untreated control plasmids. BER of ssDNA breaks in CAG repeats generated by Fenton reaction with 0.03% and 0.9% H_2O_2 led to a significant increase in the amount of repeat deletion products with $(CAG)_{10–18}$ repeats

6. pcDNA3.1/CT-GFP-TOPO-CAG_{35} Plasmid: 1 μg/μL, stored at −20 °C.
7. 100% and 70% ethanol, stored at −20 °C.
8. 3 M sodium acetate, pH 5.2, stored at 25 °C (*see* **Note 2**).
9. TE buffer: 10 mM Tris–HCl, 1 mM EDTA, pH 8.0, stored at 25 °C.
10. 10× TBE buffer: 1 M Tris base, 1 M boric acid, and 20 mM EDTA, stored at 25 °C.
11. 7.5 μg/μL linear acrylamide solution, stored at 25 °C.

2.3 Determination of ssDNA Breaks in CAG Repeats Using LM-PCR

1. Synthesized double-strand oligonucleotide DNA linkers (10 μM) (*see* Fig. 3b) and PCR primers (10 μM), stored at −20 °C.
2. 10 mM dNTP stock solution, stored at −20 °C.
3. 10 U/μL T4 polynucleotide kinase, stored at −20 °C.
4. 2 U/μL Deep Vent DNA polymerase, stored at −20 °C.
5. 10× PCR enhancer solution, stored at −20 °C.
6. 400 U/μL T4 DNA ligase, stored at −20 °C.
7. 100 mM ATP stock solution, stored at −20 °C.
8. AmpliTaq Gold 360 DNA polymerase, stored at −20 °C.
9. 10× AmpliTaq Gold 360 buffer, stored at −20 °C.
10. 3000 Ci/mmol ATP [γ-^{32}P] (10 mCi/mL), 100 μCi, stored at −20 °C.
11. 18% 8 M urea-denaturing polyacrylamide gel.
12. 2× stopping buffer: 95% deionized formamide, 0.5 mM EDTA, pH 8.0, 0.25% (w/v) bromophenol blue, 0.25% (w/v) xylene cyanol FF.

2.4 Reconstitution of BER

1. Purified recombinant human BER enzymes including DNA polymerase β (Pol β), AP endonuclease 1 (APE1), flap endonuclease 1 (FEN1), and DNA ligase I (LIG I), stored at −80 °C.
2. BER reaction buffer: 50 mM Tris–HCl, 50 mM KCl, 0.1 mg/mL BSA, 1 mM DTT, 0.1 mM EDTA store at 4 °C.
3. 10 mM dNTP stock solution, stored at −20 °C.
4. 50 mM Mg^{2+} stock solution, stored at −20 °C.
5. 10× BER reaction stopping buffer: 100 mM EDTA, pH 8.0, stored at −20 °C.

2.5 Isolation of Repaired Products

1. Pierce Avidin–Agarose, 5 mL of settled gel (10 mL total slurry volume).
2. Binding Buffer: 0.1 M phosphate, 0.15 M NaCl, 1% Nonidet P-40, pH 7.2.

3. 3 M sodium acetate, pH 5.2, stored at 25 °C.
4. 100% and 70% ethanol, stored at −20 °C.
5. TE buffer: 10 mM Tris–HCl, 1 mM EDTA, pH 8.0, stored at 25 °C.
6. 7.5 μg/μL linear acrylamide solution.

2.6 Determination of TNR Instability

1. A polyacrylamide gel purified synthesized oligonucleotide control that contains a TNR tract without a base lesion (*see* **Note 3**).
2. Purified plasmid control without hydroxyl radical treatment (*see* **Note 4**).
3. Forward PCR primers tagged by a 6-FAM group: 10 μM, stored at −20 °C.
4. Reverse PCR primers: 10 μM, stored at −20 °C.
5. AmpliTaq Gold 360 DNA polymerase, stored at −20 °C.
6. 10× AmpliTaq Gold 360 buffer, stored at −20 °C.
7. DNA size markers (MapMarker X-Rhodamine Labeled 50–1000 bp) for DNA fragment size determination through capillary electrophoresis.

3 Methods

3.1 Construction of TNR Oligonucleotide Substrates Containing a Synthesized DNA Base Lesion

1. Mix 50 pmol purified synthesized DNA oligonucleotide containing a TNR tract with a base lesion (8-oxoG or abasic site; a synthesized tetrahydrofuran residue can be used as an analogy of an abasic site), 100 pmol template strand, and 5 μL 10× annealing buffer in a 1.5 mL Eppendorf tube and add deionized water to a total volume of 50 μL.
2. Heat the annealing mixture in a heat block at 95 °C for 10 min and turn off the heat block until the temperature drops to room temperature (25 °C).
3. Dilute the substrate stock solution (1 μM) with BER reaction buffer to various concentrations for assembly of BER reactions according to the need of experiments.

3.2 Construction of a Plasmid-Based TNR Substrate Containing ssDNA Breaks at Multiple Sites in a Long CAG Repeat Tract

1. Make 30 μL DNA plasmid solution with 10 μg pcDNA3.1/CT-GFP-TOPO-CAG_{35} plasmid in Tris buffer containing 50 mM Tris–HCl, pH 7.5.
2. Mix 10 μL of thiourea (0.4 M) with 32 μL EDTA (200 mM), pH 8 to make 42 μL quench solution (95 mM thiourea and 152 mM EDTA, pH 8).
3. Make 10 μL Fenton reaction solution: 0.01 mM Fe^{2+} and 0.02 mM EDTA, pH 8.0 (*see* **Note 1**).

4. Make 10 μL H_2O_2 solution (0.03%, 0.15%, 0.3%, 0.6%, and 0.9% H_2O_2) (*see* **Note 1**).
5. Mix DNA plasmid solution with 10 μL Fenton reaction solution, 10 μL 1 mM sodium ascorbate solution. Quickly add 10 μL H_2O_2 solution with various concentrations into each of reaction mixtures and incubate them in the dark at room temperature (25 °C) for 5 min.
6. Add 30 μL quench solution to the reaction mixture (total volume: 60 μL) and incubate at 25 °C for 30 min.
7. Add 6 μL 3 M sodium acetate (pH 5.2) and cold 120 μL 100% ethanol to the reaction mixture and mix well by gently tapping the bottom of the Eppendorf tube.
8. Freeze the reaction mixture at −80 °C for 30 min.
9. Pellet plasmid DNA by centrifugation at 11,269 × *g* (Eppendorf Centrifuge 5417R) for 30 min.
10. Wash DNA plasmid pellets once with cold 70% ethanol.
11. Dry DNA plasmid pellets in the air for 5 min.
12. Dissolve DNA plasmid pellets in sterile water.
13. Measure the concentrations of DNA plasmids with Nano-Drop and store the plasmid DNA at −20 °C for further analysis.
14. Check DNA strand breaks in the plasmids by running the plasmid samples (1 μg) in 1% agarose gel in 1× TBE buffer through gel electrophoresis at 150 V for 1 h.

3.3 Identification of Site-Specific Distribution of ssDNA Breaks in a Long CAG Repeat Tract in a Plasmid Using LM-PCR

1. Phosphorylate the 5′-end of ssDNA breaks in a TNR tract in DNA plasmids (2 μg) by mixing 1 μL T4 PNK, with 2.5 μL 10× PNK buffer, 2.5 μL 10 mM ATP. Add deionized water to a total volume of 25 μL. Incubate the reaction at 37 °C for 1 h. Terminate the phosphorylation reaction and denature plasmid DNA at 75 °C for 10 min.
2. Perform primer extension by incubating phosphorylated plasmids (25 μL) with 5 μL 10× Thermolpol reaction buffer, 1 μL of Deep Vent DNA polymerase (2 U), 0.8 μL of 10 mM dNTP, 4 μL of 10× PCR Enhancer, 5 μL of 10 μM primer 1, 2 μL of 100 mM $MgSO_4$. Add deionized water to 50 μL.
3. Perform the primer extension reaction in a PCR machine with the following program.
   (a) 95 °C for 5 min.
   (b) 62 °C for 30 min.
   (c) 72 °C for 10 min.
4. Mix 9 μL (360 ng plasmid DNA) primer extension products with 1 μL of 10 μM double-strand linker DNA (*see* Fig. 3b), 2 μL of 10× ligation buffer, 1.5 μL T4 DNA ligase (400 U/μL),

and 6.5 μL water and incubate the ligation mixture at 16 °C overnight.

5. Mix 100 ng of the ligated DNA with 2.5 μL of 50 mM $MgCl_2$, 2.5 μL of 10× AmpliTaq Gold 360 buffer, 0.5 μL of AmpliTaq Gold 360 DNA polymerase (2.5 U), 0.8 μL of 10 mM dNTPs, 2.5 μL of 10× PCR Enhancer, 2.5 μL of 10 μM of Linker primer, 2.5 μL of 10 μM of primer 2. Add deionized water to make a total volume of 25 μL.
6. Perform PCR amplification with following the program for 40 cycles with a final extension at 72 °C for 10 min.
   (a) 95 °C for 5 min.
   (b) 95 °C for 30 s.
   (c) 60 °C for 30 s.
   (d) 72 °C for 2 min for 30 s.
7. Remove the PCR reaction mixture from a PCR machine and spin down the mixture at 704 × *g* for 5 min, and leave the reaction mixture on ice and the store it at −20 °C later.
8. Radiolabel the 5′-end of Primer 3 and size markers by incubating 10 pmol Primer 3 or 50 ng size markers with 1 μL ATP [γ-^{32}P] (3000 Ci/mmol 10 mCi/mL, 100 μCi), 1 μL 10× PNK buffer, 10 U PNK, add deionized water to total volume of 10 μL and incubate the mixture at 37 ° C for 30 min.
9. Mix 5 μL plasmids (25 ng DNA) with 0.5 μL of AmpliTaq Gold 360 DNA (2.5 U), 1 μL 10× AmpliTaq Gold 360 buffer, 1 μL of 1 μM 5′-labeled Primer 3, 1 μL of 10 mM dNTPs, 0.5 μL of, 1.5 μL deionized water (total volume of 10 μL).
10. Place the reaction mixture in a PCR machine and set up the labeling reaction with following the program for three cycles with one cycle of a final extension at 72 °C for 10 min.
    (a) 95 °C for 5 min.
    (b) 95 °C for 1 min.
    (c) 60 °C for 3 min.
    (d) 72 °C for 2 min and 30 s.
11. Place the reaction mixture on ice and spin it down at 704 × *g* for 5 min. Add 2.8 μL of 100 mM EDTA to the reaction mixture to terminate the labeling reaction.
12. Transfer the reaction mixture to a 1.5 mL Eppendorf tube.
13. Add 1.3 μL of 3 M sodium acetate (pH 5.2), 1.5 μL of 7.5 μg/μL linear acrylamide and 26 μL of cold 100% ethanol to precipitate DNA plasmids at −80 °C for 30 min.
14. Spin down DNA by centrifugation at 11,269 × *g* for 30 min.
15. Wash DNA pellets once with cold 70% ethanol.

16. Dissolve DNA pellets into 10 μL TE buffer. Based on the counts of radioactivity of DNA samples, add 20 μL 2× stopping buffer.
17. Run the DNA samples along with the 5′-labeled size marks in 18% 8 M urea-denature polyacrylamide gel overnight (12–14 h) at 28 W (*see* **Note 5**).
18. Dry the gel with Bio-Rad Gel Dryer Model 583 and expose the gel in a phosphoimager cassette overnight (*see* **Note 5**).
19. Scan the gel by a phosphorimager to obtain a gel image.

3.4 Reconstitution of BER

1. Mix purified BER enzymes, APE1 (10 nM), Pol β (10 nM), FEN1 (10 nM), LIG I (25 nM) with 50 nM synthesized oligonucleotide or plasmid substrates in BER reaction buffer with a total volume of 50 μL.
2. Incubate reaction mixtures at 37 °C for 30 min (*see* **Note 6**).
3. Terminate BER reaction with the addition of 5 μL 10× BER stopping buffer.

3.5 Isolation of Repaired Products

1. To isolate repaired products resulting from BER with oligonucleotide substrates, mix 50 μL reaction mixture with 30 μL avidin–agarose beads and incubate the mixture at 25 °C for 2 h on a rotator.
2. Spin down avidin–agarose beads at 1957 × *g* for 1 min, and remove the supernatant.
3. Wash the repaired products that are bound by avidin–agarose beads for three times with 50 μL binding buffer.
4. Separate the repaired strands from their template strands by denaturation through incubation with 50 μL 0.15 M NaOH for 15 min with rotation at 25 °C.
5. Spin down avidin–agarose beads that bind to the template strands at 1957 × *g* for 2 min.
6. Carefully transfer 50 μL supernatant containing the repaired strands to a new Eppendorf tube.
7. To isolate repaired products resulting from BER with plasmid substrates, skip **steps 1–6** and directly go to **step 8**.
8. Mix 50 μL repaired products from oligonucleotide or plasmid substrates with 5 μL 3 M sodium acetate (pH 5.2) and 100 μL cold 100% ethanol.
9. Freeze the DNA precipitation mixture at −80 °C for 30 min.
10. Pellet DNA repaired products by centrifugation at 11,269 × *g* for 30 min.
11. Wash the DNA repaired product pellets once with cold 70% ethanol.

12. Air-dry DNA pellets for 5 min.
13. Dissolve DNA pellets into the sterilized water and store at −20 °C.

3.6 Determination of TNR Instability

1. Mix 50 ng repaired products with 2 μL 10× AmpliTaq Gold 360 buffer, 0.5 μL of AmpliTaq Gold 360 DNA (2.5 U), 2 μL 10 μM 6-FAM-tagged forward primer and 2 μL 10 μM reverse primer, 2 μL 50 mM Mg^{2+}, 2 μL 1 mM dNTPs and add sterile water to total volume of 20 μL.
2. Perform PCR reaction with the following program for 30 cycles.
   (a) 95 °C for 5 min.
   (b) 62 °C for 30 min.
   (c) 72 °C for 10 min.
3. Collect PCR products and store at −20 °C.
4. Mix 2 μL PCR products with 1 μL MapMarker 1000 size markers and 9.5 μL Hi-Di formamide.
5. Run the products through capillary electrophoresis to separate different sizes of DNA fragments with various lengths of TNR tracts.
6. Analyze and determine the sizes of TNRs of repaired products using software GeneMapper v.5.0 (Life Technologies-ThermoFisher Scientific).

4 Notes

1. Solutions of Fenton reaction, sodium ascorbate, and H_2O_2 must be prepared freshly at the time when experiments are set up.
2. 3 M sodium acetate solution must be checked monthly to see if there is any crystal formed at the bottom of a solution bottle or tube. If any crystal is identified in the solution, fresh sodium acetate solution needs to be made.
3. Oligonucleotide control with TNRs must be carefully purified by 8 M urea-denaturing polyacrylamide gel electrophoresis through UV shadowing. It is important to use a relatively high percentage (18–20%) gel and avoid loading too much DNA in one lane of the gel.
4. It should be noted that purified pcDNA3.1/CT-GFP-TOPO-CAG_{35} plasmids contain both $(CAG)_{35}$ and $(CAG)_{34}$ repeats. This suggests that $(CAG)_{35}$ repeats were deleted during replication of the plasmids in *E. Coli* bacteria.

5. The watts and running time of polyacrylamide gel electrophoresis for separating multiple sizes of DNA fragments should be optimized based on the size of DNA products. Gel exposing time should be optimized based on the intensity of radioactivity.
6. The concentrations of BER enzymes can be varied to test the role of each individual repair enzyme in modulating TNR instability during BER.

Acknowledgments

This work was supported by NIH grant R01ES023569 to Y. Liu.

References

1. Paulson HL, Fischbeck KH (1996) Trinucleotide repeats in neurogenetic disorders. Annu Rev Neurosci 19:79–107. Epub 1996/01/01. https://doi.org/10.1146/annurev.ne.19.030196.000455
2. McMurray CT (2010) Mechanisms of trinucleotide repeat instability during human development. Nat Rev Genet 11(11):786–799. https://doi.org/10.1038/nrg2828
3. Wells RD, Dere R, Hebert ML, Napierala M, Son LS (2005) Advances in mechanisms of genetic instability related to hereditary neurological diseases. Nucleic Acids Res 33 (12):3785–3798. https://doi.org/10.1093/nar/gki697
4. McMurray CT (1999) DNA secondary structure: a common and causative factor for expansion in human disease. Proc Natl Acad Sci U S A 96(5):1823–1825
5. Wells RD (1996) Molecular basis of genetic instability of triplet repeats. J Biol Chem 271 (6):2875–2878
6. Mirkin SM (2007) Expandable DNA repeats and human disease. Nature 447 (7147):932–940. https://doi.org/10.1038/nature05977
7. Usdin K, House NC, Freudenreich CH (2015) Repeat instability during DNA repair: insights from model systems. Crit Rev Biochem Mol Biol 50(2):142–167. https://doi.org/10.3109/10409238.2014.999192
8. Freudenreich CH (2018) R-loops: targets for nuclease cleavage and repeat instability. Curr Genet 64:789–794. https://doi.org/10.1007/s00294-018-0806-z
9. Liu Y, Wilson SH (2012) DNA base excision repair: a mechanism of trinucleotide repeat expansion. Trends Biochem Sci 37 (4):162–172. https://doi.org/10.1016/j.tibs.2011.12.002
10. Xu M, Lai Y, Torner J, Zhang Y, Zhang Z, Liu Y (2014) Base excision repair of oxidative DNA damage coupled with removal of a CAG repeat hairpin attenuates trinucleotide repeat expansion. Nucleic Acids Res 42(6):3675–3691. https://doi.org/10.1093/nar/gkt1372
11. Lai Y, Beaver JM, Lorente K, Melo J, Ramjagsingh S, Agoulnik IU et al (2014) Base excision repair of chemotherapeutically-induced alkylated DNA damage predominantly causes contractions of expanded GAA repeats associated with Friedreich's ataxia. PLoS One 9 (4):e93464. https://doi.org/10.1371/journal.pone.0093464
12. Lai Y, Budworth H, Beaver JM, Chan NL, Zhang Z, McMurray CT et al (2016) Crosstalk between MSH2-MSH3 and polβ promotes trinucleotide repeat expansion during base excision repair. Nat Commun 7:12465. https://doi.org/10.1038/ncomms12465
13. Beaver JM, Lai Y, Xu M, Casin AH, Laverde EE, Liu Y (2015) AP endonuclease 1 prevents trinucleotide repeat expansion via a novel mechanism during base excision repair. Nucleic Acids Res 43(12):5948–5960. https://doi.org/10.1093/nar/gkv530
14. Beaver JM, Lai Y, Rolle SJ, Liu Y (2016) Proliferating cell nuclear antigen prevents trinucleotide repeat expansions by promoting repeat deletion and hairpin removal. DNA Repair (Amst) 48:17–29. https://doi.org/10.1016/j.dnarep.2016.10.006
15. Xu M, Lai Y, Jiang Z, Terzidis MA, Masi A, Chatgilialoglu C et al (2014) A 5′, 8-cyclo-2′-deoxypurine lesion induces trinucleotide

repeat deletion via a unique lesion bypass by DNA polymerase beta. Nucleic Acids Res 42 (22):13749–13763. https://doi.org/10.1093/nar/gku1239

16. Kovtun IV, Liu Y, Bjoras M, Klungland A, Wilson SH, McMurray CT (2007) OGG1 initiates age-dependent CAG trinucleotide expansion in somatic cells. Nature 447(7143):447–452. https://doi.org/10.1038/nature05778
17. Xu M, Gabison J, Liu Y (2013) Trinucleotide repeat deletion via a unique hairpin bypass by DNA polymerase beta and alternate flap cleavage by flap endonuclease 1. Nucleic Acids Res 41(3):1684–1697. https://doi.org/10.1093/nar/gks1306
18. Lai Y, Xu M, Zhang Z, Liu Y (2013) Instability of CTG repeats is governed by the position of a DNA base lesion through base excision repair. PLoS One 8(2):e56960. https://doi.org/10.1371/journal.pone.0056960
19. Pfeifer GP, Riggs AD (1993) Genomic footprinting by ligation mediated polymerase chain reaction. Methods Mol Biol 15:153–168. https://doi.org/10.1385/0-89603-244-2:153
20. Ren Y, Lai Y, Laverde EE, Lei R, Rein HL, Liu Y (2017) Modulation of trinucleotide repeat instability by DNA polymerase beta polymorphic variant R137Q. PLoS One 12(5): e0177299. https://doi.org/10.1371/journal.pone.0177299

Chapter 6

Extracting and Measuring dNTP Pools in *Saccharomyces cerevisiae*

Radha Subramaniam, Natalie A. Lamb, Yoonchan Hwang, Lauren Johengen, and Jennifer A. Surtees

Abstract

Regulation of dNTP pools in an intracellular environment is not only vital for DNA replication but also plays a major role in maintaining genomic stability. Ribonucleotide reductase (RNR) catalyzes the rate-limiting step in dNTP synthesis and altered regulation of RNR leads to imbalanced dNTP pools. Increased dNTP levels are mutagenic and have the potential to interfere with pathways that are involved in DNA replication, repair and DNA damage control. However, the mechanisms through which altered dNTP pools affect these pathways are poorly understood. Nonetheless, altered dNTP pools have been identified in a number of cellular contexts, including cancer. In order to interpret and analyze the effects of altered dNTP pools, we need quantitative information about dNTP pools in different genetic and environmental contexts in vivo. Here we describe a high-throughput fluorescence-based assay that uses a qPCR-based approach to quantify dNTP levels for use with *Saccharomyces cerevisiae* extracts.

Key words Deoxnucleotide triphosphate, dNTP, DNA replication, Genome instability, Mutagenesis, Quantitative PCR, *Saccharomyces cerevisiae*

1 Introduction

DNA replication requires an adequate supply of deoxyribonucleotide triphosphates (dNTPs) in order to accurately and efficient complete DNA synthesis. Ribonucleotide reductase (RNR) catalyzes the rate-limiting step in dNTP synthesis and is subject to several levels of regulation—transcriptional, posttranslational via protein–protein interactions, localization (at least in yeast), and allosteric inhibition [1–5]. The extent of regulation highlights the importance of maintaining the appropriate dNTP pools—both levels and ratios—throughout the cell cycle.

Altered dNTP pools is increasingly recognized as a hallmark of cancer [2, 6] and impacts both genomic and cellular fitness [5, 7–10]. Limiting concentrations of dNTPs induce replication stress, including fork stalls and fork collapse, potentially resulting in

Lata Balakrishnan and Jason A. Stewart (eds.), *DNA Repair: Methods and Protocols*, Methods in Molecular Biology, vol. 1999, https://doi.org/10.1007/978-1-4939-9500-4_6,

DNA double-strand breaks. Conversely, elevated or altered dNTP pools are mutagenic. There is a proposed shift in the equilibrium between DNA synthesis and DNA proofreading (exonuclease activity) by the replicative polymerases under these conditions, leading to more rapid synthesis and reduced proofreading with a net negative impact on replication fidelity. Consistent with the hypothesis that elevated dNTPs compromise replication fidelity is the observation that a twofold increase in dNTPs combined with defects in mismatch repair (MMR), that is, the spell-checker of the replisome, leads to synergistic increases in mutation rates [8–11].

dNTP pools are altered in the presence of DNA damage and in different genetic backgrounds [7]. It is therefore of interest to quantify dNTP pools in a straightforward way, in order to understand the ways in which altered dNTP pools might play a role in genome stability and cellular fitness.

The sensitive qPCR-based approach described here was adapted from a protocol for dNTP quantification in human cells developed by the Ladner lab [12]. This approach allows a fairly high throughput analysis of dNTP levels extracted from cells. We have adapted it for use with *Saccharomyces cerevisiae*, for which we have developed an effective dNTP extraction protocol. This approach uses equipment and reagents that are available to most labs and avoids the use of radiolabel in the assay.

2 Materials

2.1 Growth Media

YPD plates (1% yeast extract, 2% peptone, 2% dextrose, 2% agar).

YPD liquid media (1% yeast extract, 2% peptone and 2% dextrose).

30 °C incubator.

30 °C shaking incubator.

2.2 Quantitative dNTP Measurements Reagents

Filtered pipette tips.

Template probes, DT1 and DT2 (*see* Table 1) (*see* **Note 1**).

FAM double-quenched probes (*see* Table 1) (*see* **Notes 1** and **2**).

NDP1 primer (*see* Table 1) (*see* **Note 1**).

Omnipure sterile, nuclease-free water.

Individual dNTPs (*see* **Note 3**).

AmpliTaq Gold DNA Polymerase with 15 mM $MgCl_2$ buffer (Buffer I).

MicroAmp Optical 96-well Reaction Plates.

Microseal "B" adhesive sealing film.

Quantitative real-time thermocycler (qPCR) machine. We used theCFX96 Touch™ Real-Time PCR Detection System.

Rotor to centrifuge 96-well plates (optional).

Table 1
Oligonucleotide primers and probes used to quantify dNTP levels [12]

Primer- NDP1	**CCGCCTCCACCGCC**
Probe-FAM-dTTP	***6FAM***/AGGACCGAG/***ZEN***/GCAAGAGCGAGCGA/***IBFQ***
Template- DT1-dTTP	TCGCTCGCTCTTGCCTCGGTCCTTTATTTGGCGGTGGAGGCGG
Template- DT2-dTTP	TCGCTCGCTCTTGCCTCGGTCCTTT*A*TTT*A*TTTGGCGG TGGAGGCGG
Probe-FAM-dATP	***6FAM***/TGGTCCGTG/***ZEN***/GCTTGTGCGTGCGT/***IBFQ***
Template- DT1-dATP	ACGCACGCACAAGCCACGGACCAAATAAAGGCGGTGGAGGCGG
Template- DT2-dATP	ACGCACGCACAAGCCACGGACCAAA**T**AAA**T**AAAGGCGG TGGAGGCGG
Probe-FAM-dGTP	***6FAM***/ACCATTCAC/***ZEN***/CTCACACTCACTCC/***IBFQ***
Template- DT1-dGTP	GGAGTGAGTGTGAGGTGAATGGTTTCTTTGGCGGTGGAGGCGG
Template- DT2-dGTP	GGAGTGAGTGTGAGGTGAATGGTTT*C*TTT*C*TTTGGCGG TGGAGGCGG
Probe-FAM-dCTP	***6FAM***/AGGATTGAG/***ZEN***/GTAAGAGTGAGTGG/***IBFQ***
Template- DT1-dCTP	CCACTCACTCTTACCTCAATCCTTTGTTTGGCGGTGGAGGCGG
Template- DT2-dCTP	CCACTCACTCTTACCTCAATCCTTT*G*TTT*G*TTTGGCGG TGGAGGCGG

All primers are written 5′ to 3′

1× Master Mix for Standard Curve, cell extract, and NDC control:

1 μL dTTP DT1 template (10 μM).

1 μL NDP1 (10 μM).

1 μL dTTP FAM (10 μM).

1 μL dTTP drop out mix.

2.5 μL 10× Buffer I (15 mM $MgCl_2$).

0.175 μL *AmpliTaq Gold*.

15.825 μL nuclease-free water.

22.5 μL to each well.

1× Master Mix for Cell Extract:

1 μL dTTP DT1 template (10 μM).

1 μL NDP1 (10 μM).

1 μL dTTP FAM (10 μM).

1 μL dTTP drop out mix.

2.5 μL 10× Buffer I (15 mM $MgCl_2$).
0.175 μL *AmpliTaq Gold.*
13.325 μL nuclease-free water.
20 μL to each well.

1× No Template Control (NTC):
X dTTP DT1 template (10 μM).
1 μL NDP1 (10 μM).
1 μL dTTP FAM (10 μM).
1 μL dTTP drop out mix.
2.5 μL 10× Buffer I (15 mM $MgCl_2$).
0.175 μL *AmpliTaq Gold.*
16.825 μL nuclease-free water.
22.5 μL to each well.

1× No AmpliTaq Control (NAC):
1 μL dTTP DT1 template (10 μM).
1 μL NDP1 (10 μM).
1 μL dTTP FAM (10 μM).
1 μL dTTP drop out mix.
2.5 μL 10× Buffer I (15 mM $MgCl_2$).
X *AmpliTaq Gold.*
16 μL nuclease-free water.
22.5 μL to each well.

1× No FAM Control (NFC):
1 μL dTTP DT1 template (10 μM).
1 μL NDP1 (10 μM).
X dTTP FAM (10 μM).
1 μL dTTP drop out mix.
2.5 μL 10× Buffer I (15 mM $MgCl_2$).
0.175 μL *AmpliTaq Gold.*
16.825 μL nuclease-free water.
22.5 μL to each well.

2.3 dNTP Extraction Reagents

1.7 mL microcentrifuge tubes.
60% methanol (MeOH) (v:v).
100% methanol (MeOH).
Omnipure sterile, nuclease-free water.
Acetonitrile.

Dry ice.

Acid-washed glass beads, 425–600 μm.

Amicon Ultra 0.5 mL 3 K centrifugal units.

Ice buckets.

Plastic container for acetonitrile bath.

Centrifuge that can go down to −20 °C.

Speed-vacuum centrifuge.

Boiling water bath/heat block.

2.4 Yeast Strains

We use the *W303 rad5$^+$RNR1* and *W303 rad5$^+$ rnr1D57N* strain backgrounds, which were generated in the Chabes lab [8, 9].

3 Methods

3.1 Principle of Method

This protocol adapts a rapid, sensitive approach to quantifying individual dNTPs using a fluorescence-based assay [12]. This is fundamentally an enzymatic assay, using quantitative polymerase chain reactions (qPCR) to determine the levels of a given dNTP available for incorporation into DNA via DNA polymerase.

This assay employs a modified oligonucleotide template with three distinct regions (Fig. 1):

1. A region that anneals the NDP1 qPCR primer.
2. A region that binds 6-FAM-labeled, double-quenched (IBFQ-ZEN) probe.
3. A middle region that forms a gap to be filled in by DNA synthesis.

Thermus aquaticus (Taq) DNA polymerase is added to initiate DNA synthesis in the presence of excess dNTPs, except for the dNTP to be quantified, dTTP in this illustration. When the dNTP to be quantified is present (two dTTP molecules in this example of dTTP-DT2 template), *Taq* polymerase synthesis extends across the gap from the NDP1 primer to the downstream quenched probe.

Once the downstream probe is reached, the intrinsic 5′–3′ exonuclease activity of *Taq* polymerase excises the 5′ nucleotide of the probe, which is labeled with a 6-FAM fluorophore, releasing FAM from the dual quenched ZEN-IBFQ probe. This generates a fluorescence signal in response to excitation-induced photon energy (h*ν*). The fluorescence generated is directly proportional to the concentration of the limiting dNTP. When the limiting dNTP is at very low concentrations, *Taq* polymerase cannot extend through the mid template region and therefore does not encounter the probe and the fluorescence remains quenched.

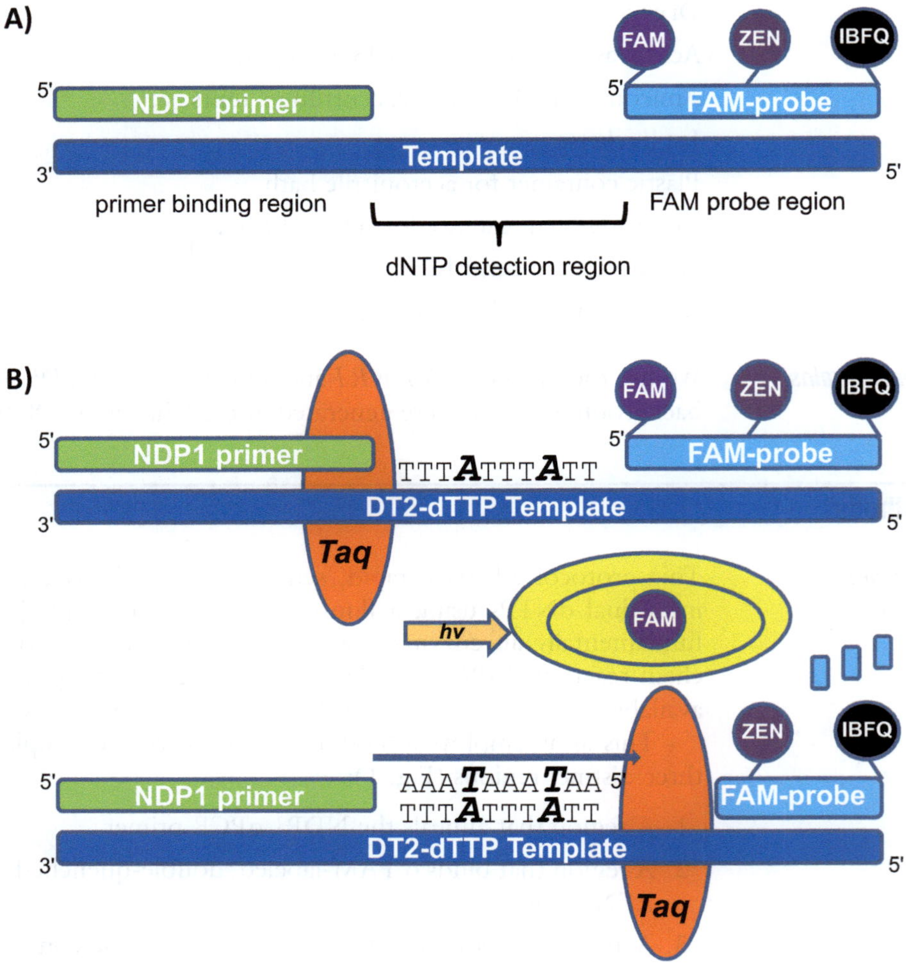

Fig. 1 Cartoon representation of qPCR-based approach for dNTP quantification. (**a**) Oligonucleotide setup for fluorescence-based assay. (**b**) Cartoon of the principle behind the quantification. *Thermus aquaticus* (*Taq*) DNA polymerase is primed from NDP1 and synthesizes across the dNTP detection region, inserting a known number of a specific dNTP. In this case, two dTTPs are incorporated to fill in the central gap. Once the gap is filled, *Taq* stops synthesizing DNA and its 5′ to 3′ exonuclease activity is activated, leading to release of FAM from the probe and a resulting fluorescent signal that is proportional to the number of dTTPs incorporated. If there are insufficient dTTPs in the reaction, *Taq* stalls in the gap, FAM is not released and no fluorescence is generated

3.2 qPCR Plate and Standard Curve Setup

1. Here we illustrate this protocol with quantification of dTTP using DT1-dTTP template as an example. The same approach would apply for the other dNTPs and templates (*see* **Notes 4** and **5**).
2. Resuspend primer NDP1 (used for all templates) to generate a 100 μM stock solution. Dilute the stock 1:10 to get working stock of 10 μM.

3. Resuspend the dTTP DT1 template and dTTP FAM to generate 100 μM stock solutions. Make 20 μL aliquots and store at −20 °C (*see* **Note 6**). When ready to use, dilute an aliquot of the stock solution 1:10 to get working stock of 10 μM (add 180 μL nuclease-free water to 20 μL frozen aliquot for a 10 μM solution).
4. Make a dTTP drop out mix containing dATP, dGTP, and dCTP **each** at a final concentration of 2.5 mM (400 μL = 100 μL dH_2O + 100 μL 10 mM dATP + 100 μL 10 mM dGTP + 100 μL 10 mM dCTP).
5. For each standard curve, quantify six known values of dTTP for a DT1-dTTP template, each in triplicate: 1.25 pmol, 2.5 pmol, 5 pmol, 7.5 pmol, 10 pmol, 15 pmol. Concentrations of the standard dNTP are adjusted so that the same volume (2.5 μL of each standard value) is added to each reaction. In a standard plate, the total reaction volume is 25 μL. For example, 2.5 μL of 0.5 μM dTTP would be added to a 25 μL reaction (final volume) for 1.25 pmol dTTP (*see* **Notes** 7 and **8**).
6. To make the appropriate stock solutions for the desired dNTP amounts, make the following standard dilutions starting from dTTP stock (100 mM). Use a new tip for each dilution and vortex well after each addition (*see* **Note 9**).
   (a) 10 mM: 10 μL 100 mM stock into 90 μL H_2O.
   (b) 1 mM: 10 μL 10 mM stock into 90 μL H_2O.
   (c) 100 μM: 10 μL 1 mM stock into 90 μL H_2O.
   (d) 10 μM: 20 μL 100 μM stock into 180 μL H_2O.
   (e) 6 μM: 120 μL 10 μM stock into 80 μL H_2O **(2.5 μL = 15 pmol).**
   (f) 4 μM: 133.3 μL 100 μM stock into 66.7 μL H_2O **(2.5 μL = 10 pmol).**
   (g) 3 μM: 150 μL 100 μM stock into 50 μL H_2O **(2.5 μL = 7.5 pmol).**
   (h) 2 μM: 133.3 μL 100 μM stock into 66.7 μL H_2O **(2.5 μL = 5 pmol).**
   (i) 1 μM: 100 μL 100 μM stock into 100 μL H_2O **(2.5 μL = 2.5 pmol).**
   (j) 0.5 μM: 100 μL 100 μM stock into 100 μL H_2O **(2.5 μL = 1.25 pmol).**
7. For each 25 μL qPCR reaction, add the following (*see* **Note 10**):
   (a) 1 μL of primer NDP1 (10 pmol/μL).
   (b) 1 μL of selected template (10 pmol/μL) (e.g., DT1 or DT2 template).

   (c) 1 μL of selected probe (10 pmol/μL).
   (d) 1 μL of dNTP dropout mix (−A/C/T/G) (2.5 pmol/μ L) lacking the dNTP to be quantified.
   (e) 2.5 μL of 10× PCR Buffer (*AmpliTaq Gold* Buffer I with 15 mM $MgCl_2$).
   (f) 0.175 μL of *AmpliTaq Gold* polymerase (5 U/μL).
   (g) Up to 10 μL of dNTP standard (*see above*) or extract (*see below*), as required to get a signal above the 0 pmol dNTP control (*see* **step 8**).
   (h) Nuclease-free H_2O to make up a final volume of 25 μL.

8. The following controls should be included with each experiment:
   (a) No dNTP control (NDC): This control omits the limiting dNTP being quantified and is used to measure any background fluorescence generated by the probe in presence of the polymerase. The raw fluorescence units obtained for all concentrations are normalized to account for background probe fluorescence using values generated from this control.
   (b) No template control (NTC): This control omits the DNA template from a reaction, and serves as a general control to detect any amplification in absence of the true template. This control detects any external source of contamination.
   (c) No amplification control (NAC): The *AmpliTaq Gold* DNA polymerase was excluded from the PCR reaction. This control detects background fluorescence that is ideally not present in a qPCR reaction and if this control amplifies, it usually indicates the fluorescent probe is degraded.
   (d) No FAM control (NFC): This control omits the dNTP-specific FAM probe and detects any contaminating fluorescence.

   Figure 2 illustrates the relative values for these controls, which are highly reproducible, with the exception of the NDC, which shows more variability between experiments.

9. Make up master mixes for the standard curve, with known dNTP amounts, controls and cell extracts, if applicable (*see* **Note 11**).

   Master Mixes

   (a) One Master Mix sufficient to set up standards, for 2.5 μL of cell extracts and NDC. Add 22.5 μL Master Mix to each well +2.5 μL additional volume for 25 μL volume per well.
   (b) One Master Mix sufficient to set up 5 μL of cell extracts and 2.5 μL of cell extracts + 2.5 μL spike. Add 20 μL

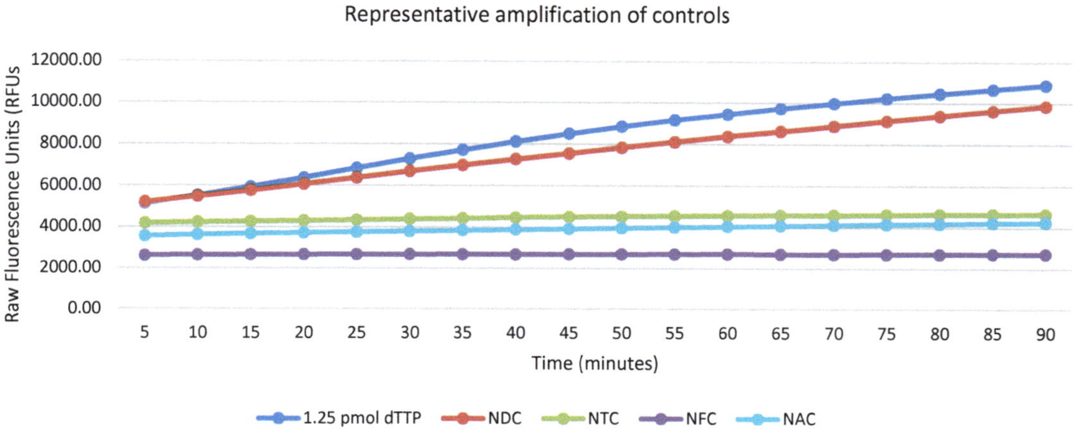

Fig. 2 dNTP quantification negative controls. This is raw data for negative controls. NTC, NFC, and NAC consistently exhibit no amplification over the course of the experiment, with NFC showing the lowest background fluorescence. The NDC control amplifies with more variability between experiments, but this is corrected for by averaging the NDC controls and subtracting these NDC values to normalize fluorescence units for experiments

Master Mix added to each well + 5 μL additional volume for 25 μL volume per well.

(c) Three separate Master Mixes for NTC, NAC, and NFC controls (22.5 μL each).

10. When setting up a qPCR plate, add Master Mix to all wells, then add water (for NDC, 0 pmol controls), standard values of dTTP and add cell extracts last. This is all done on ice.

11. Seal the top of the plate. Be careful to avoid fingerprints on the top of the film.

12. If available, spin down the plate using a 96-well plate spinner prior to quantification.

13. Set up the program for dNTP quantification of the 6-FAM fluorophore on the real-time PCR thermocycler. The thermocycler is programmed to perform three steps:

(a) The first step is a one-time 6 min step at 95 °C to provide a hot-start for the DNA polymerase (*see* **Note 12**).

(b) The second is a 5 min 60 °C incubation step during which the polymerization occurs.

(c) The third step is a fluorescence plate-read at the conclusion of each 5 min incubation step.

The incubation and plate-read steps are repeated for a minimum of 10 cycles (50 min total). This allows real-time analysis of assay progression and kinetics in 5 min cycles up to 40 min. However, the number of cycles can be extended since the assay progression might differ depending on the amount of dNTPs present in the extract or the instruments used to perform the reaction.

3.3 Quantification of Fluorescence

1. Fluorescence was detected using the CFX96 Touch™ Real-Time PCR Detection System and analyzed with Bio-Rad CFX Maestro software (*see* **Note 13**).
2. Unlike standard real-time PCR reactions, the polymerization in this method is not cyclic and the data are analyzed differently. The generation of C_t values is not applicable. Instead, the raw fluorescence spectra for each well (or triplicate wells) should be exported and analyzed. Using the Bio-Rad CFX Maestro software the "Baseline" setting is adjusted to "No Baseline Subtracted" (*see* **Note 14**).
3. Data export is achieved by exporting all data sheets as Microsoft Excel 2007 (.xls) files (Microsoft, Redmond, WA, USA). The fluorescence spectra for 6-FAM is typically quantified every 5 min, that is, after each qPCR "cycle." The "Quantification Amplification Results" sheet contains raw fluorescence values after each cycle and is used in subsequent analysis steps to quantify fluorescence units measured for the amount of each limiting dNTP.
4. Subtract raw fluorescence values for blank reactions (0 pmol average; i.e., limiting dNTP omitted—NDC) from raw fluorescence units in the presence of the limiting dNTP to give normalized fluorescence units (NFUs). This accounts for any background probe fluorescence. An example is shown in Table 2:

3.4 Statistical Analysis and Assay Validation

1. Plot the normalized data for a time course analysis for each individual sample within a triplicate to determine variance *within* an experiment. We observe minimal intraassay variance within the dynamic range of a particular DT substrate (*see* Fig. 3).
2. Plot the average normalized fluorescence units (NFUs) generated for each concentration to detect any variance in fluorescence *between* experiments for a particular concentration of dNTP (Fig. 4). We observe some variation between experiments, although the trends remain unaffected (*see* Figs. 4 and 5).
3. Determine coefficient of variance (C.V.) for a measure of interassay variation. For three different experiments measuring dTTP, we measured a % C.V. (= SD/MEAN $\times$ 100) of 5.34 $\pm$ 0.56. For this reason we recommend generating an internal calibration curve for each experiment when quantifying dNTPs from cell extracts.

Table 2
Background correction for raw fluorescence data

Time in min	2.5 pmol	NDC RFU's	Difference = NFU's
0	0	0	0
5	201.03	206.72	−5.697493731
10	399.05	225.99	173.0570352
15	501.59	172.49	329.1041714
20	744.19	157.87	586.3199753
25	851.07	198.42	652.6477418
30	1019.94	272.78	747.1531056
35	1294.93	344.18	950.750761
40	1635.77	414.51	1221.255359
45	2032.15	501.00	1531.144881
50	2495.99	602.82	1893.165638
55	3209.14	699.28	2509.855808

4. Analyze the relationship between assay response and known concentrations of dNTPs with a calibration curve and evaluate using regression coefficients (R^2) (Figs. 6 and 7). The calibration curve will be used to quantify the unknown dNTP levels from cell extracts. Note that the R^2 value improves (is closer to 1) after 15 min, or three cycles (Fig. 6).

3.5 dNTP Extraction from Yeast Cells

dNTP extracts from *Saccharomyces cerevisiae* can be challenging. It is necessary to permeabilize the cells to release the dNTPs, but we have observed that full lysis results in contaminants that inhibit the qPCR reaction. The protocol below combines protocols from a number of different sources to generate a straightforward and reproducible extraction method [12–17].

1. Two days before extraction set up a 5 mL overnight yeast culture from a single colony in media of choice (in afternoon/evening). Grow to saturation at 30 °C with shaking.
2. Approximately 24 h later, dilute the saturated culture, 1:1000, in 125 mL media, in 500 mL flasks.
3. Grow cells to mid-log phase OD_{600} ~ 0.6 A. This will take approximately 12–13 h, depending on the strain background.

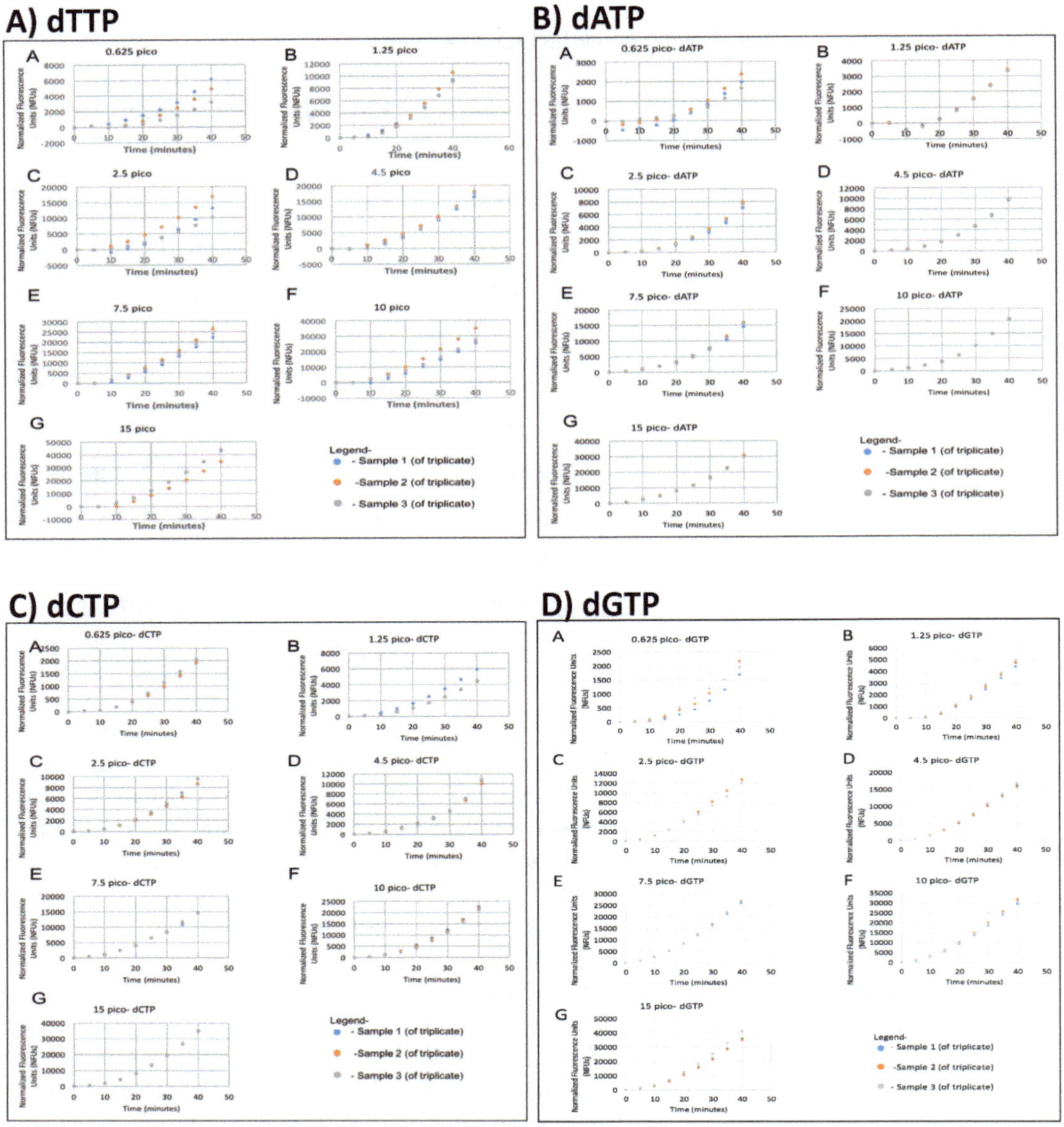

Fig. 3 Intraassay variation. Normalized fluorescence data are shown for known quantities of (**a**) dTTP, (**b**) dATP, (**c**) dCTP, and (**d**) dGTP in triplicate

4. The day of the extraction, prepare and chill 60% methanol. Dispense 20 mL per extraction into 50 mL conical tubes. Typically four extractions are performed per strain.
5. Incubate 60% methanol on ice to maintain ~1–2 °C.
6. Dispense 100% methanol in a 15 mL conical tube—200 μL is required per extraction.
7. Incubate 100% methanol on ice to maintain ~1–2 °C.

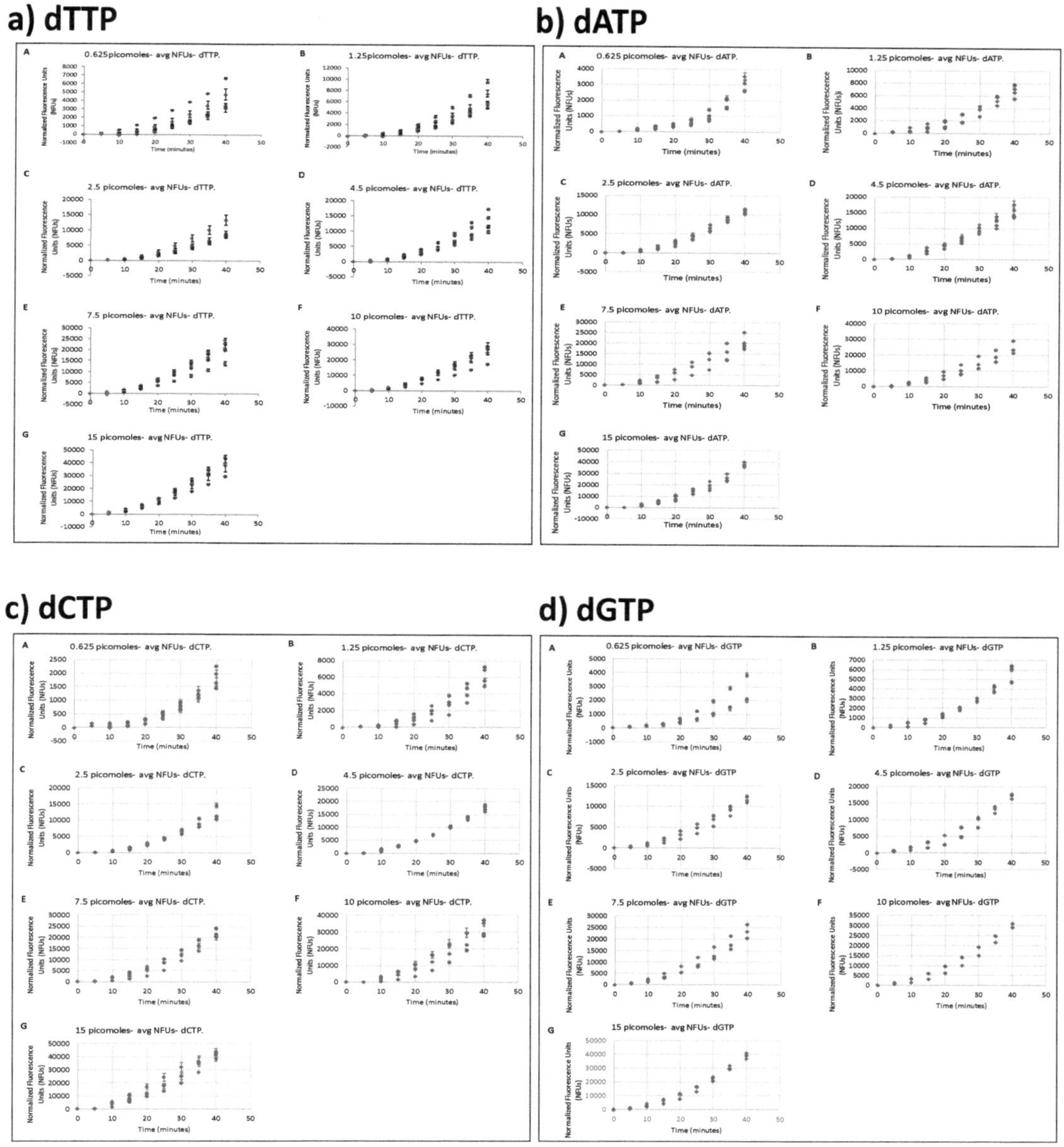

Fig. 4 Interassay variation. Average normalized fluorescence units are shown from multiple independent experiments for know quantities of (**a**) dTTP, (**b**) dATP, (**c**) dCTP, and (**d**) dGTP

8. Set up an acetonitrile–dry ice bath. Slowly add dry ice pellets to 100% acetonitrile in a plastic container set inside an ice bucket. The bath should be ~−40 °C and about 2.5 cm deep.
9. Turn on Beckman centrifuge to get it down to −20 °C (for **step 14**) (*see* **Note 15**).
10. Take 4 × 25 mL samples from a single culture. Transfer these samples into 4 × 50 mL conical tubes for centrifugation.

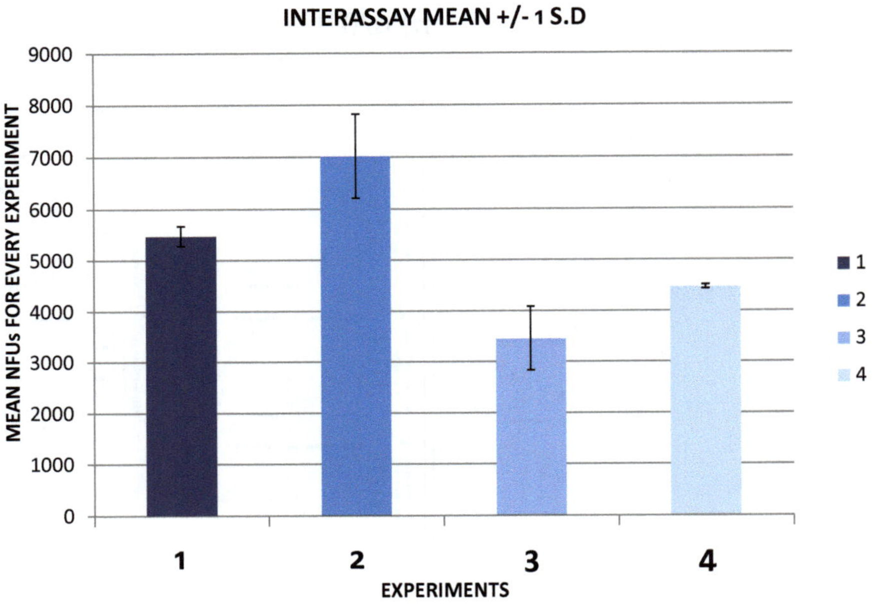

Fig. 5 Interassay means for four independent experiments with dTTP. There is sufficient variability to warrant generating calibration curves with each experiment

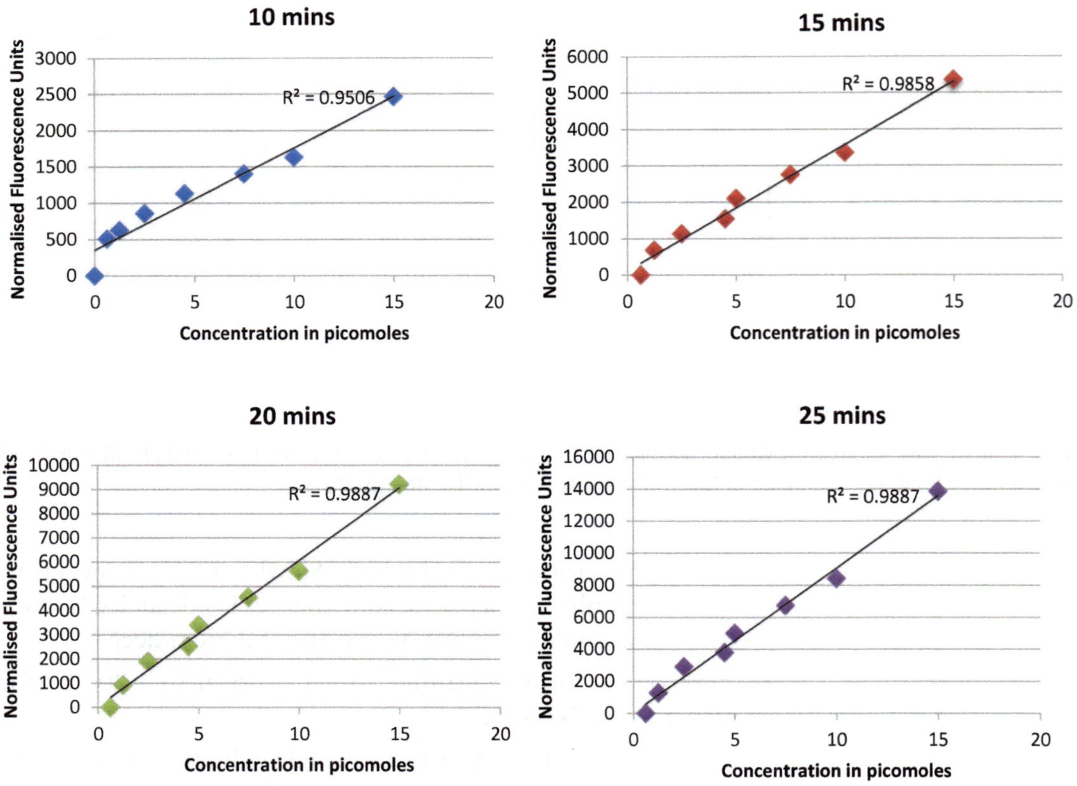

Fig. 6 Calibration curves for dTTP. The fluorescence generated with increasing amounts (pmol) of dTTP is plotted after 2, 3, 4, and 5 qPCR cycles. Note the increased signal (and different *y*-axes) as the time course proceeds

11. Pellet the cells, four conical tubes per strain, at 3500 rpm (2,800 × *g*) at room temperature for 5 min.
12. Discard the supernatant and resuspend the pellet up to 5 mL volume by adding sterile nuclease-free water.
13. Add entire suspension (now ~5 mL) to the center of 20 mL of ice cold 60% methanol in a 50 mL conical tube on ice, using a pipette. This is the quench step.
14. Pellet the cells at 4000 rpm (5,000 × *g*) at −20 °C for 5 min.
15. Discard supernatant and resuspend pellet in residual liquid.
16. The cells will be frozen at this point and will need to defrost a little before transferring with a pipette.

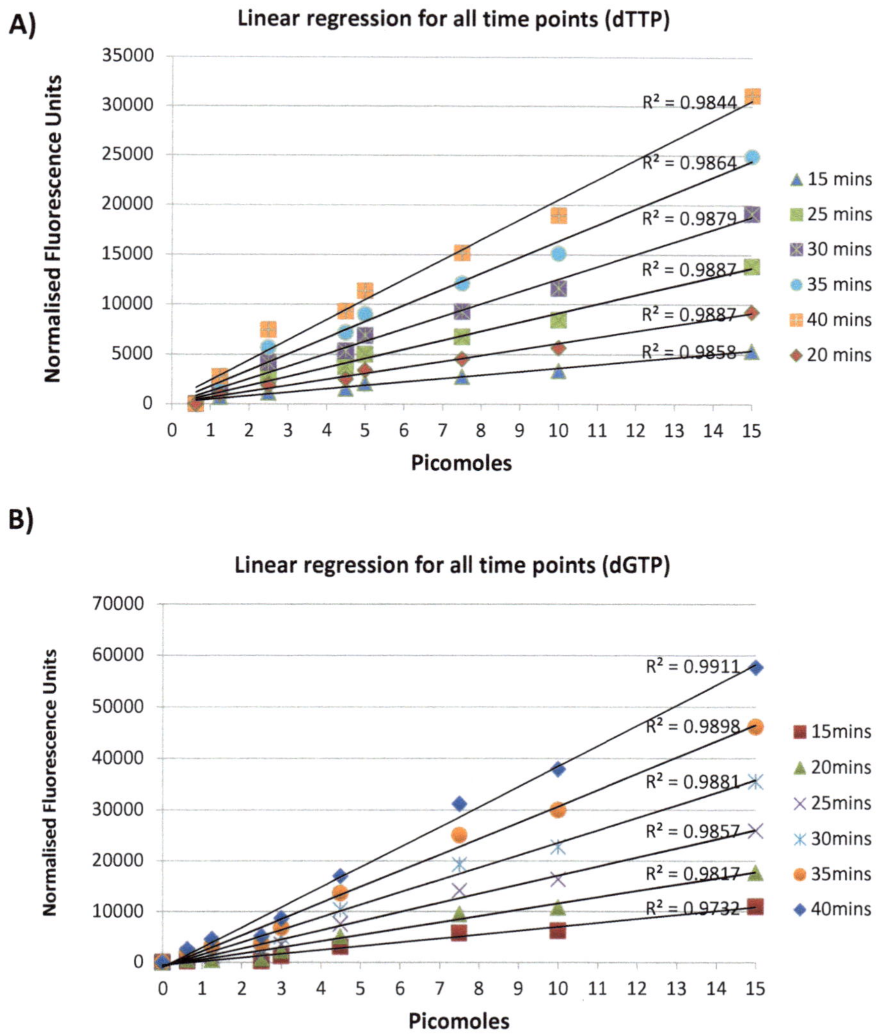

Fig. 7 Calibration curves for (**a**) dTTP, (**b**) dGTP, (**c**) dATP, and (**d**) dCTP

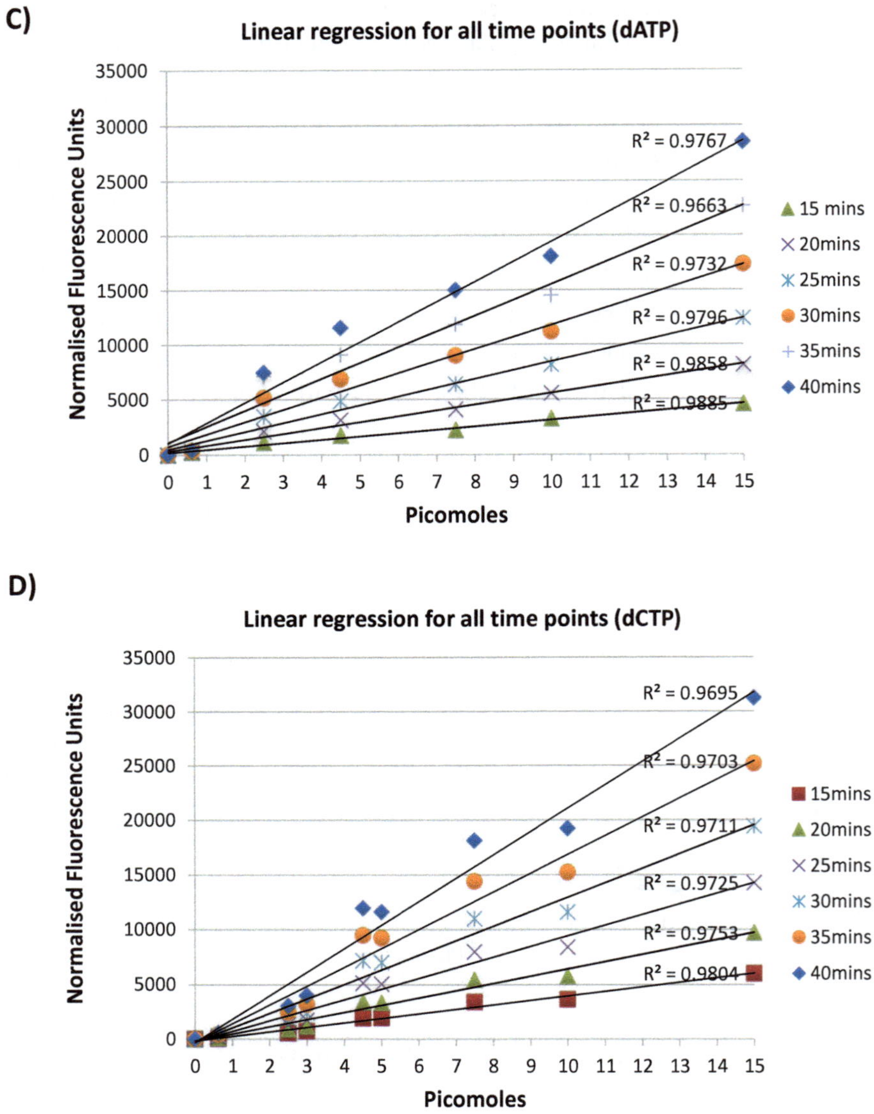

Fig. 7 (continued)

17. Transfer suspension (~250 μL) to labeled microcentrifuge tube and add 250 μL ice cold 100% methanol.
18. Add 0.3 g acid-washed beads and hand vortex in 10 s cycles at 4 °C transferring to ice bucket for 2 min between cycles to keep extracts cold. Repeat six times (*see* **Note 16**).
19. Flip tubes upside down in between cycles to ensure even mixing for efficient lysis.
20. Transfer microcentrifuge tubes to −40 °C acetonitrile bath for 30 min (*see* **Note 17**).

21. Transfer tubes to ice bath to thaw, 3 min.
22. Spin down at 14,000 rpm (18,407 × *g*) for 10 min at room temperature.
23. At the same time equilibrate an Amicon column with 400 μL sterile diH_2O.
24. Transfer cell extract supernatant to new microcentrifuge tube.
25. Pass extracts through preequilibrated Amicon ultra 0.5 mL 3 K centrifugal units (14,000 rpm for 10 min) (*see* **Note 18**).
26. Transfer filtrate to new microcentrifuge tube.
27. Heat at 98 °C for 3 min to denature any protein.
28. Spin down briefly to get condensation off lid.
29. Evaporate filtrate under centrifugal vacuum at 72 °C for ~1–1.5 h for 500 μL of cell extract.
30. Once all residual liquid has evaporated (pellets are translucent and orange in color), resuspend the pellet in room temperature nuclease-free water (100 μL typically). Incubate at room temperature for 5–10 min, flicking gently to fully resuspend the dNTP pellet. Do not vortex.
31. Store extracts at −20 °C (*see* **Note 19**).

3.6 Quantification of dNTPs from Cell Extracts

1. *See* Fig. 8 for a suggested qPCR plate layout that includes a standard curve and multiple cell extract samples in triplicate. A typical plate has the following:
   (a) Standard curve with known values of dNTP, including 0 pmol control.
   (b) Control master mixes (NTC, NFC, and NAC) with known values of dNTP added.
   (c) Cell extracts.
   (d) Cell extracts + known value of dNTP (spiked controls) (*see* **Note 20**).
   (e) Control master mixes with cell extracts added.
2. To quantify the dNTPs in the cell extracts, perform qPCR with 2.5 μL of cell extracts in triplicate, 5 μL of cell extracts in triplicate, and a spiked control with cell extract + 2.5 pmol of the relevant dNTP (*see* **Note 10**). This will be done with all four dNTPs. The volumes of cell extract to be added can be adjusted as needed, depending on fluorescence signals.
3. Create a scatter plot of your time course data for each dNTP to determine whether you have quantifiable amounts of dNTP. These curves (Fig. 9) will also determine whether there are linear increases in fluorescence with increases in cell extract.
4. Generate a calibration curve (scatter plot) based on your known standards.

	1	2	3	4	5	6	7	8	9	10	11	12	
A	15 pico	15 pico	15 pico	10 pico	10 pico	10 pico	7.5 pico	7.5 pico	7.5 pico	5 pico	5 pico	5 pico	dNTP DT1 Standard
B	2.5 pico	2.5 pico	2.5 pico	1.25 pico	1.25 pico	1.25 pico	0 pico	0 pico	0 pico	NTC 2.5 pico	No FAM 2.5 pico	No Taq 2.5 pico	
C	2.5 uL CE	2.5 uL CE	2.5 uL CE	5 uL CE	5 uL CE	5 uL CE	2.5 uL CE 2.5 ul 1 uM (2.5 pico)	2.5 uL CE 2.5 ul 1 uM (2.5 pico)	2.5 uL CE 2.5 ul 1 uM (2.5 pico)	NTC 2.5 ul CE	No FAM 2.5 ul CE	No Taq 2.5 ul CE	Cell extract #1
D	2.5 uL CE	2.5 uL CE	2.5 uL CE	5 uL CE	5 uL CE	5 uL CE	2.5 uL CE 2.5 ul 1 uM (2.5 pico)	2.5 uL CE 2.5 ul 1 uM (2.5 pico)	2.5 uL CE 2.5 ul 1 uM (2.5 pico)	NTC 2.5 ul CE	No FAM 2.5 ul CE	No Taq 2.5 ul CE	Cell extract #2
E	2.5 uL CE	2.5 uL CE	2.5 uL CE	5 uL CE	5 uL CE	5 uL CE	2.5 uL CE 2.5 ul 1 uM (2.5 pico)	2.5 uL CE 2.5 ul 1 uM (2.5 pico)	2.5 uL CE 2.5 ul 1 uM (2.5 pico)	NTC 2.5 ul CE	No FAM 2.5 ul CE	No Taq 2.5 ul CE	Cell extract #3
F	2.5 uL CE	2.5 uL CE	2.5 uL CE	5 uL CE	5 uL CE	5 uL CE	2.5 uL CE 2.5 ul 1 uM (2.5 pico)	2.5 uL CE 2.5 ul 1 uM (2.5 pico)	2.5 uL CE 2.5 ul 1 uM (2.5 pico)	NTC 2.5 ul CE	No FAM 2.5 ul CE	No Taq 2.5 ul CE	Cell extract #4
G	2.5 uL CE	2.5 uL CE	2.5 uL CE	5 uL CE	5 uL CE	5 uL CE	2.5 uL CE 2.5 ul 1 uM (2.5 pico)	2.5 uL CE 2.5 ul 1 uM (2.5 pico)	2.5 uL CE 2.5 ul 1 uM (2.5 pico)	NTC 2.5 ul CE	No FAM 2.5 ul CE	No Taq 2.5 ul CE	Cell extract #5
H	2.5 uL CE	2.5 uL CE	2.5 uL CE	5 uL CE	5 uL CE	5 uL CE	2.5 uL CE 2.5 ul 1 uM (2.5 pico)	2.5 uL CE 2.5 ul 1 uM (2.5 pico)	2.5 uL CE 2.5 ul 1 uM (2.5 pico)	NTC 2.5 ul CE	No FAM 2.5 ul CE	No Taq 2.5 ul CE	Cell extract #6

Fig. 8 Standard qPCR plate setup for performing dNTP extract quantification reactions. Each cell represents a well in a qPCR plate. Written on each cell is the amount of sample (or standard) added to each master mix. It serves as a useful guide when setting up a large number of reactions

5. Generate a trendline or line of best fit from your calibration curve. Use the time point that generates a trendline with the highest R^2 value and determine the equation of this line ($y = mx + b$). Figure 10a shows the trendline for the calibration curve of dTTP at the 40 min time point.

6. Take the average NFUs for the same time point of your unknown cell extract sample. Insert this number into the equation of the trendline determined above *(NFUs = y)*, solving for *x* to determine the number of picomoles of dNTP in your unknown sample (Fig. 10b). Repeat for all the quantification experiments. Since experiments were repeated at least three times, the calculated concentrations for a particular dNTP are pooled and the average concentration is calculated and plotted.

7. You can then calculate the number of picomoles/10^8 cells to normalize your calculated dNTP levels to the culture size (e.g., Fig. 11). Figure 11 plots dNTP quantification from two different yeast strains: *RNR1*, or wild-type, and *rnr1D57N*, which has been demonstrated to exhibit a 1.5- to twofold increase in dNTP levels [9] as a result of the loss of allosteric regulation of

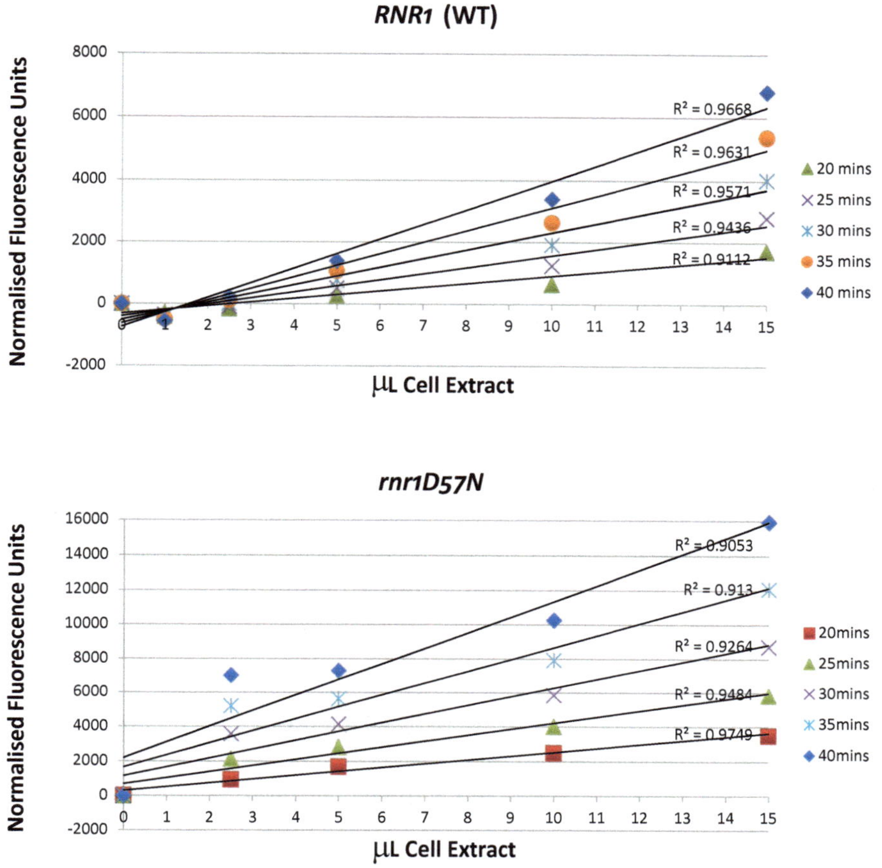

Fig. 9 Quantification of dTTP from wild-type (*RNR1*) and mutant (*rnr1D57N*) cell extracts. Note the differences in scale on the *y*-axis

ribonucleotide reductase, the enzyme that catalyzes the rate-limiting step in dNTP synthesis [9, 18, 19]. We observe a similar increase in dNTP pools, and similar overall dNTP levels, using this qPCR-based fluorescence protocol [9].

4 Notes

1. All oligonucleotides were synthesized by Integrated DNA Technologies (Coralville, IA, USA). The primers are subjected to polyacrylamide gel electrophoresis purification and reconstituted in Omnipure sterile nuclease-free water (EMD Chemicals USA, Gibbstown, NJ, USA) at a stock concentration of 100 μmol/L. Oligonucleotides were then diluted to a stock of 10 μmol/L and aliquoted to avoid repeated freeze–thaw cycles. Store oligonucleotides at −20 °C and thaw on ice when needed.

A)

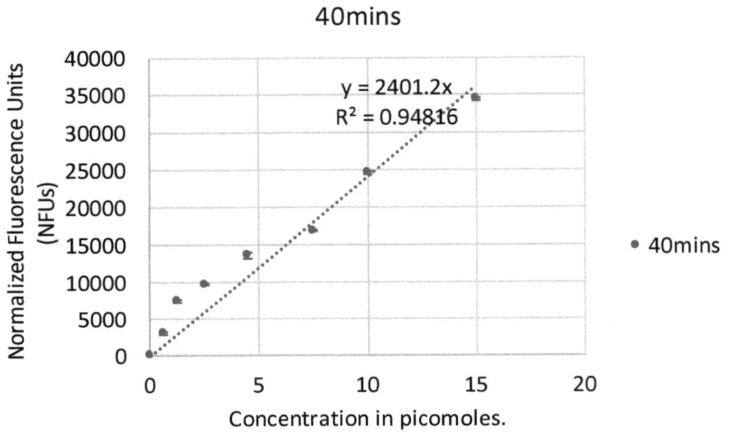

B)

From time course (**Figure 8**):

Time/Volume	**NFUs- 10ul- *RNR1***
NFUs @ 40 minutes/10 L	8357.9

From the standard calibration curve above:

$$y = 2401.2x$$
$$8357.9 = 2401.2x$$
$$x = 8357.9/2401$$
$$x = 3.48 \text{ pmol dTTP}/10\ \mu\text{L extract}$$

Fig. 10 Calculation of unknown quantities of dNTP from cell extract based on standard calibration curves. (**a**) Calibration curve for dTTP quantities measured after 40 min (eight cycles). (**b**) Calculation of dTTP levels in dNTP extracts from wild-type (*RNR1*) cells based on fluorescence counts (Fig. 8) and the calibration curve in (**a**)

2. The qPCR reaction will measure the fluorescence generated by release of 6-FAM from the FAM probe. The ZEN and IBFQ quenchers reduce the background fluorescence and prevent a significant signal from being detected until FAM is released from the probe by *AmpliTaq* exonuclease activity.
3. A dNTP dropout mix was prepared, which *excludes* the dNTP to be assayed and a separate dNTP mix is required for the detection of each of the dNTPs: For example, to detect dTTP a dNTP mix consisting of 100 μL of stock concentrations of 10 mmol/L dATP, 100 μL of dCTP, 100 μL of dGTP, and 100 μL of sterile nuclease-free water will yield 400 μL of a

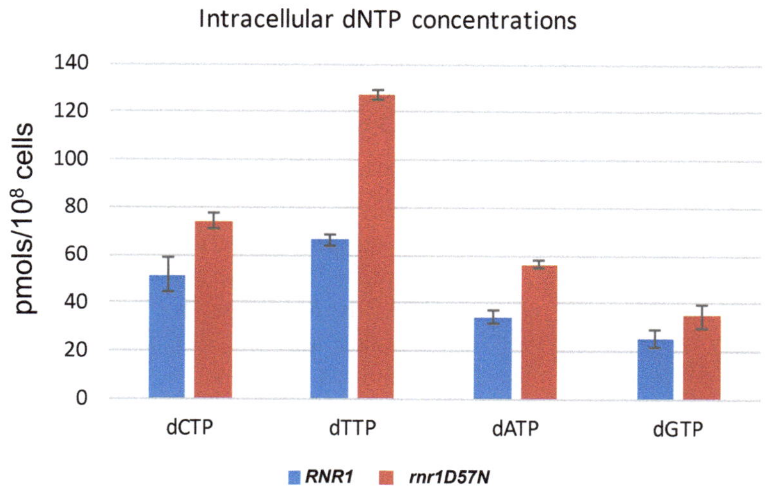

Fig. 11 Quantification of each dNTP from cell extracts from two yeast strains. Both wild-type (*RNR1*) and *rnr1D57N* exhibit similar relative (balanced) dNTP pools, but *rnr1D57N* has elevated levels, with a 1.5- to twofold increase in absolute dNTP levels

dC/G/ATP dNTP mix at 2.5 mmol/L ready for use in the assay to detect dTTP.

4. In order to determine dNTP levels from cellular extracts, standard curves for each dNTP are generated with known dNTP concentrations. For each dNTP to be quantified, the other three dNTPs are in excess and the known amounts of the dNTP of interest are titrated into the reactions. All dilutions are made using nuclease-free water and filtered tips.
5. All reagents should be thawed on ice prior to use. The *AmpliTaq Gold* should be kept at −20 °C until immediately prior to use.
6. The FAM, double-quenched probe is stable if stored at −20 °C. The probes are light-sensitive and so light exposure should be minimized. If they will be exposed to sustained light exposure, the tubes should be covered with foil or brown-colored tubes should be considered.
7. For the dTTP Standard curve, use a range of dTTP amounts in combination with the dTTP dropout mix. For a DT1 template (any dNTP), the linear range of detection is 1.25 pmol to 10–15 pmol of standard dNTP. At lower concentrations (<1.25 pmol), measurements are typically below 0 pmol control. At higher concentrations (>20 pmol), the reactions are saturated and little separation is observed between the different dNTP levels. For the DT2 template (any dNTP), the linear range of detection is 5 pmol to ~30 pmol of standard dNTP,

thereby allowing accurate quantification of higher dNTP levels, but reducing the sensitivity at the other end of the spectrum.

8. Increasing the $MgCl_2$ concentration narrows this range and also lowers the values that can be quantified.
9. For each qPCR experiment, we recommend making a new set of dNTP standards to quantify from 100 mM stock solutions.
10. The template and the FAM probe must match. For instance, to quantify dTTP, use the DT1-(or DT2) dTTP template and the dTTP-FAM probe; to quantify dGTP use the DT1-dGTP template and the dGTP-FAM probe. The NDP1 primer is used in all reactions.
11. When making master mixes, add reagents in the following order: water, DT template, NDP1, dropout mix, buffer. The FAM probe and Taq are added last. Keep the master mix on ice after adding *AmpliTaq* and vortex briefly to thoroughly mix.
12. Longer "hot start" times produced more variable results in our hands.
13. Real-time PCR platforms vary, but all commercially available platforms allow for detection of the 6-FAM fluorophore and facilitate the export of raw fluorescence spectra to excel sheets.
14. The baseline is typically adjusted based on the initial cycles of qPCR. In a standard qPCR reaction there is little observed change in fluorescence during the first few cycles. With our approach, especially for samples with higher starting fluorescence values, the baseline is inappropriately adjusted when subtracted automatically. We have noted that the baseline is consistent with the No FAM control and there is little variability across a plate; therefore, it has negligible effects on quantified fluorescence values and is not appropriate for this qPCR application.
15. Ensure centrifuge rotor is cold.
16. Hand vortexing works best for us. An automatic vortexer such as a "MixMate" is not sufficiently vigorous, resulting in separate cell and bead phases. It is important to be timely with these steps. As soon as the dNTPs are quenched, they are very unstable and degrade quickly off the ice.
17. Be sure to keep a map of where tubes are located in ice bucket as acetonitrile will take off all sharpie labeling (red sharpie is best).
18. Amicon Ultra column capacity is 500 μL. Therefore, multiple spins may be necessary, depending on volume of extract. The same column can be reloaded with additional extract followed by centrifugation. Typically two passes are necessary per extraction.

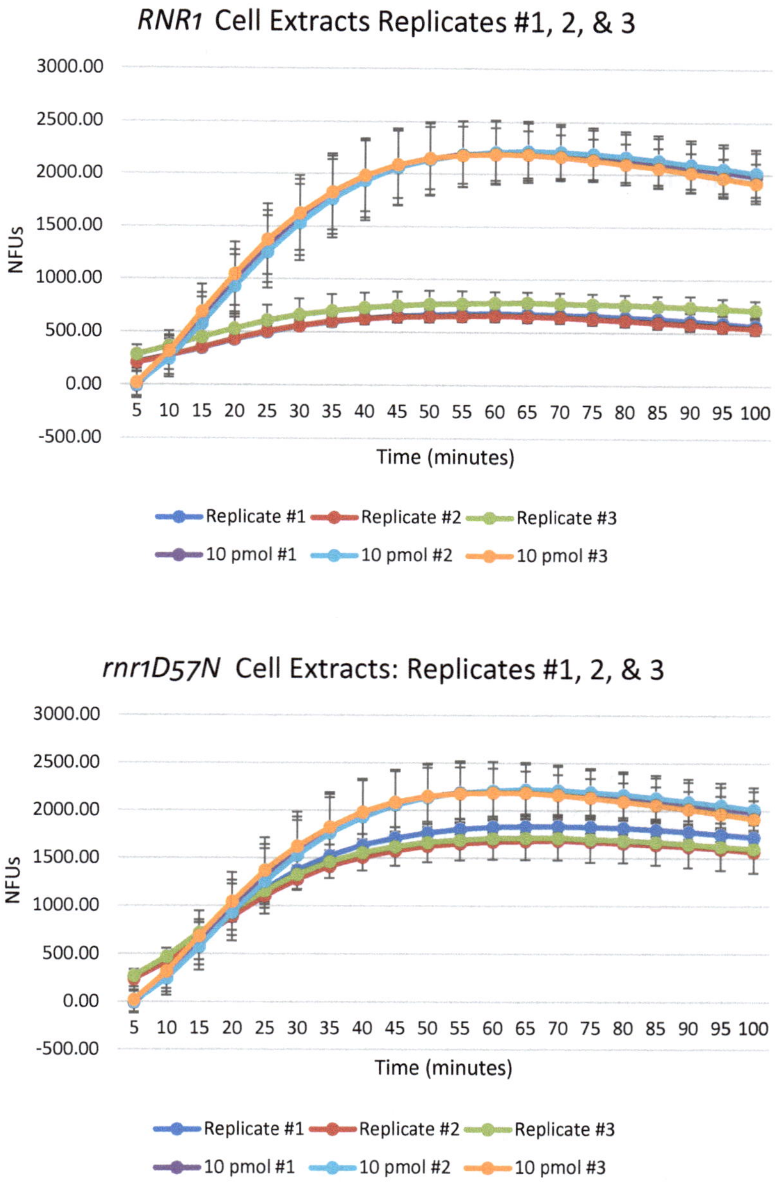

Fig. 12 Quantification of dTTP from cell extracts from *RNR1* and *rnr1D57N* is stable over three days. dTTP was measured in extracts from (**a**) *RNR1* and (**b**) *rnr1D57N* the day following extraction (Replicate 1) and two subsequent days (Replicates 2 and 3) to determine the stability of the extracted dNTP. The dTTP was stored at −20 °C between assays and thawed on ice for each experiment. Normalized fluorescence units are shown for 5 μL of each extract as well as a 10 pmol dTTP standard quantified alongside the extracts

19. We observe minimal degradation after freeze–thaw cycles and quantification of the same extracts over 3 days resulted in similar results (Fig. 12). Nonetheless, it is best to keep extracts on ice or frozen at all times.

20. The spiked controls help determine whether there is inhibition of qPCR in the presence of the cell extract. They are also useful in quantifying low levels of dNTPs.

Acknowledgments

We are grateful to members of the Surtees lab (past and present) for helpful discussions and input. We thank Jumana Badar for her help in working out calibration curves. N.A.L. is a University at Buffalo Presidential Scholar. This work was supported by the American Cancer Society (RSG-14-235-01 to J.A.S.). J.A.S. is an ACS Research Scholar.

References

1. Mathews CK (2006) DNA precursor metabolism and genomic stability. FASEB J 20:1300–1314
2. Mathews CK (2014) Deoxyribonucleotides as genetic and metabolic regulators. FASEB J 28:3832–3840
3. Mathews CK (2015) Deoxyribonucleotide metabolism, mutagenesis and cancer. Nat Rev Cancer 15:528
4. Johansson R, Jonna VR, Kumar R, Nayeri N, Lundin D, Sjöberg B-M, Hofer A, Logan DT (2016) Structural mechanism of allosteric activity regulation in a ribonucleotide reductase with double ATP cones. Structure 24:906–917
5. Pai CC, Kearsey S (2017) A critical balance: dNTPs and the maintenance of genome stability. Genes 8:57
6. Aye Y, Li M, Long MJC, Weiss RS (2015) Ribonucleotide reductase and cancer: biological mechanisms and targeted therapies. Oncogene 34:2011–2021
7. Schmidt TT, Reyes G, Gries K, Ceylan CÜ, Sharma S, Meurer M, Knop M, Chabes A, Hombauer H (2017) Alterations in cellular metabolism triggered by URA7 or GLN3 inactivation cause imbalanced dNTP pools and increased mutagenesis. Proc Natl Acad Sci 114:E4442–E4451
8. Xu X, Page JL, Surtees JA, Liu H, Lagedrost S, Lu Y, Bronson R, Alani E, Nikitin AY, Weiss RS (2008) Broad overexpression of ribonucleotide reductase genes in mice specifically induces lung neoplasms. Cancer Res 68:2652–2660
9. Chabes A, Georgieva B, Domkin V, Zhao X, Rothstein R, Thelander L (2003) Survival of DNA damage in yeast directly depends on increased dNTP levels allowed by relaxed feedback inhibition of ribonucleotide reductase. Cell 112:391–401
10. Watt DL, Buckland RJ, Lujan SA, Kunkel TA, Chabes A (2016) Genome-wide analysis of the specificity and mechanisms of replication infidelity driven by imbalanced dNTP pools. Nucleic Acids Res 44:1669–1680
11. Buckland RJ, Watt DL, Chittoor B, Nilsson AK, Kunkel TA, Chabes A (2014) Increased and imbalanced dNTP pools symmetrically promote both leading and lagging strand replication infidelity. PLoS Genet 10:e1004846
12. Wilson PM, LaBonte MJ, Russell J, Louie S, Ghobrial AA, Ladner RD (2011) A novel fluorescence-based assay for the rapid detection and quantification of cellular deoxyribonucleoside triphosphates. Nucleic Acids Res 39:e112
13. Koc A, Mathews CK, Wheeler LJ, Gross MK, Merrill GF (2006) Thioredoxin is required for deoxyribonucleotide pool maintenance during S phase. J Biol Chem 281:15058–15063. https://doi.org/10.1074/jbc.M601968200
14. Wheeler LJ, Mathews CK (2012) Effects of a mitochondrial mutator mutation in yeast POS5 NADH kinase on mitochondrial nucleotides. J Biol Chem 287:31218–31222
15. Soriano-Carot M, Bañó MC, Igual JC (2012) The yeast mitogen-activated protein kinase Slt2 is involved in the cellular response to genotoxic stress. Cell Div 7:1
16. Prasad Maharjan R, Ferenci T (2003) Global metabolite analysis: the influence of extraction methodology on metabolome profiles of *Escherichia coli*. Anal Biochem 313:145–154
17. Villas-Bôas SG, Højer-Pedersen J, Akesson M, Smedsgaard J, Nielsen J (2005) Global metabolite analysis of yeast: evaluation of sample preparation methods. Yeast 22:1155–1169

18. Eriksson S, Gudas LJ, Ullman B, Clift SM, Martin DW (1981) DeoxyATP-resistant ribonucleotide reductase of mutant mouse lymphoma cells. Evidence for heterozygosity for the protein M1 subunits. J Biol Chem 256:10184–10188

19. Reichard P, Eliasson R, Ingemarson R, Thelander L (2000) Cross-talk between the allosteric effector-binding sites in mouse ribonucleotide reductase. J Biol Chem 275:33021–33026

Part III

Cellular and Cell Extract-Based Assays to Measure DNA Repair

Chapter 7

Conversion Tract Analysis of Homology-Directed Genome Editing Using Oligonucleotide Donors

Yinan Kan and Eric A. Hendrickson

Abstract

Homology-directed genome editing is the intentional alteration of an endogenous genetic locus using information from an exogenous homology donor. A conversion tract is defined as the amount of genetic information that is converted from the homology donor to a given strand of the targeted chromosomal locus. Because of this, conversion tract analysis retrospectively not only elucidates the mechanism of homology-directed genome editing but also provides valuable insights on the conversion efficiency of every nucleotide in the homology donor. Here we describe a blue fluorescent protein-to-green fluorescent protein conversion system that can be conveniently used to measure the efficiency and analyze the lengths of conversion tracts of homology-directed genome editing using oligonucleotide donors in mammalian cells.

Key words Homology-directed repair (HDR), Clustered regularly interspaced short palindromic repeats (CRISPR)/CRISPR-associated 9 (Cas9) {CRISPR-Cas9}, Gene conversion, Conversion tract, Oligonucleotide, Gene editing

1 Introduction

Somatic gene targeting has two general applications of importance and broad interest. One application is the inactivation of genes ("knockouts"), a process utilized mostly by basic researchers to delineate the loss-of-function phenotype(s) of a particular gene [1]. This approach routinely involves the introduction of a DNA double-strand break (DSB) within a genomic locus and the reliance on a cellular DNA repair process termed non-homologous end joining (NHEJ) that frequently introduces small insertions and deletions (aka "indels") during the repair process ([2]; Fig. 1b). If these indels occur in the gene's open reading frame they invariably lead to frameshift mutations that functionally inactivate the gene.

The second application ("knockins"), which is of significantly more clinical and pharmaceutical interest, is the conversion of a preexisting endogenous genomic sequence to a new sequence specified by the researcher [1]. The latter application is best

Lata Balakrishnan and Jason A. Stewart (eds.), *DNA Repair: Methods and Protocols*, Methods in Molecular Biology, vol. 1999, https://doi.org/10.1007/978-1-4939-9500-4_7, © Springer Science+Business Media, LLC, part of Springer Nature 2019

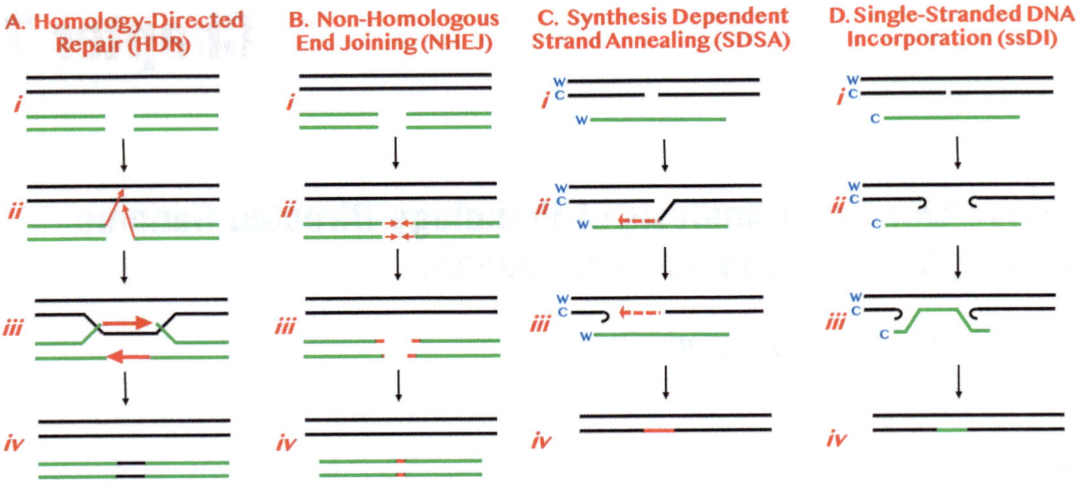

Fig. 1 Models of DNA repair pathways relevant to gene editing. (**a**) Homology directed repair (HDR). (*i*) A chromosome suffers a DNA DSB (shown as nonconnected double green lines). An undamaged homolog or sister chromatid is shown as double black lines. (*ii*) The 3′-ends of the DSB attack (thin red arrows) the undamaged homolog or sister chromatid. (*iii*) The 3′-ends are used to copy the missing genetic information via DNA replication (thick red arrows). (*iv*) The DSB is repaired in an almost error-free fashion such that the repaired chromosome now contains some of the genetic information from the homolog or sister chromatid (green and black double lines). (**b**) Nonhomologous end joining (NHEJ). (*i*) A chromosome suffers a DNA DSB (shown as nonconnected double green lines). An undamaged homolog or sister chromatid is shown as double black lines. (*ii*) The ends of the DSB are "reamed" back together (thin red arrows) irrespective of the presence of the undamaged homolog or sister chromatid. (*iii*) The resultant DNA ends often contain variable, small insertions or deletions (i.e., indels; small red lines). (*iv*) The DSB is repaired in an ~error-prone fashion such that the repaired chromosome now contains an indel at the site where the DSB occurred (green and red double lines). (**c**) Synthesis Dependent Strand Annealing (SDSA). (*i*) A chromosome (double black line) is shown with the strands arbitrarily labeled as Watson (W) or Crick (C). The chromosome suffers a DNA nick (shown as nonconnected black lines). The nick can occur on either strand, but here it is arbitrarily shown as occurring on the C strand. In addition, a donor single-stranded DNA oligonucleotide of W polarity is shown (green line). (*ii*) The 3′-end of the nick attacks the donor oligonucleotide and DNA synthesis (red dashed line) occurs. (*iii*) The extended 3′-end of the nick now reanneals to the chromosome, likely creating flap structures in the process. (*iv*) The flaps are removed and the nick is religated to complete repair which results in a chromosome with new genetic information that is copied off of the donor oligonucleotide on one of its strands (black and red line). (**d**) Single-Stranded DNA Incorporation (ssDI). (*i*) A chromosome (double black line) is shown with the strands arbitrarily labeled as Watson (W) or Crick (C). The chromosome suffers a DNA nick (shown as nonconnected black lines). The nick can occur on either strand, but here it is arbitrarily shown as occurring on the C strand. In addition, a donor single-stranded DNA oligonucleotide of C polarity is shown (green line). (*ii*) The nick on the chromosome is enlarged to a gap. (*iii*) The donor oligonucleotide can now anneal to the chromosome, likely creating flap structures in the process. (*iv*) The flaps are removed and the nick(s) are religated to complete repair which results in a chromosome with new genetic information that is derived from the donor oligonucleotide on one of its strands (black and green line)

exemplified by the concept of gene therapy where an offending, pathology-generating DNA sequence of an individual is replaced with a wild-type version to restore the function of the gene and the health of the patient. This approach routinely also involves the introduction of a DSB within a genomic locus but differs from a knockout in (1) that an exogenous donor DNA is supplied and (2) that the repair relies on a different cellular DNA repair process termed homology-directed repair (HDR; Fig. 1a). In HDR, a portion of the chromosome is "swapped out" for similar (but nonidentical) genetic information on the donor DNA [3]. Since HDR is otherwise essentially error-free, the repaired chromosome is generally functional and contains precisely the novel DNA sequence that the investigator specified.

The HDR-dependent ability to perform precision genome editing has been available to the scientific community since the late 1980s/early 1990s but involved technologies that were cumbersome, labor-intensive, inefficient, and slow. Thus, for about the subsequent almost three decades, precision genome engineering was by and large an esoteric field of science practiced by only the most obstinate of researchers. The discovery of clustered regularly interspaced short palindromic repeats (CRISPR) and the CRISPR-associated 9 (Cas9) protein and their adaptation to eukaryotic cell systems brought electrifying life to this moribund field [4–9]. Thus, the high frequency and efficiency with which CRISPR/Cas9 is capable of generating a DNA DSB at a desired locus has made precision genome editing practical in organisms as disparate as otherwise genetically intractable microorganisms [10], commercially important crops [11], and animals [12] as well as humans [13, 14]. These applications have, in turn, resulted in a veritable plethora of new genome-modified research models, agricultural products, and therapeutic reagents.

Precision genome editing takes advantage of the endogenous HDR pathways that exist in most organisms. During an HDR-mediated repair event, genetic information is transferred from an exogenous homology donor to a given strand of the targeted chromosomal locus. Technically, the portion of transferred genetic information is called a conversion tract. Importantly, conversion tract analysis not only elucidates the mechanism of homology-directed genome editing [15–17], but also provides valuable insights on the conversion efficiency of every nucleotide in the homology donor. The precise size of a conversion tract depends upon many variables but is strongly influenced by the type of genomic lesion that is used to initiate the event, the pathway utilized to repair the lesion and the type of DNA that is used as the donor.

DNA DSBs are the standard lesion used to initiate genome engineering. When this occurs, the conversion tracts associated with the gene editing event are quite large and often extend for hundreds of nucleotides in either direction from the site of the DSB [17]. Because of this feature, the use of DSBs coupled with dsDNA donors are the gold standard for introducing large modifications onto a chromosome. However, DNA DSBs have the serious disadvantage that they can be repaired by either HDR (desired) or NHEJ (highly undesirable) with the daunting empirical observation that the NHEJ pathway usually predominates in most cells/organisms. Because of this significant drawback, a variety of investigators sought out potentially safer ways to carry out gene editing. DNA nicks were known to be recombinogenic [18], and thus investigators engineered the Cas9 protein such that it would make only nicks and not DSBs [19]. Importantly, nicks seldom, if ever, engage the NHEJ repair pathway and so this significant drawback is obviated [20]. The downside of utilizing nicks is that they are less recombinogenic than DSBs, and thus the frequency of precision gene editing is decreased. Although not necessarily a drawback, an additional caveat of nick-mediated gene editing is that it utilizes different subpathways of HDR to carry out the repair event. Thus, the repair of DSBs with a dsDNA donor template utilizes the canonical DSB repair (DSBR) HDR pathway [15, 16]. In contrast, nick-initiated repair events often engage either the synthesis-dependent strand annealing (SDSA; Fig. 1c) or single-strand DNA incorporation (ssDI; Fig. 1d) pathways of HDR [17, 21, 22].

One additional advantage of nick-mediated precise genome engineering (besides the lack of interference from NHEJ) is that single-stranded oligonucleotides (ODNs) can be used as donors. ODNs are usually only 30–200 nt in length although the recent description of "megamers"—ODNs that can be upward of 2 kb long—have significantly increased the potential utility of this approach [23]. Importantly, ODNs can be conveniently chemically synthesized and efficiently convert genetic information into the chromosome in the presence of either DSBs or nicks. One limitation of this approach is that it is accompanied by small conversion tracts. Thus conversion tract analyses demonstrated that ODNs are only efficient in converting genetic information within ~15 nucleotides of the targeted genomic lesion [17].

In summary, our laboratory (along with others) has demonstrated that when large amounts of genetic information must be introduced onto a chromosome then the use of a DSB as the initiating lesion accompanied by a dsDNA donor is the preferred approach. When smaller amounts of genetic information (especially changes such as single nucleotide alterations) must be introduced then the use of a nick as the initiating lesion accompanied by an ODN donor will often work best, or at least as well.

In the course of carrying out these projects, our laboratory has needed to develop a variety of assays to measure the efficiency of gene editing and the length of conversion tracts. Here we describe a blue fluorescent protein (BFP)—to—enhanced green fluorescent protein (EGFP) conversion system that can be conveniently used to measure the efficiency and analyze the conversion tracts of homology-directed genome editing using ODN donors in mammalian cells. Please note that this is a modified protocol from our previous publication where the selection strategy was inverted, that is, EGFP to BFP [17].

2 Materials

1. Cell line: the human colorectal carcinoma HCT116 (ATCC) cell line is used here as an example. A multitude of additional human cell lines have, however, been used with great success. These include cancer cell lines such as U2OS, HeLa, HEK293, MCF7, and MCF10A; the immortalized, but nontransformed RPE + hTERT cell line; and the primary cell lines, RPE and BJ fibroblasts. Each of these cell lines requires different culture and handling conditions from HCT116 cells and thus the reader is recommended to inform themselves in regard to the culture requirements for their particular cell line.
2. Cell culture related materials: the cells are cultured in either 100 mm, 6-well or 96-well tissue culture dishes in a humidified 37 °C incubator with 5% CO_2 in McCoy's 5A medium supplemented with 10% fetal bovine serum (FBS), 1% L-glutamine, and 1% penicillin/streptomycin. TrypLE Express enzyme (Fisher) is utilized according to the manufacturer's instructions for trypsinization and sub-cloning and all cell washing is done with a phosphate buffered saline (PBS) solution, pH 7.4. A Neon™ transfection system (Fisher) using the 10 μL kit is recommended for the introduction of reagents and donor ODNs into HCT116 and for most other cell lines.
3. The BFP reporter: an expression vector for BFP (*see* **Notes 1** and **2**), which was specially derived from EGFP-N2 by introducing an H66Y mutation into the EGFP coding sequence was used [24]. However, many different BFP (and/or EGFP) expression vectors can be reversibly adapted for this step as long as they contain a nucleotide sequence with the CAC (histidine; H) > TAC (tyrosine; Y) mutation at amino acid H66. One such vector is pLenti CMV GFP Puro, which is available through Addgene (plasmid #17448). In this example, the reporter needs to be modified by introducing a single nucleotide TAC > CAC mutation into the EGFP open reading frame to convert it into BFP. The relevant sequence is shown below with the Y66H mutation highlighted in blue (*see* **Note 2**).

CGAGGGCGATGCCACCTACGGCAAGCTGACCCTGAAGTTCATCT

GCACCACCGGCAAGCTGCCCGTGCCCTGGCCCACCCTCGTGAC

CACCCTGACCCACGGCGTGCAGTGCTTCAGCCGCTACCCCGAC

CACATGAAGCAGCACGACTTCTTCAAGTCCGCCATGCCCGAAGG

CTACGTCCAGGAGCGCACCATC

We introduced this single nucleotide modification into our EGFP expression vector using a QuikChange II Site-Directed Mutagenesis kit from Agilent, but many similar kits are available from a variety of commercial sources, and all of them should work equally effectively for this simple modification. In our published experiments, we have integrated the reporter specifically at the hypoxanthine phosphoribosyl transferase (HPRT) locus on the X chromosome [24], but in practice the reporter can be integrated into any chromosome either via random integration (using, for example, a standard lentiviral-based vector and lentiviral transduction) or via homology-directed genome editing (such as CRISPR/Cas9-mediated approaches) as the investigator wishes.

4. Cas9 and gRNAs: TrueCut™ Cas9 Protein v2 and TrueGuide™ 1-piece modified synthetic gRNA with the following protospacer sequences are recommended. The nickase versions of Cas9 (i.e., the D10A and H840A variants) are available from Applied Biological Materials.

   BFP_S gRNA: CTCGTGACCACCCTGACC

   BFP_AS gRNA: GCACTGCACGCCGTGGGTCA

5. ODN donors: the central GFP mutation is highlighted in green and flanking synonymous single nucleotide polymorphisms (SNPs) (which can be used for conversion tract analyses) highlighted in red (*see* **Note 3**).

   GFP_S0:

ACGGCAAGCTGACCCTGAAGTTCATCTGCACCACCGGCAAGCTG

CCCGTGCCCTGGCCCACCCTCGTGACCACCCTGACCTACGGCG

TGCAGTGCTTCAGCCGCTACCCCGACCACATGAAGCAGCACGAC

TTCTTCAAGTCCGCCATGCCCGAAGGCTA

GFP_S1:

ACGGCAAGCTGACCCTGAAATTCATCTGCACCACCGGCAAGCTG

CCCGTGCCCTGGCCCACACTCGTGACCACCCTGACCTACGGCG

TGCAGTGCTTCAGTCGCTACCCCGACCACATGAAGCAGCACGAC

TTCTTCAAGTCTGCCATGCCCGAAGGCTA

GFP_S2:

ACGGCAAGCTGACCCTGAAGTTCATCTGCACCACCGGCAAACTG

CCCGTGCCCTGGCCCACCCTCGTCACCACCCTGACCTACGGCG

TGCAGTGTTTCAGCCGCTACCCCGACCACATGAAACAGCACGAC

TTCTTCAAGTCCGCCATGCCCGAAGGCTA

GFP_S3:

ACGGCAAGCTGACCCTGAAGTTCATCTGCACCACCGGCAAGCTG

CCCGTCCCCTGGCCCACCCTCGTGACCACGCTGACCTACGGCG

TCCAGTGCTTCAGCCGCTACCCTGACCACATGAAGCAGCACGAC

TTCTTCAAGTCCGCCATGCCCGAAGGCTA

GFP_AS0:

TAGCCTTCGGGCATGGCGGACTTGAAGAAGTCGTGCTGCTTCAT

GTGGTCGGGGTAGCGGCTGAAGCACTGCACGCCGTAGGTCAGG

GTGGTCACGAGGGTGGGCCAGGGCACGGGCAGCTTGCCGGTG

GTGCAGATGAACTTCAGGGTCAGCTTGCCGT

GFP_AS1:

TAGCCTTCGGGCATGGCAGACTTGAAGAAGTCGTGCTGCTTCAT

GTGGTCGGGGTAGCGACTGAAGCACTGCACGCCGTAGGTCAGG

GTGGTCACGAGTGTGGGCCAGGGCACGGGCAGCTTGCCGGTG

GTGCAGATGAATTTCAGGGTCAGCTTGCCGT

GFP_AS2:

TAGCCTTCGGGCATGGCGGACTTGAAGAAGTCGTGCTGTTTCAT

GTGGTCGGGGTAGCGGCTGAAACACTGCACGCCGTAGGTCAGG

GTGGTGACGAGGGTGGGCCAGGGCACGGGCAGTTTGCCGGTG

GTGCAGATGAACTTCAGGGTCAGCTTGCCGT

GFP_AS3:

TAGCCTTCGGGCATGGCGGACTTGAAGAAGTCGTGCTGCTTCAT

GTGGTCAGGGTAGCGGCTGAAGCACTGGACGCCGTAGGTCAGC

GTGGTCACGAGGGTGGGCCAGGGGACGGGCAGCTTGCCGGTG

GTGCAGATGAACTTCAGGGTCAGCTTGCCGT

6. Miscellaneous reagents: FACS sorter: any FACS sorter with 405 nm and 488 nm lasers and corresponding filters for BFP and GFP. We do not recommend colinear lasers. Genomic DNA is prepared using a KAPA Mouse Genotyping Kit. PCR was performed using as a master mix the KAPA HiFi HotStart ReadyMixPCR Kit.
7. Primers (DNA ODNs):

   For Sanger sequencing:

   GFP_SF: CGAGGGCGATGCCACC

   GFP_SR: GATGGTGCGCTCCTGGA

   For next generation sequencing (NGS) (NGS adaptors highlighted in orange, *see* **Note 4**):

   GFP_NGSF:

ACACTCTTTCCCTACACGACGCTCTTCCGATCTCGAGGGCGATG

CCACC

GFP_NGSR:

GACTGGAGTTCAGACGTGTGCTCTTCCGATCTGATGGTGCGCTC

CTGGA

8. ODN purification: Agencourt AMPure Solid Phase Reversible Immobilization (SPRI).

3 Methods

3.1 Generation of the BFP Reporter Cell Line

1. Construct the BFP reporter cassette described in Subheading 2. The reporter can be integrated into the chromosome via plasmid or lentiviral random integration. Below (in Subheading 3.1, **step 2**) we briefly describe the generation of a BFP reporter using pLenti-CMV-GFP-Puro and lentiviral integration. Alternatively, a method of site-specific integration of the BFP expression cassette via homology-directed genome editing using recombinant adeno-associated virus (rAAV) has been published [17] and the plasmid is available upon request.
2. Introduce the TAC > CAC mutation into the pLenti-CMV-GFP-Puro vector (described in Subheading 3) using site-directed mutagenesis. Package the resultant BFP expression plasmid into a lentivirus using a standard second or third generation lentiviral packaging protocol. Infect the desired target cell line using the BFP lentivirus. Culture the infected cells for 7 days. A stably integrated cell line with bright BFP expression can be isolated via FACS sorting or serial dilution.

3.2 Perform the Transfection

1. Seed the BFP reporter cell line onto a 100 mm plate(s) 1–2 days before the transfection. Approximately 1×10^5 cells will be needed per transfection. Moreover, the cells should be approximately 70% confluent (i.e., still proliferating) at the time of the transfection.
2. Add 1 mL of fresh culture medium into one well of a 6-well plate for each gRNA and ODN combination that will be used and prewarm the plate in a 37 °C incubator. The ODNs GFP_S0 and GFP_AS0 are used to determine the efficiency of homology-directed genome editing (Subheading 3.3), whereas the other six ODNs (GFP_S1, GFP_S2, GFP_S3, GFP_AS1, GFP_AS2, or GFP_AS3) are used to map the conversion tracts (Subheading 3.4). A control transfection without any gRNA is also recommended to determine the baseline green fluorescence in the absence of DSBs.
3. Add 1.25 μg TrueCut Cas9 Protein v2 (or nickase variants) and 0.25 μg TrueGuide 1-piece modified Synthetic gRNA into 6 μL of Resuspension Buffer R (provided in the Neon transfection system 10 μL kit) into a PCR tube for every transfection. Gently mix and incubate at room temperature for 15 min.
4. Harvest 1×10^5 cells per transfection using TrypLE. Neutralize the TrypLE with three volumes of culture medium and centrifuge the cells at 400 rcf for 5 min.
5. Wash the cell with PBS. Centrifuge again at 600 rcf for 5 min.

6. In the meantime, add 1 μg of the corresponding ODN donor (GFP_S1, GFP_S2, GFP_S3, GFP_AS1, GFP_AS2, or GFP_AS3) to each PCR tube.
7. Aspirate the PBS and resuspend the cells with 6 μL Resuspension Buffer R per transfection.
8. Aliquot 6 μL of the cell suspension into each PCR tube containing Cas9, gRNA and ODNs. Gently mix.
9. Electroporate the mixture using 10 μL Neon tips (provided in the Neon transfection system 10 μL kit) using optimized conditions for your cell line. For most cell lines, 1450 V/10 ms/3 pulses (for high viability) or 1600 V/10 ms/3 pulses (for optimal transfection) efficiency are used.
10. Immediately transfer the cells into one well of the 6-well plate with prewarmed culture medium. Culture the cells for 2 days in the 37 °C incubator.

3.3 Measure the Conversion Efficiency

1. Harvest all adherent cells using TrypLE approximately 2 days after transfection. Neutralize the TrypLE using 3 volumes of medium. Spin down the cells at 400 rcf for 5 min and resuspend the cells in 350 μL of culture medium.
2. Determine the efficiency of homology-directed genome editing as the percentage of GFP-positive cells using a flow cytometry or a FACS sorter. Gate the single cell cluster out of the debris and doublets and then draw a BFP^+ gate that is above the BFP negative cells in the no gRNA control. Use the same BFP gate to quantify and separate the BFP-positive cells (Fig. 2; *see also* **Note 5**).

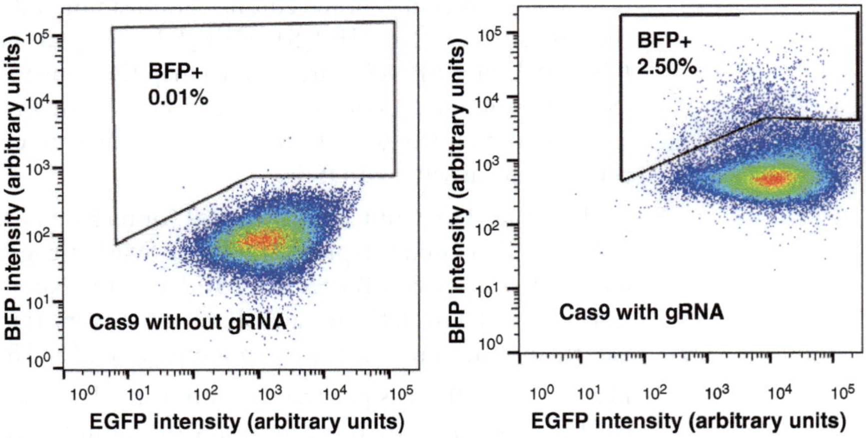

Fig. 2 Homology-directed genome editing measured by the EGFP-to-BFP reporter. The percentage of BFP-positive cells in the absence and presence of gRNA are analyzed using flow cytometry. Although an EGFP-to BFP conversion is shown here, we recommend (as described in the text) carrying out this experiment in the reverse direction (i.e., BFP-to-EGFP) because it gives a better separation of the converted cells

3.4 Mapping the Conversion Tracts

1. Harvest the transfected cells using TrypLE approximately 2 days after transfection and collect the GFP-positive cells using a FACS sorter (as in Subheading 3.3, **step 1**, above; *see* **Note 6**).
2. For Next Generation sequencing (NGS), collect at least 10,000 GFP-positive cells in an Eppendorf tube.
3. Centrifuge the cells at 1000 rcf for 5 min. Prepare the genomic DNA using a KAPA Mouse Genotyping Kit.
4. PCR amplify the targeted GFP fragment using primers GFP_NGSF × GFP_NGSR and the KAPA HiFi HotStart ReadyMixPCR Kit according to the manufacturer's instructions. To minimize PCR bias, we recommend using the minimal number of cycles which results in a visible band on a gel.
5. Barcode each transfection using designated barcoding primers recommended by the NGS service provider by subjecting them to 15 PCR cycles using a KAPA HiFi HotStart ReadyMixPCR kit.
6. Pool all the PCR products and purify them using Agencourt AMPure SPRI beads according to the manufacturer's instructions. Send the library to the NGS facility.
7. For Sanger sequencing, place individual GFP-positive cells in a total of 10 prewarmed 96-well plates with culture medium.
8. Culture the cells in the 37 °C incubator for up to 2 weeks until the individual colonies fill about one-third area of the well.
9. Harvest 100 individual colonies for each transfection using TrypLE. Move the trypsinized cells into 96-well PCR plates with medium. Centrifuge the cells at 1000 rcf for 5 min. Prepare the genomic DNA using a KAPA Mouse Genotyping Kit as described above.
10. PCR amplify the targeted GFP fragment using primers GFP_SF × GFP_SR and KAPA HiFi HotStart ReadyMixPCR Kit for 30–35 PCR cycles.
11. Send all individual unpurified PCR products to a Sanger sequencing facility.
12. Compile all sequencing reads containing the central TAC nucleotides highlighted in green (Materials, Subheading 2, **item 5**). Calculate the coconversion frequency of each flanking SNP highlighted in red (Materials, Subheading 2, **item 5**). Plot the coconversion frequency of all flanking SNPs against their distance to the targeted genomic lesion to generate a conversion tract map for each transfection. Example conversion tracts are shown below (Fig. 3). Usually the SDSA pathway generates unidirectional conversion tracts starting near the genomic lesion, whereas the ssDI pathway generates bidirectional conversion flanking the genomic lesion (*see* **Note 7**).

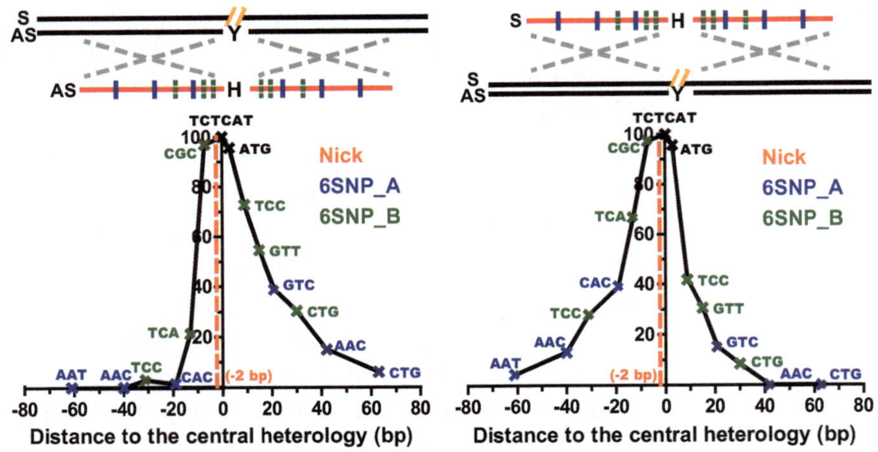

Fig. 3 Example conversion tracts of Cas9 nickase-induced, homology-directed genome editing. Coconversion efficiency of flanking SNPs is mapped against their distance to the nicking sites. The left and right panels show conversion maps using ODNs donors with the same sense as the complementary strand and the nicked strand, respectively. Schematic elements are colored as follows: genomic DNA, black; genomic lesions, orange; ODNs, red; SNPs on different ODNs, blue and green; homology regions, dashed silver crosses

4 Notes

1. Although the reporters can theoretically be used in either direction, the BFP-to-EGFP conversion direction is recommended as the resultant EGFP fluorescence is much brighter than the original BFP. Thus, the EGFP conversion products can be cleanly separated from BFP in spite of minor spectral overlap, whereas an EGFP-to-BFP experiment (which was the direction we initially described [17]) does not yield as clean a separation.
2. We minimized the changes to convert BFP to EGFP to one nucleotide (as opposed to several nucleotides as originally described [17]) to avoid a bias toward longer conversion tracts containing the full BFP-to-EGFP mutations.
3. ODNs with too many flanking SNPs will reduce targeting efficiency. ODNs with SNPs too close to each other may favor the coconversion of adjacent SNPs. The ODNs GFP_S1-3 and GFP_AS1-3 are designed to generate smooth conversion tracts with minimal bias from coconversion.
4. Please consult your NGS service provider for NGS adaptors and barcode information.
5. Compared to the EGFP-to-BFP reporter, the GFP cells generated by the BFP-to-EGFP reporter are much brighter and more separable from the BFP-positive background. *See* **Note 1**. Compensation is optional.

6. NGS is preferred over Sanger sequencing for the ease of operation and high coverage. A library grade PCR enzyme and minimal PCR cycles are required to minimize the skew during library construction.
7. The conversion tracts can be affected by other factors. Minimal 3′ resection may affect the starting position of the conversion tracts. Also, clustered SNPs on the ODNs favor coconversion and thus generate longer conversion tracts. The modified ODN designs help to generate sharper and more characteristic conversion tracts compared to our previous publication [17].

References

1. Hendrickson EA (2008) Gene targeting in human somatic cells. In: Conn PM (ed) Source book of models for biomedical research. Humana Press, Inc., Totowa, NJ, pp 509–525
2. Chang HHY, Pannunzio NR, Adachi N, Lieber MR (2017) Non-homologous DNA end joining and alternative pathways to double-strand break repair. Nat Rev Mol Cell Biol 18 (8):495–506. https://doi.org/10.1038/nrm.2017.48
3. Jasin M, Rothstein R (2013) Repair of strand breaks by homologous recombination. Cold Spring Harb Perspect Biol 5(11):a012740. https://doi.org/10.1101/cshperspect.a012740
4. Mali P, Esvelt KM, Church GM (2013) Cas9 as a versatile tool for engineering biology. Nat Methods 10(10):957–963. https://doi.org/10.1038/nmeth.2649
5. Hsu PD, Lander ES, Zhang F (2014) Development and applications of CRISPR-Cas9 for genome engineering. Cell 157(6):1262–1278. https://doi.org/10.1016/j.cell.2014.05.010
6. Doudna JA, Charpentier E (2014) Genome editing. The new frontier of genome engineering with CRISPR-Cas9. Science 346 (6213):1258096. https://doi.org/10.1126/science.1258096
7. Pennisi E (2013) The CRISPR craze. Science 341(6148):833–836. https://doi.org/10.1126/science.341.6148.833
8. Baker M (2014) Gene editing at CRISPR speed. Nat Biotechnol 32(4):309–312. https://doi.org/10.1038/nbt.2863
9. Ledford H (2015) CRISPR, the disruptor. Nature 522(7554):20–24. https://doi.org/10.1038/522020a
10. Shapiro RS, Chavez A, Collins JJ (2018) CRISPR-based genomic tools for the manipulation of genetically intractable microorganisms. Nat Rev Microbiol 16:333–339. https://doi.org/10.1038/s41579-018-0002-7
11. Arora L, Narula A (2017) Gene editing and crop improvement using CRISPR-Cas9 system. Front Plant Sci 8:1932. https://doi.org/10.3389/fpls.2017.01932
12. Lamas-Toranzo I, Guerrero-Sanchez J, Miralles-Bover H, Alegre-Cid G, Pericuesta E, Bermejo-Alvarez P (2017) CRISPR is knocking on barn door. Reprod Domest Anim 52 (Suppl 4):39–47. https://doi.org/10.1111/rda.13047
13. Pankowicz FP, Jarrett KE, Lagor WR, Bissig KD (2017) CRISPR/Cas9: at the cutting edge of hepatology. Gut 66(7):1329–1340. https://doi.org/10.1136/gutjnl-2016-313565
14. Smith AJ, Carter SP, Kennedy BN (2017) Genome editing: the breakthrough technology for inherited retinal disease? Expert Opin Biol Ther 17(10):1245–1254. https://doi.org/10.1080/14712598.2017.1347629
15. Langston LD, Symington LS (2004) Gene targeting in yeast is initiated by two independent strand invasions. Proc Natl Acad Sci U S A 101 (43):15392–15397. https://doi.org/10.1073/pnas.0403748101
16. Kan Y, Ruis B, Lin S, Hendrickson EA (2014) The mechanism of gene targeting in human somatic cells. PLoS Genet 10(4):e1004251. https://doi.org/10.1371/journal.pgen.1004251
17. Kan Y, Ruis B, Takasugi T, Hendrickson EA (2017) Mechanisms of precise genome editing using oligonucleotide donors. Genome Res 27 (7):1099–1111. https://doi.org/10.1101/gr.214775.116
18. Lee GS, Neiditch MB, Salus SS, Roth DB (2004) RAG proteins shepherd double-strand breaks to a specific pathway, suppressing error-

prone repair, but RAG nicking initiates homologous recombination. Cell 117(2):171–184

19. Ran FA, Hsu PD, Lin CY, Gootenberg JS, Konermann S, Trevino AE, Scott DA, Inoue A, Matoba S, Zhang Y, Zhang F (2013) Double nicking by RNA-guided CRISPR Cas9 for enhanced genome editing specificity. Cell 154(6):1380–1389. https://doi.org/10.1016/j.cell.2013.08.021
20. Vriend LE, Krawczyk PM (2017) Nick-initiated homologous recombination: protecting the genome, one strand at a time. DNA Repair (Amst) 50:1–13. https://doi.org/10.1016/j.dnarep.2016.12.005
21. Davis L, Maizels N (2014) Homology-directed repair of DNA nicks via pathways distinct from canonical double-strand break repair. Proc Natl Acad Sci U S A 111(10):E924–E932. https://doi.org/10.1073/pnas.1400236111
22. Davis L, Maizels N (2011) DNA nicks promote efficient and safe targeted gene correction. PLoS One 6(9):e23981. https://doi.org/10.1371/journal.pone.0023981
23. Quadros RM, Miura H, Harms DW, Akatsuka H, Sato T, Aida T, Redder R, Richardson GP, Inagaki Y, Sakai D, Buckley SM, Seshacharyulu P, Batra SK, Behlke MA, Zeiner SA, Jacobi AM, Izu Y, Thoreson WB, Urness LD, Mansour SL, Ohtsuka M, Gurumurthy CB (2017) Easi-CRISPR: a robust method for one-step generation of mice carrying conditional and insertion alleles using long ssDNA donors and CRISPR ribonucleoproteins. Genome Biol 18(1):92. https://doi.org/10.1186/s13059-017-1220-4
24. Heim R, Tsien RY (1996) Engineering green fluorescent protein for improved brightness, longer wavelengths and fluorescence resonance energy transfer. Curr Biol 6(2):178–182

Chapter 8

Reporter Assays for BER Pathway

Dorota Piekna-Przybylska

Abstract

Base excision repair (BER) is one of the most active DNA repair pathways in cells correcting DNA damage from oxidation, deamination, alkylation, and damages induced by free radicals and ionizing radiation. Deregulation or deficiencies in BER mechanisms increase the level of mutations leading to carcinogenesis, and single-strand DNA break formation, which may be converted to double-strand breaks and induce apoptosis. BER deficiency is associated with development of diseases causing neurodegenerative disorders, such as Alzheimer's disease (AD) and amyotrophic lateral sclerosis (ALS). In addition, BER mechanisms can be affected by viral infections, such as HPV, HTLV-1, and HIV-1. Deficiencies in DNA repair in cells can be analyzed using a very convenient and effective approach, where mammalian cells are transfected with plasmids carrying a reporter gene of fluorescent protein that contain inactivating damages. The repair of DNA damages depends on the cellular machinery and is reflected by expression of the reporter gene measured by flow cytometry. In this chapter, we describe this plasmid-based reporter gene system to investigate in cell the repairs of DNA damages involving BER mechanisms.

Key words Base excision repair, BER, DNA repair, Reporter assay, Flow cytometry

1 Introduction

The DNA lesions repaired by BER pathways are mostly base modifications, abasic sites, and single-strand DNA breaks [1]. The major source of DNA damage is reactive oxygen species (ROS) produced as a normal product of cellular metabolism and also generated from exogenous sources, such as ionizing radiation. There are more than 100 types of oxidative base modifications described in mammalian DNA, with guanine the most susceptible to oxidation by ROS [2–4]. The 8-hydroxyguanine (8-oxoG) lesion, which is an oxidized derivative of guanine, is estimated to form in 2800 residues in DNA per day in a single human cell [5, 6]. Another common lesion repaired by BER is hydrolytic deamination of cytosine (100–500 per day per cell), which results in the formation of uracil [7]. Both, 8-oxoG and uracil are premutagenic lesions, because if left unrepaired they would mispair with adenine during DNA replication, and this mispairing will

Lata Balakrishnan and Jason A. Stewart (eds.), *DNA Repair: Methods and Protocols*, Methods in Molecular Biology, vol. 1999, https://doi.org/10.1007/978-1-4939-9500-4_8,

result in G → T (for oxidized G) and C → T (for deaminated C) transversion mutations. However, the most frequent damage repaired by BER is the hydrolysis of the glycosidic bonds in DNA, which results in a loss of about 10,000 bases per day and formation of abasic sites (AP; apurinic and apyrimidinic sites) [7, 8]. AP sites are also generated by DNA glycosylases at the first step of BER.

The BER pathway is initiated by a damage-specific DNA glycosylase, which removes the flipped out damaged base leaving an abasic site [9]. The AP site is subsequently processed by AP endonuclease, APE1 to produce a single-strand break. There are currently described 11 distinct DNA glycosylases in mammalian cells, and most of them excise a wide variety of damaged bases [10]. Few of them are bifunctional and perform also cleavage of phosphodiester bonds using its AP lyase activity [11]. After the strand cleavage, the 1 nucleotide gap intermediate can be repaired either through short patch BER (SP-BER), which is a single nucleotide polymerization by DNA polymerase β, or long patch BER (LP-BER) involving replacement of several (2–15) nucleotides by DNA polymerase δ/ε. What subpathway of BER will be used to repair the damage base is determined by the 5′ dRPase activity of DNA polymerase β, which may remove the 5′ dRP group generated by APE1 [12, 13]. If the 5′ dRP is removed, than SP-BER is finalized by sealing a nick in ligation reaction carried out by DNA ligase III (LIG3) in cooperation with XRCC1, or by DNA ligase I (LIG1). If the 5′ dRP is not removed, than the repair is continued through LP-BER subpathway, where polymerization is initiated by DNA polymerase δ/ε displacing the strand containing the 5′ dRP (known as the "flap") in cooperation with the clamp loading factor RFC and the processivity factor PCNA. The flap structure is subsequently removed by the flap endonuclease FEN1, and a nick is sealed by LIG1.

Our DNA repair assay monitoring BER mechanism in cells relies on transfecting cells with plasmids carrying the enhanced green fluorescent protein (eGFP) gene that contains inactivating damages. The efficiency of DNA repairs is assessed using flow cytometry, which allows for analyzing quantitatively the level and fluorescent intensity of eGFP (Fig. 1). The assay described in this chapter focuses on the construction of plasmids bearing damages subsequently repaired by BER mechanisms, but we also used this method to investigate DNA mismatch repair (MMR) pathway [14]. Such a plasmid system can be used, for example, in infected cells to monitor virus interference in DNA repair, or in cells with inactivated genes to determine whether loss of particular protein affects the repair of damages in reporter gene [14]. The system may also be used in combination with inhibitors or drugs to measure their effect on DNA repair in cells [14].

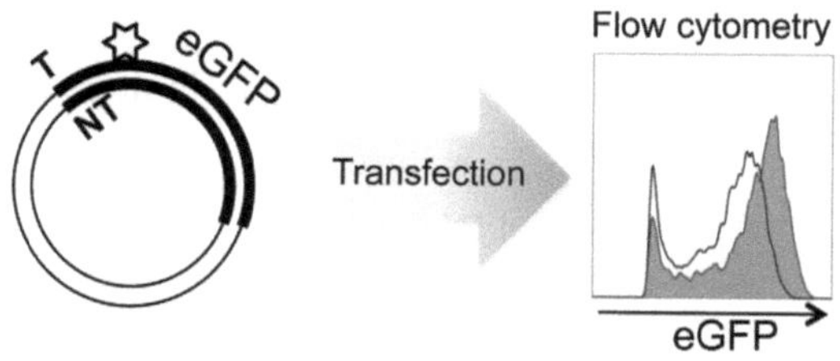

Fig. 1 Reporter assay to analyze BER in cells. Mammalian cells are transfected with plasmids expressing a reporter gene, such as eGFP that contain inactivating damages (*star*) within the open reading frame (*thick line*) in the transcribed (*T*) DNA strand only. The efficiency of DNA repair is determined by flow cytometry analysis of eGFP expression in cells transfected with undamaged construct (*in histogram gray profile*) and with damaged construct (*white*). *NT* nontranscribed DNA strand

2 Materials

2.1 Preparation of Lesion-Specific BER Plasmid Substrates

1. Plasmid constructs pD59 or pD61; or other plasmids expressing reporter gene designed to incorporate synthetic oligonucleotide bearing DNA damage.
2. Nicking enzyme Nb.BbvCI and CutSmart Buffer (New England BioLabs); or other nicking enzyme, for which two restriction sites were incorporated into the reporter gene.
3. Synthetic unmodified and damage containing oligonucleotides (available from Eurofins MWG Operon, Sigma, Integrated DNA Technologies and Bio-Synthesis).
4. Restriction enzyme HindIII, or any enzyme for which only one restriction site is present in the plasmid construct.
5. T4 polynucleotide kinase (PNK).
6. T4 DNA Ligase and 10× Reaction Buffer: 500 mM Tris–HCl, 100 mM $MgCl_2$, 100 mM DTT, 10 mM ATP, pH 7.5.
7. Agarose for gel electrophoresis.
8. 10 mg/mL ethidium bromide (EtBr) solution in TE buffer. EtBr is a DNA intercalating agent and potential carcinogen. Handle only with gloves and proper precautions. Store at room temperature, protected from light.
9. DNA size marker.
10. DNA loading buffer.
11. Cesium chloride (CsCl).
12. 5 M NaCl.
13. Isopropanol.
14. 70% and 100% ethanol.
15. 3 M sodium acetate (NaOAc), pH 5.2.

16. TE buffer: 10 mM Tris–HCl, pH 7.5, 1 mM EDTA.
17. 18- and 21-gaude needles.
18. 5 mL syringe.
19. UV lamp.
20. PCR thermal cycler accommodating 0.5 mL tubes.
21. Beckman Coulter Quick-Seal Ultra-Clear tubes with Spacers.
22. Beckman Quick-Seal Tube Sealer.
23. Beckman 70.1 Ti rotor.
24. Beckman Coulter ultracentrifuge.
25. Beckman Coulter high-speed centrifuge.
26. Microcentrifuge.
27. Heating block.
28. GeneQuant II RNA/DNA Calculator (Pharmacia Biotech).

2.2 Transfection of Cells with Lesion-Specific BER Plasmids

1. Human colon adenocarcinoma cell line HCT116 (Sigma).
2. HCT116 growth medium: McCoy's 5A medium (Sigma) with 10% fetal bovine serum (FBS), 2 mM glutamine, and 1% penicillin–streptomycin.
3. Trypsin–EDTA solution: 0.25% trypsin–1 mM EDTA (Gibco™).
4. Lipofectamine® LTX with PLUS reagent (Invitrogen™).
5. Opti-MEM® I Reduced Serum Media (Gibco™).
6. Phosphate buffered saline (PBS) without Mg^{2+}/Ca^{2+}, pH 7.2 (Gibco™).
7. Tissue culture plates (6-well) and other supplies for cell culture work.
8. Sorvall centrifuge.
9. Vi-CELL XR cell counter (Beckman Coulter).
10. CO_2 incubator (37 °C, 5% CO_2 and maximal humidity).

2.3 Flow Cytometry Analysis

1. BD Cytofix™ Fixation buffer (BD Biosciences), or PBS with 4% formaldehyde.
2. BD Pharmingen™ Stain Buffer (FBS) (BD Biosciences), or PBS with 5–15% of FBS.
3. Sorvall centrifuge.
4. Flow cytometer LSR II 18-Color flow cytometer (Becton Dickinson) or any flow cytometer emitting the blue-green 488 nm wavelength for excitation of eGFP and emission filters for detection wavelength of 509 nm.
5. FlowJo software for analyzing flow cytometry data.

3 Methods

3.1 Principle of the BER Assay In Vivo

Two of the most common damages repaired by BER are 8-oxoG and deamination of C resulting in the formation of U. The principle of the assay monitoring the repair of these two damages relies on the generation of stop codons in mRNA of reporter gene, only if 8-oxoG or uracil are left unrepaired. The stop codons in mRNA would result from transcriptional mutagenesis. Transcription through 8-oxoG is not efficiently blocked and mostly results in erroneous incorporation of adenine instead of cytosine by RNA polymerase II [15, 16]. The presence of uracil in the transcribed strand of DNA may also lead to insertion of adenine opposite to uracil during transcription [17–19]. Thus, in designing plasmid constructs 8-oxoG and uracil should be placed in the open reading frame (ORF) of reporter gene in codons that would be converted to stop codons during transcription, if left unrepaired. For example, in the transcribed strand the anticodon AGC, which is complementary to codon TCG for Serine, will generate stop codon UAG in RNA, if oxidized G in AGC is not repaired (Fig. 2a). The same design is applied to analyze the repair of deaminated cytosine resulting in the formation of uracil. For example, ACC in the transcribed strand is complementary to TGG codon for tryptophan, but it will generate stop codon UAG or UGA in RNA, if one of the C residues is replaced by U in ACC, and left unrepaired (Fig. 2a). In consequence, reductions in a reporter gene expression from the constructs with 8-oxoG or uracil present in the transcribed strand will indicate that the repairs of these damages in cells are not efficient.

Using similar scheme, the repair of 8-oxoG and uracil can also be analyzed with plasmid construct containing stop codon (e.g., TAA) within the ORF of reporter gene to inactivate its translation (Fig. 2b). However, if damaged base is placed in the transcribed strand (8-oxoGTT or UTT), and left unrepaired, then during transcription the RNA polymerase II erroneously incorporate adenine creating in RNA a sense codon (AAA—lysine) allowing for expression of reporter gene. The same design can be used to monitor the repair of abasic sites, which are transcribed by mammalian polymerase II as cytosine (Fig. 2b) [19]. In consequence, the rescue of reporter gene expression from constructs with stop codon in the ORF will indicate that repairs of 8-oxoG, uracil or abasic site are not efficient, since reporter gene would be translated only if transcriptional mutagenesis occurred on unrepaired sites.

The plasmid constructs containing a damaged base in defined location and generating, or eliminating a stop codon, if left unrepaired, can monitor efficiency of BER in cells, but will not indicate if damage was repaired through SP- or LP-BER pathway. For investigating LP-BER mechanism, the colon carcinoma cell line

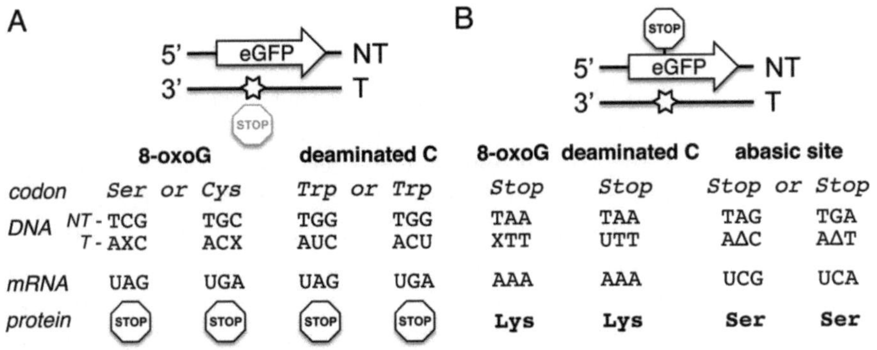

Fig. 2 Strategy in designing plasmid constructs to analyze DNA repairs through BER mechanisms. A damaged base 8-oxoG (indicated as *X*), uracil (*U*), or abasic site (Δ) are placed in the defined location in transcribed strand (*T*) of DNA to generate (**a**), or eliminate (**b**) a stop codon during transcription (*mRNA*), if damages are left unrepaired. Stop codons and amino acids arising from transcription of unrepaired templates are indicated in the last line (*protein*). *NT* nontranscribed strand

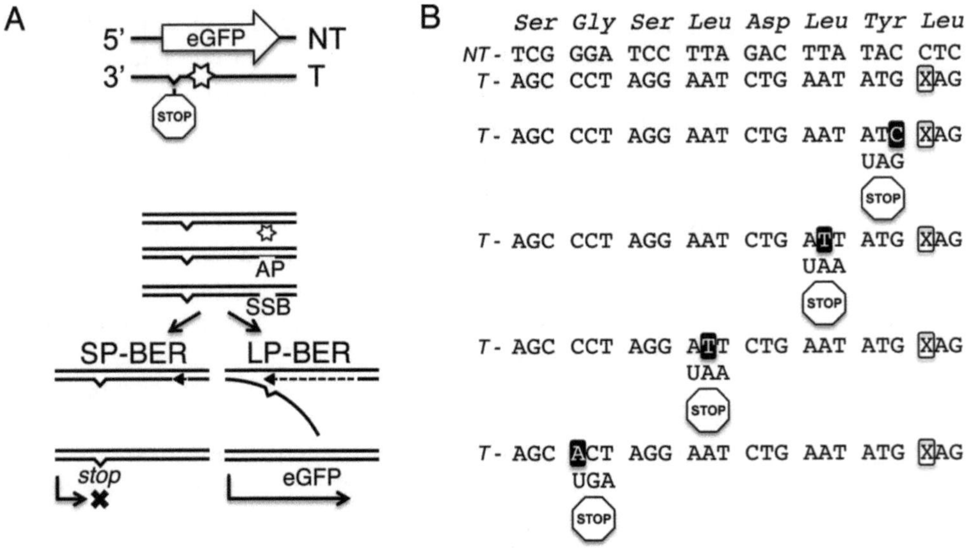

Fig. 3 Strategy to analyze DNA repairs through LP-BER pathway. (**a**) Diagram illustrating the principle of the assay to exclusively monitor LP-BER. (**b**) Illustration of plasmid substrates designed to analyze the repair of damaged base, for example 8-oxoG (indicated as *X*), through LP-BER pathway by application of mismatches creating stop codon (*on a black background*) placed at varying distances from the lesion. *NT* nontranscribed strand, *T* transcribed strand

HCT116 is particularly useful, since the cells are deficient in DNA mismatch repair [20]. This allows using a mismatch together with a damaged base to monitor exclusively a repair through LP-BER pathway (Fig. 3a). A mismatch is introduced in the transcribed strand only in a way to create a stop codon in mRNA during transcription. The mismatch is positioned at various locations from the damage base on its 3′ side, so its removal depends on a synthesis-driven displacement of a DNA segment with the damage (Fig. 3b). A plasmid construct with a mismatch-creating stop codon, but without the lesion is used as additional reference sample. For these settings, the reporter gene cannot be expressed unless mismatch-creating stop codon is repaired through LP-BER mechanism initiated at the damaged base.

3.2 Design of Plasmid Construct for Site-Specific Incorporation of DNA Damages

Plasmid constructs with modifications placed at desired location are produced by annealing of a synthetic oligonucleotide (with modification) to a circular plasmid DNA with a precreated complementary single-stranded gap (Fig. 4a). To generate the gap, the plasmid substrates contain two sites recognized by a sequence-specific nicking enzyme. The cleavage sites are placed so that only a selected region of the transcribed strand of the reporter gene is excised. Excised fragment is removed by heat–cool cycle performed in the presence of competitor oligonucleotide (with modification), which anneals to the gap.

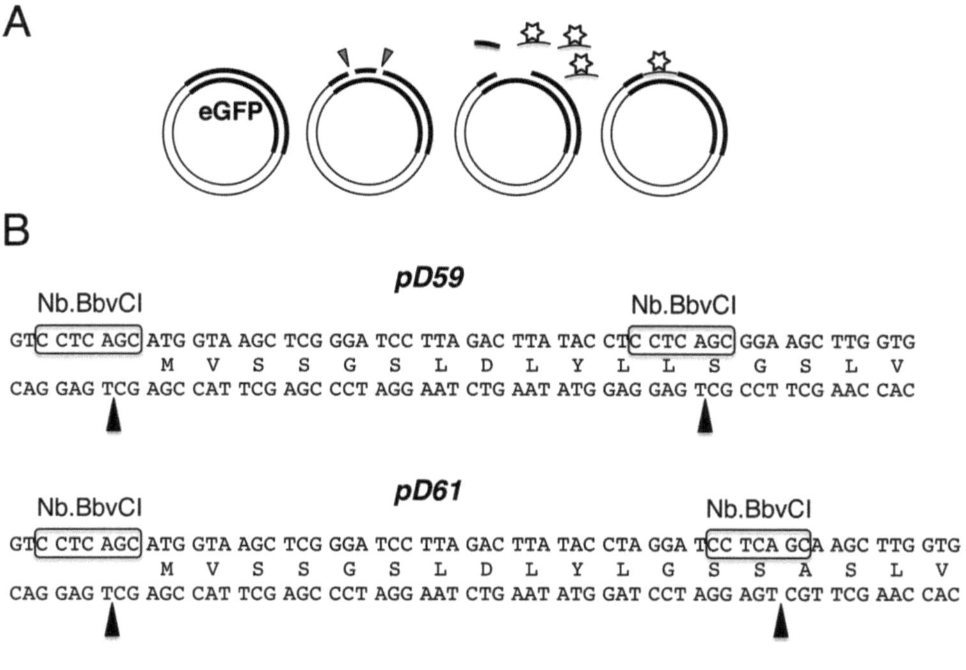

Fig. 4 Plasmids constructs for site-specific incorporation of DNA damages. (**a**) Outline of the strategy used for incorporation of DNA damages into eGFP gene. Two cleavage sites recognized by the nicking enzyme were inserted into the eGFP ORF to efficiently remove the native single-stranded DNA fragment in the transcribed DNA strand only. The single-stranded DNA fragment excised by nicking endonuclease is sequentially exchanged for the synthetic oligonucleotide containing a damaged base. (**b**) Sequences of the cassettes in pD59 and pD61 plasmid constructs placed at the beginning of the eGFP ORF and containing sites recognized by Nb.BbvCI to efficiently remove the native single-stranded DNA fragment. The cut positions of Nb.BbvCI are indicated by *black triangle*

In our system, plasmid constructs are based on a vector backbone of pCMV-kEGFPluc, containing N-terminal-modified eGFP-luciferase fusion protein under the control of a CMV promoter [21]. At the beginning of the eGFP ORF we placed a short cassette with two restriction sites recognized by the sequence-specific nicking enzyme Nb.BbvCI [14]. Nb.BbvCI is one of the altered restriction enzymes produced by New England BioLabs that hydrolyzes only one strand on a double-stranded DNA substrate, to produce DNA molecules that are only nicked but not cleaved. When constructing the plasmid for site-specific incorporation of damaged bases, other than Nb.BbvCI nicking enzymes can be used. Currently, ten different nicking enzymes are available from New England BioLabs. In our plasmids, one Nb.BbvCI restriction site is placed just upstream of the eGFP ORF, and the second site is placed within the ORF about 30 nt and 35 nt downstream in pD59 and pD61, respectively (Fig. 4b) [14]. The cleavage sites are placed so that nicks are made in the transcribed strand of the eGFP gene. After nicking reaction, the original single-stranded DNA between the two Nb.BbvCI sites is melted out, the gap is annealed with the synthetic oligonucleotide (with modification), and finally the nicks are sealed in ligation reaction.

3.3 Preparation of Lesion-Specific BER Plasmid Substrates

Preparation of plasmid constructs containing damaged bases should be scaled up to at least 150 μg of the plasmid used in nicking reaction (Subheading 3.3.1) (*see* **Note 1**).

3.3.1 Preparation of Double-Nicked Plasmids

Conditions of nicking reaction should be first adjusted experimentally to avoid star activity of nicking enzyme (Fig. 5; *see* **Note 2**).

1. Mix 60 μg of plasmid DNA with 13 U of Nb.BbvCI in 120 μL of 1× CutSmart Buffer and incubate for 1.5 h at 37 °C.
2. In order to inactivate the enzyme, transfer the tube to heat block with 80 °C, and incubate for 25 min. Cool sample at room temperature. If the efficiency of nicking reaction has already been verified, proceed to Subheading 3.3.3, or store in −20 °C.

3.3.2 Verification of Nicking Reaction

The following protocol is used to determine whether nicking of plasmid DNA occurred at both Nb.BbvCI sites. Efficiency of nicking reaction is verified by testing whether original single-stranded DNA can be replaced with synthetic oligonucleotide not phosphorylated at the 5′ end, and subsequent ligation reaction in the absence and in the presence of PNK.

1. Prepare two microcentrifuge tubes, T (test) and C (control); in each tube mix 1 μg of nicked plasmid DNA with 90 pmol of synthetic oligonucleotide in a total volume of 10 μL of 1× T4 DNA ligase buffer. Add 1 μL of PNK only to tube C to allow phosphorylation of competitor oligonucleotide at the 5′ end.

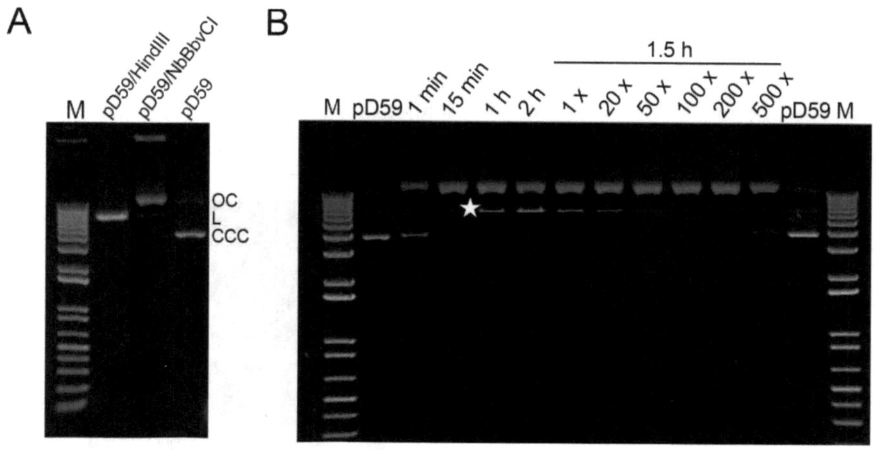

Fig. 5 Optimization of nicking reaction. (**a**) Nicking of plasmid results in the formation of an open circular (OC) form (*pD59/Nb.BbvCI, lane 3*), which migrates in the ethidium bromide agarose gel slower than covalently closed (CCC) untreated plasmid (*pD59, lane 4*), and linear (L) form (*lane 2*) of plasmid from treatment with HindIII. (**b**) Example of titration of nicking enzyme activity to determine conditions when OC form is generated, while CCC (nicks were not made) or linear form (star activity) are not observed. Nicking reactions of plasmid DNA were performed for different time periods, from 1 min to 2 h with undiluted enzyme; and for 1.5 h with diluted enzyme. The dilutions of enzyme stock were from onefold (*1*×) to 500-fold (*500*×). Star indicates linear form resulting from star activity of the nicking enzyme. Untreated plasmid (*p59*) was used as a reference sample of CCC form. *M* DNA size marker

2. Using a thermal cycler, incubate samples at 37 °C for 30 min (5′ end phosphorylation of competitor oligonucleotide), then at 80 °C for 10 min (inactivation of PNK and melting out of the excised oligomer in the double-nicked plasmid), and cool slowly to 50 °C at a temperature transition rate of $-0.02\ ^\circ\mathrm{C} \times \mathrm{s}^{-1}$ (annealing of the competitor oligonucleotide with the gapped plasmid).
3. Chill samples to 4 °C and then add 1 Weiss U of T4 DNA ligase.
4. Incubate samples for 18 h at 16 °C.
5. Analyze samples on 0.8% agarose gel containing 0.5 mg/L EtBr together with untreated plasmid and plasmid treated with nicking enzyme (Fig. 6). Complete inhibition of ligation reaction in tube T will result in the formation of an open circular (OC) form, and will indicate that two nicks were made by nicking enzyme in the same DNA strand, and the original single-strand DNA fragment was successful replaced with synthetic oligonucleotide (*see* **Note 3**).

3.3.3 Preparation of Plasmids with Inserted Altered Synthetic Oligonucleotide

The following reactions are carried out in a thermal cycler using 0.5 mL PCR thermal cycler tubes to accommodate larger reaction volumes.

1. For one 0.5 mL tube, mix 30 μg of double-nicked plasmid (from Subheading 3.3.1, **step 2**) with 1.36 nmol of synthetic

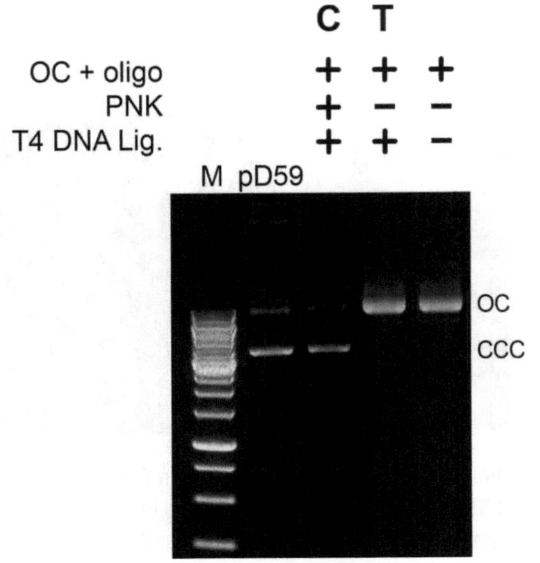

Fig. 6 Verification of nicking reaction and incorporation of oligonucleotide into gapped plasmid. Efficiency of generating two nicks in plasmid construct is verified by annealing unphosphorylated synthetic oligonucleotide (*oligo*) with gapped plasmid (*OC*) and ligation reaction in samples treated (*C*) and not treated (*T*) with PNK responsible for phosphorylation of the 5′ end of synthetic oligonucleotide. Reactions performed in the presence of PNK prior ligations should result in covalently closed plasmid (*CCC*), whereas reactions performed in the absence of PNK should result in the formation only the OC form of the plasmid. Untreated plasmid (*p59*) was used as a reference sample of CCC form. *M* DNA size marker

oligonucleotide, and 150 U of PNK in a total volume of 300 μL of 1× T4 DNA ligase buffer.

2. Incubate at 37 °C for 30 min.
3. Incubate at 80 °C for 10 min.
4. Cool slowly to 50 °C at a temperature transition rate of $-0.02\ ^\circ C \times s^{-1}$, and then chill samples to 4 °C.
5. Add 30 Weiss U of T4 DNA ligase and incubate samples for 18 h at 16 °C.

3.3.4 Purification of Supercoiled Plasmids by CsCl Density Centrifugation

The covalently closed circular forms (CCC, or supercoiled DNA) of plasmid constructs are separated from not sealed forms (open circle, OC) by equilibrium centrifugation in CsCl-EtBr gradients (*see* **Note 4**).

1. Combine ligation reactions from Subheading 3.3.3, **step 5** to get at least 150 μg of double-nicked plasmid used in ligation reaction.

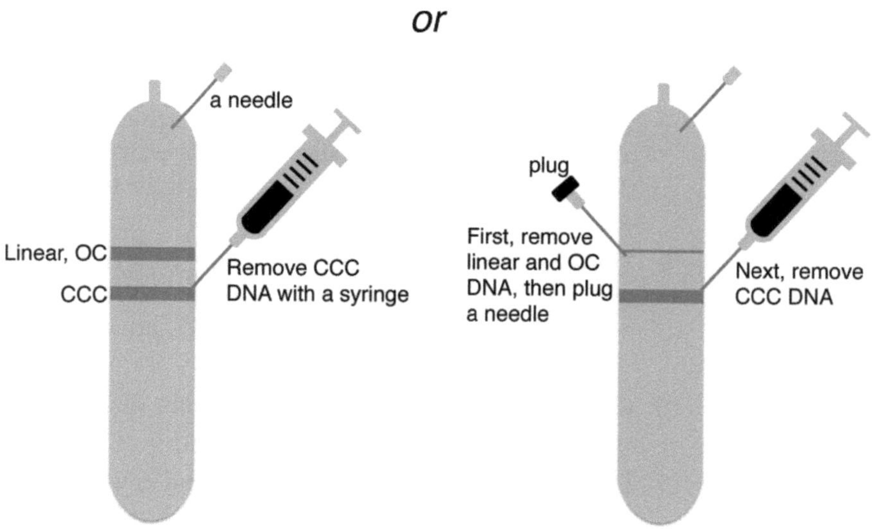

Fig. 7 Approaches to collect the supercoiled plasmid DNA after CsCl-EtBr density gradient centrifugation

2. Mix ligation reactions with TE buffer to a final volume of 7.8 mL. Add 400 μL of EtBr (10 mg/mL) and 8 g of solid CsCl, and then mix gently until all salt is dissolved.
3. Transfer the CsCl-EtBr-DNA solution into Beckman Coulter Quick-Seal Ultra-Clear tubes. If necessary, fill up the tubes with CsCl solution 1 mg/mL in TE and seal.
4. Centrifuge the density gradients in Beckman 70.1 Ti rotor at 329738.4 × *g* at 20 °C for 20 h using slow acceleration and deceleration.
5. Use long-wavelength UV light (360 nm) to visualize the OC and CCC DNA bands (Fig. 7). Two bands of DNA should be visible at the center of the gradient. The upper band is linear and OC form of nicked plasmid DNA, and the lower band is CCC form of closed circular plasmid DNA.
6. Release the vacuum in a tube by gently inserting a 21-gaude needle into the top of the tube.
7. Collect the lower band (supercoiled plasmid DNA) by inserting just below the band an 18-gaude needle attached to a 5 mL disposable syringe (Fig. 7). Alternatively, to prevent contamination with OC form, collect first the upper band. Before removing syringe, plug a 21-gaude needle at the top of the tube with modeling clay. Then remove syringe, but leave the 18-gaude needle in place, and plug it with modeling clay (Fig. 7). Now, insert below the CCC band another 18-gaude needle with syringe, then unplug a 21-gaude needle at the top of the tube and collect the CCC band.

8. Remove EtBr from collected DNA sample by extraction with an equal volume of NaCl-saturated isopropanol (mix equal parts of 5 M NaCl and isopropanol, wait until the salt settles to the bottom and two phases separate, then use the top phase for extraction). Mix and allow the phases to separate. Discard the top phase and repeat extraction until the top phase becomes colorless.
9. Transfer solution with plasmid DNA to a new tube and mix with 2 volumes of water. To precipitate DNA, mix the solution with 2.5–3 volumes of ice-cold 100% ethanol, and place at −20 °C for overnight.
10. Centrifuge at 10,000 × *g* for 20 min at 0–4 °C. Remove supernatant and let tubes open to dry.
11. Dissolve a pellet in 200 μL of TE, and transfer to microcentrifuge tube. Add 1/10 volume of 3 M NaOAc, pH 5, mix well and add 2 volumes of the −20 °C cold 100% ethanol. Mix and leave at room temperature for 10 min.
12. Recover the precipitated DNA by centrifugation at full speed in a microcentrifuge for 15–20 min.
13. Wash the pellet with the −20 °C cold 70% ethanol, and spin at full speed in a microcentrifuge for 5 min. Remove supernatant and dry pellet.
14. Resuspend the pellet in 100 μL of TE and determine the DNA plasmid concentration at A260 (1 A260 unit = 50 μg/mL) using GeneQuant II RNA/DNA Calculator.

3.4 Transfection of Cells with Lesion-Specific BER Plasmids

3.4.1 Preparation of Cells for Transfection

In every experiment, the analysis of damaged plasmid construct and control (undamaged plasmid construct) is done in triplicate, that is, in three wells. Additional three wells are required for negative control (no transfection). A minimum of nine wells is therefore required for each host cell type to be assayed for DNA repair efficiency of particular DNA damage. The cells should be seeded on plates 1 day before transfection. The following protocol is for transfection of HCT116 cells.

1. HCT116 cells are maintained in McCoy's 5A medium with supplements. One day before transfection trypsinize the cells and count using Vi-CELL XR cell counter.
2. Calculate appropriate number of cells to be plated in 6-well culture dishes. Cells should be 50–80% confluent on a day of transfection. For transfection of HCT116, 1.5–2 × 10^5 cells in 2 mL of normal growth media per transfection (one well) are seeded into six-well dishes (*see* **Note 5**).

3.4.2 Transient Transfection (Lipofection)

Undamaged plasmid (pD59 or pD61) is used as a control of transfection efficiency, and as a reference sample to assess reductions in eGFP expression for plasmid constructs bearing particular damage. For negative control, mock transfections are prepared by omitting the Lipofectamine agent. For each transfection sample prepare mixes as follows:

1. Plasmid solution: at least a total of 1 μg of DNA is required for single transfection (one well), in 250 μL of Opti-MEM® I Reduced Serum Medium mixed with 4 μL of PLUS reagent (Lipofectamine LTX Plus). This solution can be prepared in bulk for triplicate transfections.
2. Lipofectamine solution: mix Lipofectamine LTX gently before use, then dilute 5 μL in 250 μL of Opti-MEM® I Reduced Serum Medium per well. Proceed immediately to **step 3**. Diluted Lipofectamine should sit no longer than 3–5 min at room temperature. This dilution can also be prepared in bulk for multiple wells.
3. Combine the diluted DNA (**step 1**) with diluted Lipofectamine (**step 2**), mix gently and incubate for 30 min. Total volume is 500 μL for transfection per well.
4. For each transfection (one well), add 500 μL of Opti-MEM® I Reduced Serum Medium to a tube containing DNA–Lipofectamine complexes (**step 3**). Proceed immediately to **step 5**.
5. Remove the media from cells seeded on plate 1 day before transfection (Subheading 3.4.1, **step 2**) and gently add 1 mL of diluted complexes (**step 4**) to each well. Mix gently by rocking the plate.
6. Incubate the cells at 37 °C in a CO_2 incubator for 2 h.
7. Add 1 mL of growth medium containing 2× the normal concentration of serum without removing the transfection mixture (*see* **Note 6**).

3.5 Flow Cytometry Analysis

Flow cytometry analysis can be done 24–48 h after transfection. Efficiency of BER repair is analyzed by the level and fluorescence intensity of eGFP that can be measured quantitatively in any flow cytometer emitting the blue-green 488 nm wavelength for excitation of eGFP and emission filters for detection wavelength of 509 nm.

1. 24 h after transfection wash cells with PBS and recover by scraping in an ice-cold PBS buffer with 15% FBS. Alternatively, cells can be detached from surface by trypsinization. Tubes with cells should be kept on ice in the dark and analyzed by flow cytometry within 30 min.
2. Optionally, cells can be fixed before flow cytometry analysis. Spin cells at 300 x *g*, remove supernatant and resuspend cells in 1 mL of BD Cytofix™ Fixation Buffer, or 4% formaldehyde in

PBS. Incubate for 10 min at room temperature, then spin cells at 300 × *g*, remove supernatant and resuspend cells in BD Pharmingen™ Stain Buffer (FBS), or in PBS with 5–15% FBS. Tubes with cells should be kept on ice in the dark and analyzed by flow cytometry within 2 h.

3. For each transfection at least 10,000 events are counted and green fluorescence intensity is measured using LSR II 18-Color flow cytometer.

4. Average transfection efficiency measured as percentage of eGFP positive cells transfected with undamaged plasmid ranges from 50% to 80%.

5. Analyze data using FlowJo software (Fig. 8). Relative eGFP expression (RE) is calculated with the following formula

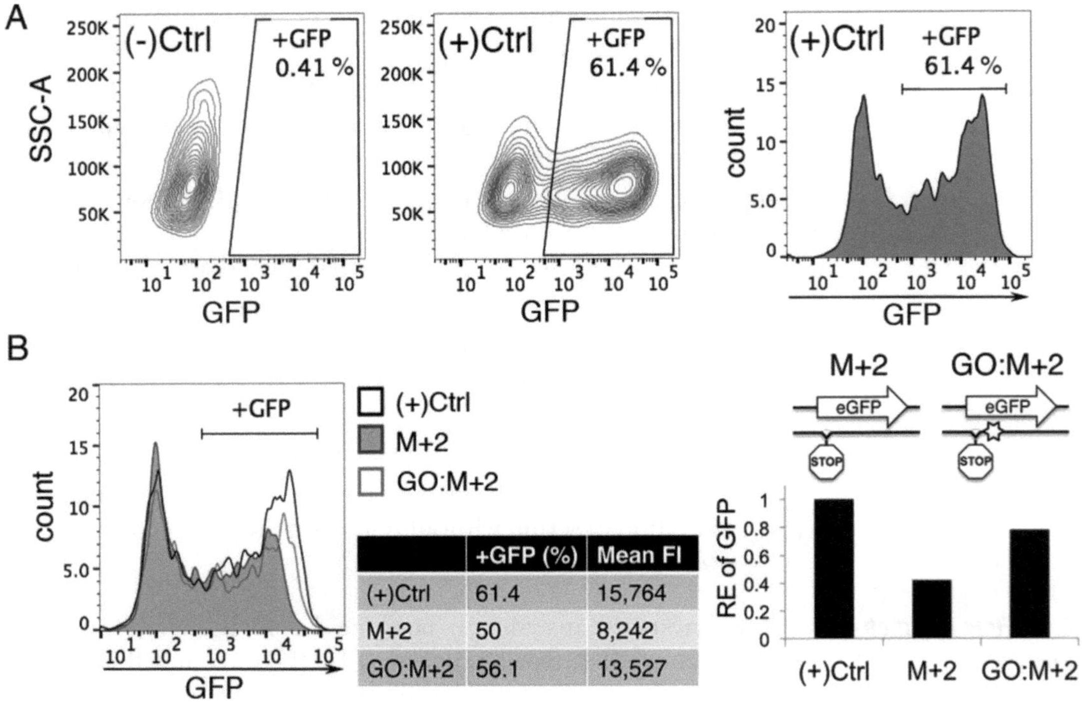

	+GFP (%)	Mean FI
(+)Ctrl	61.4	15,764
M+2	50	8,242
GO:M+2	56.1	13,527

Fig. 8 Flow cytometry analysis of BER using HCT116 cells. (**a**) Representative contour plots and histogram from flow cytometry analysis showing detection of eGFP fluorescence in cells transfected with positive control, (+)Ctrl (undamaged plasmid). Positive detection of eGFP fluorescence (+GFP) is determined by comparison to negative control, (−)Ctrl. (**b**) Representative histograms showing shift of fluorescence intensities within +GFP population reflecting repairs through LP-BER mechanisms. The M + 2 sample is plasmid construct bearing a mismatch creating a stop codon in transcribed strand within the eGFP gene. A partial rescue of the eGFP expression is observed when 8-oxoG is located in proximity (2 nt from a mismatch) (GO:M + 2). The repair of 8-oxoG occurred through LP-BER, where strand displacement synthesis allowed correction of the mismatch and expression of eGFP. In the table are indicated the percentage of +GFP population and mean fluorescence intensity for each analyzed sample, which are used to calculate relative expression (RE) of eGFP (*see* Subheading 3.5, **step 5**). Bar graph shows the RE of eGFP for analyzed samples

$RE = (M \cdot I_M)/(C \cdot I_C)$; where M and C are the percentages of eGFP positive cells transfected with damaged plasmid and undamaged control plasmid, respectively. I_M and I_C are mean fluorescence intensities of eGFP positive cells for M and C, respectively.

4 Notes

1. At least 150 μg of the plasmid should be used to prepare plasmid construct containing a damaged base. This is necessary to ensure positive detection of covalently closed plasmids (CCC form) purified by centrifugation in CsCl-EtBr gradient (Subheading 3.3.4) after insertion of altered synthetic oligonucleotide and ligation reaction (Subheading 3.3.3).
2. The high efficiency of nicking reaction is very important for successful replacement of excised region with a new sequence bearing site-specific base damage. Conditions of nicking reaction should be first adjusted experimentally to avoid star activity of nicking enzyme. First, plasmid DNA treated with nicking enzyme is analyzed on 0.8% agarose gel together with not treated sample, and sample digested with restriction enzyme, such as HindIII (Fig. 5a). Nicking of plasmid results in the formation of an open circular (OC) form, which migrates in agarose gel electrophoresis slower than covalently closed (CCC) untreated plasmid, and linear form of plasmid from treatment with HindIII. If nicking reaction results in generation of linear form in addition to OC form, then in the second step, titration of enzyme activity should be performed, and a time of nicking reaction should be adjusted to avoid star activity of nicking enzyme (Fig. 5b).
3. If ligation reaction in tube T is not completely inhibited, then nicking of plasmid DNA likely occurred only at one Nb.BbvCI site, preventing complete removal of the original single-stranded DNA fragment between the two Nb.BbvCI sites, thus resulting in its religation.
4. Although T4 DNA ligase efficiently repairs single-stranded nicks in duplex DNA, the presence of modified bases, or mismatches in incorporated synthetic oligonucleotide might reduce the efficiency of the ligation reaction.
5. The number of cells will vary for different types of cells or cell lines, and it will depend on the size of cells and their grow rate.
6. Depending on the cells used for transfection, transfection solution can be left for up to 24 h. However, if toxicity is observed after transfection, determine the optimal time of incubation and replace transfection solution with normal growth media.

References

1. Krokan HE, Standal R, Slupphaug G (1997) DNA glycosylases in the base excision repair of DNA. Biochem J 325(Pt 1):1–16
2. Croteau DL, Bohr VA (1997) Repair of oxidative damage to nuclear and mitochondrial DNA in mammalian cells. J Biol Chem 272 (41):25409–25412
3. Dizdaroglu M (1991) Chemical determination of free radical-induced damage to DNA. Free Radic Biol Med 10(3–4):225–242
4. Ames BN (1989) Endogenous oxidative DNA damage, aging, and cancer. Free Radic Res Commun 7(3–6):121–128
5. Tubbs A, Nussenzweig A (2017) Endogenous DNA damage as a source of genomic instability in cancer. Cell 168(4):644–656. https://doi.org/10.1016/j.cell.2017.01.002
6. Swenberg JA, Lu K, Moeller BC, Gao L, Upton PB, Nakamura J, Starr TB (2011) Endogenous versus exogenous DNA adducts: their role in carcinogenesis, epidemiology, and risk assessment. Toxicol Sci 120(Suppl 1): S130–S145. https://doi.org/10.1093/toxsci/kfq371
7. Lindahl T (1993) Instability and decay of the primary structure of DNA. Nature 362 (6422):709–715. https://doi.org/10.1038/362709a0
8. Nakamura J, Walker VE, Upton PB, Chiang SY, Kow YW, Swenberg JA (1998) Highly sensitive apurinic/apyrimidinic site assay can detect spontaneous and chemically induced depurination under physiological conditions. Cancer Res 58(2):222–225
9. Krokan HE, Bjoras M (2013) Base excision repair. Cold Spring Harb Perspect Biol 5(4): a012583. https://doi.org/10.1101/cshperspect.a012583
10. Jacobs AL, Schar P (2012) DNA glycosylases: in DNA repair and beyond. Chromosoma 121 (1):1–20. https://doi.org/10.1007/s00412-011-0347-4
11. Ide H, Kotera M (2004) Human DNA glycosylases involved in the repair of oxidatively damaged DNA. Biol Pharm Bull 27 (4):480–485
12. Matsumoto Y, Kim K (1995) Excision of deoxyribose phosphate residues by DNA polymerase beta during DNA repair. Science 269 (5224):699–702
13. Podlutsky AJ, Dianova II, Wilson SH, Bohr VA, Dianov GL (2001) DNA synthesis and dRPase activities of polymerase beta are both essential for single-nucleotide patch base excision repair in mammalian cell extracts. Biochemistry 40(3):809–813
14. Piekna-Przybylska D, Bambara RA, Balakrishnan L (2016) Acetylation regulates DNA repair mechanisms in human cells. Cell Cycle 15 (11):1506–1517. https://doi.org/10.1080/15384101.2016.1176815
15. Saxowsky TT, Meadows KL, Klungland A, Doetsch PW (2008) 8-Oxoguanine-mediated transcriptional mutagenesis causes Ras activation in mammalian cells. Proc Natl Acad Sci U S A 105(48):18877–18882. https://doi.org/10.1073/pnas.0806464105
16. Bregeon D, Peignon PA, Sarasin A (2009) Transcriptional mutagenesis induced by 8-oxoguanine in mammalian cells. PLoS Genet 5(7):e1000577. https://doi.org/10.1371/journal.pgen.1000577
17. Luhnsdorf B, Epe B, Khobta A (2014) Excision of uracil from transcribed DNA negatively affects gene expression. J Biol Chem 289 (32):22008–22018. https://doi.org/10.1074/jbc.M113.521807
18. Charlet-Berguerand N, Feuerhahn S, Kong SE, Ziserman H, Conaway JW, Conaway R, Egly JM (2006) RNA polymerase II bypass of oxidative DNA damage is regulated by transcription elongation factors. EMBO J 25 (23):5481–5491. https://doi.org/10.1038/sj.emboj.7601403
19. Kuraoka I, Endou M, Yamaguchi Y, Wada T, Handa H, Tanaka K (2003) Effects of endogenous DNA base lesions on transcription elongation by mammalian RNA polymerase II. Implications for transcription-coupled DNA repair and transcriptional mutagenesis. J Biol Chem 278(9):7294–7299. https://doi.org/10.1074/jbc.M208102200
20. Lei X, Zhu Y, Tomkinson A, Sun L (2004) Measurement of DNA mismatch repair activity in live cells. Nucleic Acids Res 32(12):e100. https://doi.org/10.1093/nar/gnh098
21. Sattler U, Frit P, Salles B, Calsou P (2003) Long-patch DNA repair synthesis during base excision repair in mammalian cells. EMBO Rep 4(4):363–367. https://doi.org/10.1038/sj.embor.embor796

Chapter 9

Methods for Studying DNA Single-Strand Break Repair and Signaling in *Xenopus laevis* Egg Extracts

Yunfeng Lin, Anh Ha, and Shan Yan

Abstract

DNA single-strand breaks (SSBs) are the most common type of DNA lesions as they are generated approximately 10,000 times per mammalian cell each day. Unrepaired SSBs compromise DNA replication and transcription programs, leading to genome instability, and have been implicated in many diseases including cancer. In this chapter, we introduce methods to study the ATR-Chk1 DNA damage response (DDR) pathway and DNA repair pathway in response to a site-specific, defined SSB plasmid in *Xenopus laevis* egg extracts. This experimental system can be applied in future studies to reveal many aspects of the molecular mechanisms of SSB repair and signaling in eukaryotes.

Key words DNA damage response, SSB end resection, SSB repair, SSB signaling, *Xenopus* egg extracts

1 Introduction

DNA single-strand breaks (SSBs) are derived from oxidative stress, intermediate DNA repair products, and abortive cellular enzymes [1, 2]. It has been estimated that approximately 10,000 times of SSBs are generated per mammalian cell each day; therefore, SSBs are considered as the most frequent type of DNA lesions [3, 4]. If not repaired appropriately or promptly, SSBs will compromise DNA replication and transcription programs within cells, leading to genome instability [1, 5]. Notably, accumulation of unrepaired SSBs has been implicated in cancer, aging, and neurodegenerative diseases [1, 2, 6]. However, the molecular mechanisms of SSB repair and signaling remain incomplete, largely because of the lack of a tractable experimental system [5].

Xenopus laevis egg extract systems have been developed and widely utilized in various studies of DNA metabolisms, such as DNA replication, DNA repair, and DNA damage response (DDR) pathways [7–14]. In particular, mechanistic studies using *Xenopus* system have provided critical insights into the molecular

Lata Balakrishnan and Jason A. Stewart (eds.), *DNA Repair: Methods and Protocols*, Methods in Molecular Biology, vol. 1999, https://doi.org/10.1007/978-1-4939-9500-4_9,

details of DNA repair and DDR pathways in response to DNA double-strand breaks (DSBs) and interstrand cross-links (ICLs) [15–19]. There are several different types of *Xenopus* egg extracts: low-speed supernatant (LSS), high-speed supernatant (HSS), and nucleoplasmic extracts (NPE) [20, 21]. The detailed approaches to prepare these *Xenopus* egg extracts have been described previously [7, 20, 22].

In this chapter, taking advantage of *Xenopus* HSS system, we introduce methods to study SSB repair, SSB-induced DDR pathway, and the innovative 3′–5′ SSB end resection process [5]. In particular, we will describe step-by-step methods including: generating a defined SSB plasmid, analyzing DNA repair of the defined SSB plasmid in the HSS system, elucidating activation and regulation of SSB signaling in the HSS system, isolating DNA-bound fractions from HSS to examine protein recruitment onto SSB site and dissecting the 3′–5′ SSB end resection in the HSS system. Future studies using these methods will help to dissect molecular aspects of SSB repair and signaling pathways in eukaryotes.

2 Materials

2.1 Preparation of SSB Plasmid and FAM-SSB

1. pS plasmid: mutant pUC19 by QuikChange II XL site-directed mutagenesis kit.
2. *Nt.BstNBI* site specific endonuclease.
3. 10× NEB Buffer 3.1.
4. 10× NEB CutSmart Buffer.
5. TE: 1 mM EDTA, 10 mM Tris, pH 8.0.
6. Alkaline phosphatase, calf intestine (CIP).
7. Plasmid midi or maxi kit.
8. Gel extraction kit.
9. KOD hot start DNA polymerase kit.
10. Forward primer: 6-FAM-5′-TCGGTACCCGGGGATCCTCTAG-3′.
11. Reverse primer: 5′-ACAGCTATGACCATGATTACG-3′.
12. M1-F primer: 5′-GGTAACTATCGTCTTGAATCCAACCCGGTAAGACACG-3′.
13. M1-R primer: 5′-CGTGTCTTACCGGGTTGGATTCAAGACGATAGTTACC-3′.
14. M2-F primer: 5′-CGTTCATCCATAGTTGCCTGGCTCCCCGTCGTGTAGATAAC-3′.
15. M2-R primer: 5′-GTTATCTACACGACGGGGAGCCAGGCAACTATGGATGAACG-3′.

16. M3-F primer: 5′-CCGCTTCCTCGCTCACTGGCTCGC TGCGCTCGGTCGTTC-3′.
17. M3-R primer: 5′-GAACGACCGAGCGCAGCGAGCCAGT-GAGCGAGGAAGCGG-3′.
18. 1% and 2% agarose gel.

2.2 DNA Purification

1. Agarose gel with ethidium bromide staining.
2. Nuclease-free water.
3. 3 M sodium acetate (pH 5.0).
4. Phenol–chloroform.
5. 1 mg/mL glycogen.
6. Ethanol: 100% and cold 70%.

2.3 Isolation of DNA-Bound Fractions from the HSS System

1. High-speed supernatant (HSS).
2. 10× ELB (Egg lysis buffer) salt: 25 mM $MgCl_2$; 500 mM KCl; 100 mM HEPES–KOH, pH 7.7.
3. 1 M sucrose in water.
4. ELB: 1× ELB salt, 0.25 M sucrose, pH 7.7 (*see* **Note 1**).
5. Sucrose cushion: 0.9 M sucrose in ELB.
6. 2× Laemmli Sample Buffer: 4% SDS, 20% glycerol, 10% 2-mercaptoethanol, 0.004% bromophenol blue and 0.125 M Tris–HCl, pH 6.8.

2.4 Fluorescence Analysis

1. 10× TBE running buffer: dissolve 108 g Tris base and 55 g Boric acid in 800 mL of water. Add 40 mL of 0.5 M EDTA (pH 8.0) and bring volume to 1 L (*see* **Note 2**).
2. 40% acrylamide–bis solution: 19:1 (i.e., 5% cross-linker).
3. 10% ammonium persulfate in water.
4. TEMED (tetramethylethylenediamine).
5. 2× TBE–urea sample buffer: 90 mM Tris, 90 mM boric acid, 2 mM EDTA, 12% Ficoll, 7 M urea, 0.03% bromophenol blue, 0.03% xylene cyanol.
6. Typhoon 8600 imager.
7. ImageJ software.

3 Methods

3.1 Preparation of a Defined Site-Specific SSB Plasmid

We will introduce methods to prepare control (CTL) plasmid pS, SSB plasmid, and DSB plasmid in this section (Fig. 1). To ensure the quality of SSB plasmid, steps or practices leading to unnecessary shearing of the DNA (e.g., vortex and frequent freeze–thaw) should be avoided or minimized, if possible.

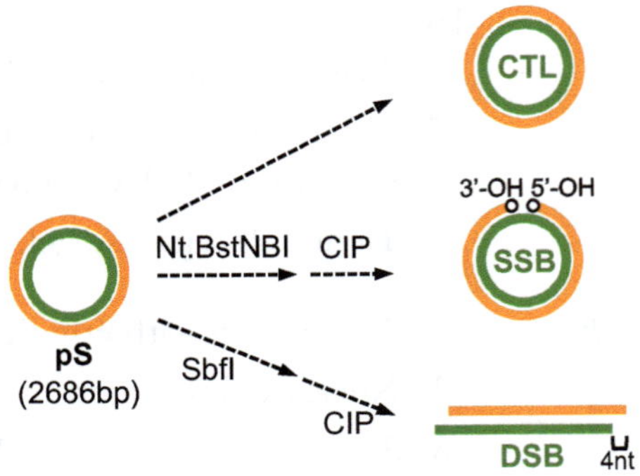

Fig. 1 Schematic diagram of the generation of defined SSB and DSB structures. This figure is adapted from Lin et al., 2018, *Nucleic Acids Research* [5]

1. To generate plasmid pS, mutant plasmid pUC19 on three sites (nt 1177–1181 on "+" strand, nt 706–710 on "−" strand, and nt 1694–1698 on "−" strand) sequentially with three pairs of primers (i.e., M1-F and M1-R primers, M2-F and M2-R primers, and M3-F and M3-R primers) using QuikChange II XL site-directed mutagenesis kit.
2. Sequence plasmid pS to verify mutations on the target sequence sites. Three point mutations (i.e., G1179A, A707G, A1695G) are created in pS. There is only one recognition site on pS for *Nt.BstNBI* (i.e., nt 427–431 on "+" strand).
3. Prepare a large amount of plasmid pS (e.g., several milligrams) using plasmid midi/maxi kit following vendor's procedure. Utilize untreated pS as control (CTL) plasmid in experiments.
4. To generate a defined SSB plasmid between C435 and T436, treat the pS with *Nt.BstNBI* (10 U/μg) in 1× NEB Buffer 3.1 for 2 h at 55 °C.
5. To generate the DSB plasmid, treat the pS with *SbfI-HF* (10 U/μg) with 1× NEB CutSmart Buffer for 2 h at 37 °C.
6. Add CIP (10 U/μg) to the enzyme-treated pS mixture for 1 h at 37 °C to catalyze the dephosphorylation at the 5′ side of SSB or DSB (Fig. 1) (*see* **Note 3**).
7. Run samples on a 1% agarose gel and then purify the SSB plasmid via the gel extraction kit, followed by an optional phenol–chloroform extraction (*see* **Note 4**).
8. Dilute the SSB plasmid to a final concentration of 375 ng/μL with TE and save at −20 °C or −80 °C in small aliquots (e.g., 20 μL) to avoid frequent freeze–thaw cycles.

3.2 Analysis of DNA Repair of the Defined SSB Plasmid in the HSS System

We have established an assay to analyze DNA repair of the defined SSB plasmid in the HSS system [5]. Here is the detailed method for SSB repair analysis.

1. Add 4 μL of the defined SSB plasmid (e.g., 375 ng/μL) to 16 μL of HSS in microcentrifuge tube.
2. After incubation at room temperature for different time points (e.g., 1, 5, and 30 min), add nuclease-free water to each reaction of HSS that was incubated with SSB plasmid to a total volume of 300 μL (*see* **Note 5**).
3. Add equal volume of phenol–chloroform (i.e., 300 μL) to the reaction mixture and mix up and down at least five times.
4. Spin the mixture for 5 min at 16,000 rcf at room temperature.
5. Transfer the top aqueous layer to a new tube and repeat the phenol–chloroform extraction another three to five times until no obvious protein is left between the boundary of phenol–-chloroform and water layers (*see* **Note 6**).
6. Supplement the final aqueous solution with sodium acetate (a final concentration of 0.3 M), glycogen (a final concentration of 1 μg/μL), and ethanol (100%, 2.5-fold volume/volume), followed by incubation at −80 °C for at least 30 min (*see* **Note 7**).
7. Spin the mixture for 15 min at 16,000 rcf at room temperature. Wash the pellet with 1 mL of cold 70% ethanol and air-dry for 30 min at room temperature before resuspension with 30–40 μL of nuclease-free water.
8. Quantify the purified DNA repair products via NanoDrop and examine by 1% agarose gel electrophoresis stained with ethidium bromide (*see* **Note 8**). Gradual repair will occur in the HSS system (Fig. 2).

3.3 Analysis of SSB Signaling in the HSS System

The defined SSB plasmid but not CTL plasmid triggers Chk1 phosphorylation in the HSS [5]. As shown below, we have optimized the experimental conditions to establish the defined SSB induced ATR-Chk1 DDR pathway in the HSS and to analyze the recruitment of checkpoint proteins to SSB site in the HSS system.

3.3.1 Establishment of the Defined SSB-Induced DDR Pathway in the HSS System

1. In microcentrifuge tube, add 2 μL of CTL or SSB plasmid to 8 μL of HSS to different final concentrations (e.g., 1, 50, 75, and 100 ng/μL), followed by incubation for different times (e.g., 1, 5, 30, or 60 min) at room temperature with gentle inversion every 10 min (*see* **Note 9**).
2. Add 50 μL of 2× Laemmli Sample Buffer to each 10 μL of reaction mixture followed by immunoblotting analysis for Chk1 phosphorylation at the Ser 344 (total Chk1 as loading control). Figure 3 shows example results. The dose-dependent

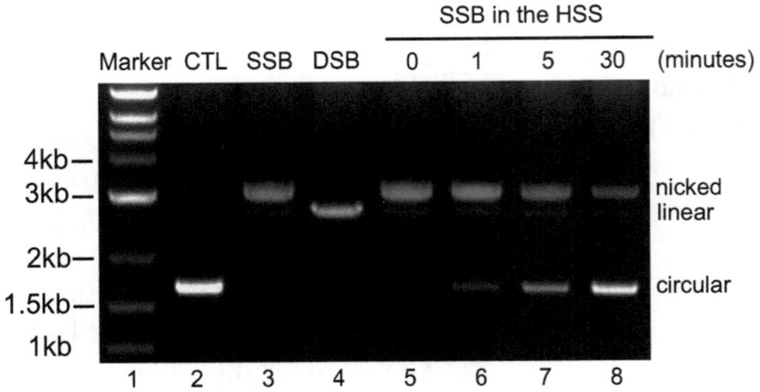

Fig. 2 The defined SSB structure is gradually repaired in HSS system after incubation at different times (0, 1, 5, or 30 min) in the HSS system. Marker represents the 1 kb DNA ladder. The result from this figure is consistent with Fig. 1b from Lin et al., 2018, *Nucleic Acids Research* [5]

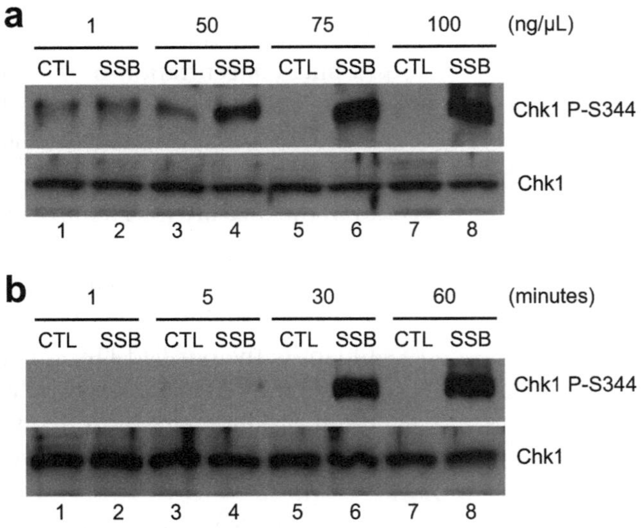

Fig. 3 SSB structure induces a novel ATR-Chk1 DDR pathway in the HSS system in a dose- and time-dependent manner. (**a**) CTL or SSB plasmid was added to HSS at different concentrations as indicated. After a 30-min incubation, extracts were examined for Chk1 phosphorylation via immunoblotting analysis. (**b**) CTL or SSB plasmid was added to HSS at a final concentration of 75 ng/μL for indicated time. Extracts were examined for Chk1 phosphorylation via immunoblotting analysis. The result from this figure is consistent with Fig. 2a, b from Lin et al., 2018, *Nucleic Acids Research* [5]

experiment indicates that 75 ng/μL of the defined SSB structure, but not CTL plasmid, triggered a robust Chk1 phosphorylation at S344 after a 30-min incubation in the HSS system (Fig. 3a) (*see* **Notes 10** and **11**). The time-dependent experiment shows that the SSB-induced Chk1 phosphorylation

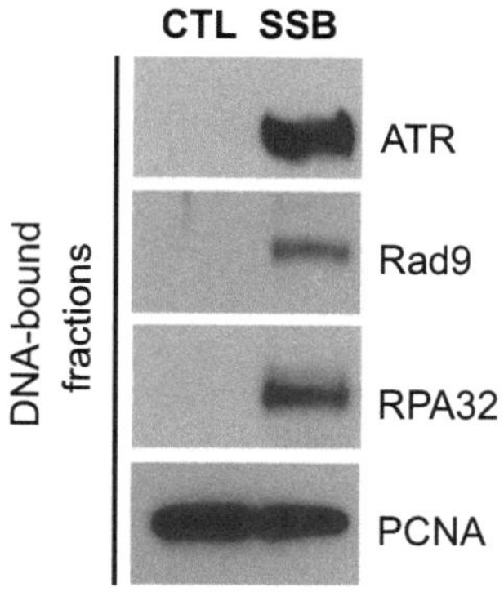

Fig. 4 Checkpoint proteins are recruited to the SSB site in the HSS system. CTL or SSB plasmid was added to HSS followed by incubation at room temperature. Then the DNA-bound fractions from HSS were isolated, and examined via immunoblotting analysis

appeared at 5 min, and peaked at 30 min after incubation in the HSS (Fig. 3b) (*see* **Note 12**). This SSB signaling technology can be tailored to future mechanistic studies using immunodepletion approach and/or small molecule inhibitors (*see* **Note 13**).

3.3.2 Analysis of Protein Recruitment to SSB Site in the HSS System

1. Incubate 10 μL of the SSB plasmid or CTL plasmid with 40 μL of HSS (to a final concentration of 75 ng/μL) for 30 min at room temperature (*see* **Note 14**).
2. Then, dilute the 50 μL of reaction mixture with 200 μL of ELB buffer and spin through a 1 mL of sucrose cushion at 9000 rcf for 15 min at 4 °C in a swinging bucket.
3. After centrifugation, remove the supernatant and resuspend the DNA-bound protein factions with 30 μL of 2× Laemmli Sample Buffer. Then examine the DNA-bound fractions via immunoblotting analysis (Fig. 4) (*see* **Note 15**).

3.4 SSB End Resection in the 3′–5′ Direction in the HSS System

To detect the SSB end resection in the 3′–5′ direction in the HSS system, we have developed a method to prepare the FAM-labeled SSB structure (designated as FAM-SSB, shown in Fig. 5a). The purified FAM-SSB structure can be utilized in the HSS system to examine the 3′–5′ SSB end resection (Fig. 5b).

3.4.1 Preparation of FAM-SSB Structure

1. First, generate a FAM-labeled dsDNA (i.e., 70 bp dsDNA with FAM-labeled at the 5′-side of plus strand) by PCR using the forward and reverse primers with the plasmid pS as template. PCR conditions include denaturation at 95 °C for 30 s, annealing at 60 °C for 10 s, and extension at 70 °C for 30 s for a total of 30 cycles.
2. Then purify the FAM-dsDNA PCR product via gel extraction.

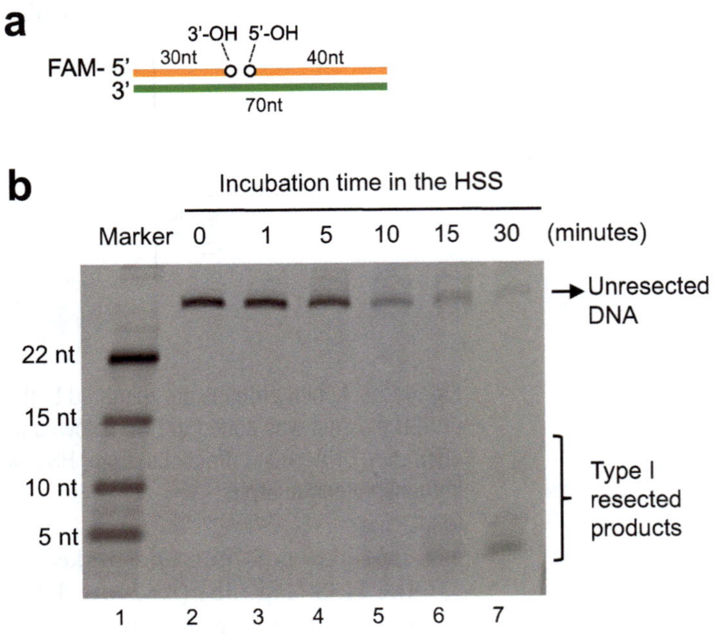

Fig. 5 The FAM-SSB structure is resected in the 3′–5′ direction in the HSS system. (**a**) Schematic presentation of the defined FAM labeled SSB structure (i.e., FAM-SSB). (**b**) FAM-SSB structure was added to HSS for different time as indicated. Then samples were examined via 16% TBE–urea denaturing gel and visualized via Typhoon imager. The result from this figure is consistent with Fig. 6b in Lin et al., 2018, *Nucleic Acids Research* [5]

3. Treat the purified FAM-dsDNA with *Nt.BstNBI* (10 U/μg) for 2 h at 55 °C.
4. Then cool the tube down to ~37 °C, and supplement with CIP (10 U/μg) for 1 h at 37 °C to dephosphorylate at the 5′ side of the SSB structure.
5. Purify the FAM-labeled SSB structure on a 2% agarose gel. After gel extraction, save the FAM-SSB structure at −20 °C in small aliquots for in vitro exonuclease assay.

3.4.2 SSB End Resection of FAM-SSB Structure in the HSS System

1. To examine how SSB is resected, incubate 8 μL of the FAM-SSB substrate (50 nM) with 112 μL of HSS at room temperature.
2. After incubation at different time points (i.e., 0, 1, 5, 10, 15, or 30 min), take out 12 μL of sample and mix with 12 μL of 2× TBE–urea sample buffer (i.e., equal volume). Then denature the samples at 95 °C for 5 min.
3. Without cooling down, load and resolve the hot samples on a 16% TBE–urea denaturing polyacrylamide gel (*see* **Note 16**). Add "Marker" including four FAM-labeled different lengths of

ssDNA to label the positions of DNA migration on TBE–urea gel.

4. To prepare a 16% polyacrylamide gel, mix 4 mL of 10× TBE running buffer, 16 mL of 40% acrylamide mixture and 8 mL of water in 100 mL flask. Add 16.8 g urea into the flask and keep stirring by magnetic bar (*see* **Note 17**). Add 300 μL of 10% ammonium persulfate and 30 μL of TEMED and cast gel within an 18.3 cm × 20 cm × 3 mm gel cassette. Insert a 15-well gel comb immediately without introducing air bubbles. Once the gel is solidified, prerun the TBE–urea gel for at least 30 min at 20 W.
5. Once the samples are loaded, run the TBE–urea gel in 1× TBE running buffer at 25 W for ~1–2 h (*see* **Note 18**).
6. Scan the TBE–urea gel using a Typhoon 8600 imager (PMT at 600 V, excitation at 532 nm and emission at 526 nm). Use ImageQuant 5.2 to visualize gels, and quantify DNA products with ImageJ, if needed. As shown in Fig. 5b, the FAM-SSB can be resected in the 3′–5′ direction into Type I resected products in the HSS system (*see* **Note 19**).

4 Notes

1. The egg lysis buffer (ELB) should be freshly made and can continue to be used for ~4 h on ice.
2. It is better to use the TBE running buffer within 2 weeks, or the boric acid will precipitate, which may or may not effect to the results. EDTA needs to be heated and the pH is adjusted to facilitate solubility.
3. After the treatment by *Nt.BstNBI*, the sample should be cooled down to 37 °C or below prior to the CIP treatment. CIP may lose its activity if added to the tube at 55 °C.
4. The recovery of purified plasmid from a gel extraction kit is around 50% in our hands. Loading the dissolved gel slices through the columns one to two more times can increase the binding of plasmid DNA to the columns, leading to higher overall yield using the gel extraction kit. Phenol–chloroform extraction can increase the quality of the purified plasmid preparation but may decrease the overall yield of the plasmid.
5. Two reasons to bring up the mixture volume up to 300 μL with nuclease-free water: one is to dilute the proteins in the HSS for better protein extraction and the other is to increase the overall yield of DNA. If the overall DNA yield is low, consider the option of resuspending the mixture with nuclease-free water to a total volume of 600 μL.

6. A lot of DNA binding proteins can be bound with plasmid from the HSS. The additional three to five repeats of phenol--chloroform extraction is quite important to get rid of DNA-bound proteins from the repair products. If proteins are not removed completely or still bound with DNA, it is difficult to analyze DNA repair products appropriately via subsequent agarose gel electrophoresis. Because additional phenol–chloroform extraction also means less recovery of DNA, a balance in the number of phenol–chloroform extractions and yield of DNA should be considered and optimized.
7. Longer incubation and lower temperature provide better recovery of DNA.
8. A lower percentage agarose gel will offer better separation of DNA repair products, but the gel is more easily broken. 1% agarose is a starting percentage and can be optimized.
9. Different batches of HSS preparation may have different quality. Thus, it is a good practice to optimize the concentration of SSB plasmid and incubation times in the HSS system. Nevertheless, ensuring high quality of HSS is very important for the SSB signaling experiments.
10. Frequent freeze–thaw cycles may generate more unwanted DNA breaks in plasmid DNA. We typically make 375 ng/μL of the defined SSB in stock. For every use, we can then just dilute the plasmid in HSS (i.e., 1 volume: 4 volume) to set up the reactions.
11. The defined SSB structure (75 ng/μL), but not CTL plasmid, triggered a robust Chk1 phosphorylation at S344 after a 30-min incubation in the HSS system (Fig. 3a). Higher concentration of SSB plasmid than 75 ng/μL in the HSS had no noticeable effect on Chk1 phosphorylation (Fig. 3a). These observations suggest that the defined SSB structure induces a unique ATR-Chk1 DDR pathway in the HSS in a dose-dependent manner.
12. Incubation periods of longer than 30 min at room temperature had no noticeable effect on Chk1 phosphorylation (Fig. 3b). These observations suggest that the SSB structure induces a unique checkpoint signaling in the HSS in a time-dependent manner.
13. This SSB signaling system is applicable for studies testing whether other checkpoint proteins play a direct role for the SSB signaling, especially proteins required for DNA replication initiation and/or elongation. The protein of interest (e.g., APE2, [5]) can be removed via immunodepletion using antibodies against the target protein. Alternatively, small molecule inhibitor specific for target protein (e.g., ATR specific inhibitor VE-822, [5]) can be added to the HSS system to test whether

such small molecule affects the SSB-induced ATR-Chk1 DDR pathway in the HSS system.

14. This is scaled up in comparison to the experimental setup described in Subheading 3.3.1 for detectable yield of DNA-bound fractions.
15. PCNA can be used as a loading control, since it binds to both CTL and SSB plasmid under these experimental conditions. The representative checkpoint proteins such as ATR, RPA32, and Rad9 are shown in Fig. 4.
16. The chosen percentage of gel is based on the size of the FAM-DNA structure, and can be optimized.
17. It may take up to 10 min for urea to dissolve. The final volume is around 40 mL.
18. The running time for TBE–urea gel can be optimized depending on the size of ssDNA products.
19. Because the resected products are arranged from ~4 nt to ~12 nt, it is estimated that the SSB structure is resected from ~18 nt to ~26 nt in the 3′ to 5′ direction in the HSS system.

Acknowledgments

The Yan lab was supported, in part, by funds from University of North Carolina at Charlotte (Duke Energy Endowment Special Initiatives Fund and Faculty Research Grants) and grants from the National Institute of General Medical Sciences (R15GM101571 and R15GM114713) and the National Cancer Institute (R01CA225637) in the National Institutes of Health.

References

1. Caldecott KW (2008) Single-strand break repair and genetic disease. Nat Rev Genet 9 (8):619–631. https://doi.org/10.1038/nrg2380
2. Yan S, Sorrell M, Berman Z (2014) Functional interplay between ATM/ATR-mediated DNA damage response and DNA repair pathways in oxidative stress. Cell Mol Life Sci 71 (20):3951–3967. https://doi.org/10.1007/s00018-014-1666-4
3. Ciccia A, Elledge SJ (2010) The DNA damage response: making it safe to play with knives. Mol Cell 40(2):179–204. https://doi.org/10.1016/j.molcel.2010.09.019
4. Tubbs A, Nussenzweig A (2017) Endogenous DNA damage as a source of genomic instability in cancer. Cell 168(4):644–656. https://doi.org/10.1016/j.cell.2017.01.002
5. Lin Y, Bai L, Cupello S, Hossain MA, Deem B, McLeod M, Raj J, Yan S (2018) APE2 promotes DNA damage response pathway from a single-strand break. Nucleic Acids Res 46 (5):2479–2494. https://doi.org/10.1093/nar/gky020
6. Nassour J, Martien S, Martin N, Deruy E, Tomellini E, Malaquin N, Bouali F, Sabatier L, Wernert N, Pinte S, Gilson E, Pourtier A, Pluquet O, Abbadie C (2016) Defective DNA single-strand break repair is responsible for senescence and neoplastic escape of epithelial cells. Nat Commun 7:10399. https://doi.org/10.1038/ncomms10399

7. Walter J, Sun L, Newport J (1998) Regulated chromosomal DNA replication in the absence of a nucleus. Mol Cell 1(4):519–529
8. Van C, Yan S, Michael WM, Waga S, Cimprich KA (2010) Continued primer synthesis at stalled replication forks contributes to checkpoint activation. J Cell Biol 189(2):233–246. https://doi.org/10.1083/jcb.200909105
9. Kumagai A, Shevchenko A, Shevchenko A, Dunphy WG (2010) Treslin collaborates with TopBP1 in triggering the initiation of DNA replication. Cell 140(3):349–359. https://doi.org/10.1016/j.cell.2009.12.049
10. Wallace BD, Berman Z, Mueller GA, Lin Y, Chang T, Andres SN, Wojtaszek JL, DeRose EF, Appel CD, London RE, Yan S, Williams RS (2017) APE2 Zf-GRF facilitates 3′–5′ resection of DNA damage following oxidative stress. Proc Natl Acad Sci U S A 114(2):304–309. https://doi.org/10.1073/pnas.1610011114
11. DeStephanis D, McLeod M, Yan S (2015) REV1 is important for the ATR-Chk1 DNA damage response pathway in Xenopus egg extracts. Biochem Biophys Res Commun 460(3):609–615. https://doi.org/10.1016/j.bbrc.2015.03.077
12. Yan S, Willis J (2013) WD40-repeat protein WDR18 collaborates with TopBP1 to facilitate DNA damage checkpoint signaling. Biochem Biophys Res Commun 431(3):466–471. https://doi.org/10.1016/j.bbrc.2012.12.144
13. Bai L, Michael WM, Yan S (2014) Importin beta-dependent nuclear import of TopBP1 in ATR-Chk1 checkpoint in Xenopus egg extracts. Cell Signal 26(5):857–867. https://doi.org/10.1016/j.cellsig.2014.01.006
14. Yan S (2015) Teaching and learning in a Xenopus research lab. Lab Anim (NY) 44(8):327. https://doi.org/10.1038/laban.817
15. Costanzo V, Robertson K, Bibikova M, Kim E, Grieco D, Gottesman M, Carroll D, Gautier J (2001) Mre11 protein complex prevents double-strand break accumulation during chromosomal DNA replication. Mol Cell 8(1):137–147
16. Dupre A, Boyer-Chatenet L, Gautier J (2006) Two-step activation of ATM by DNA and the Mre11-Rad50-Nbs1 complex. Nat Struct Mol Biol 13(5):451–457. https://doi.org/10.1038/nsmb1090
17. You Z, Shi LZ, Zhu Q, Wu P, Zhang YW, Basilio A, Tonnu N, Verma IM, Berns MW, Hunter T (2009) CtIP links DNA double-strand break sensing to resection. Mol Cell 36(6):954–969. https://doi.org/10.1016/j.molcel.2009.12.002
18. Raschle M, Knipscheer P, Enoiu M, Angelov T, Sun J, Griffith JD, Ellenberger TE, Scharer OD, Walter JC (2008) Mechanism of replication-coupled DNA interstrand crosslink repair. Cell 134(6):969–980. https://doi.org/10.1016/j.cell.2008.08.030
19. Ben-Yehoyada M, Wang LC, Kozekov ID, Rizzo CJ, Gottesman ME, Gautier J (2009) Checkpoint signaling from a single DNA interstrand crosslink. Mol Cell 35(5):704–715. https://doi.org/10.1016/j.molcel.2009.08.014
20. Lebofsky R, Takahashi T, Walter JC (2009) DNA replication in nucleus-free Xenopus egg extracts. Methods Mol Biol 521:229–252. https://doi.org/10.1007/978-1-60327-815-7_13
21. Cupello S, Richardson C, Yan S (2016) Cell-free Xenopus egg extracts for studying DNA damage response pathways. Int J Dev Biol 60(7–8–9):229–236. https://doi.org/10.1387/ijdb.160113sy
22. Willis J, DeStephanis D, Patel Y, Gowda V, Yan S (2012) Study of the DNA damage checkpoint using Xenopus egg extracts. J Vis Exp (69):e4449. https://doi.org/10.3791/4449

Chapter 10

Chromatin Immunoprecipitation (ChIP) of Plasmid-Bound Proteins in *Xenopus* Egg Extracts

Kelly B. Wolfe and David T. Long

Abstract

Xenopus egg extracts provide a cell-free system to analyze various aspects of chromatin biology. Here we describe a modified method of chromatin immunoprecipitation (ChIP) to detect the interaction of proteins with plasmid DNA incubated in extract. The combination of ChIP and *Xenopus* egg extracts provides a highly versatile and tractable approach to analyze dynamic protein–DNA interactions with great spatial and temporal detail.

Key words Xenopus egg extract, Nucleoplasmic extract, Chromatin immunoprecipitation (ChIP), DNA repair, Cell-free system

1 Introduction

Gilmour and Lis first described a method for detecting protein–DNA interactions in vivo in 1984 [1, 2]. As a proof-of-concept experiment, UV light was used to permanently couple proteins to DNA in *Escherichia coli*. The relative enrichment of RNA polymerase was then determined by immunoprecipitating the protein and hybridizing the associated DNA fragments to different DNA sequences. Current methods now utilize formaldehyde to form reversible cross-links, allowing DNA fragments to be analyzed by quantitative PCR. Over time, chromatin immunoprecipitation (ChIP) has become an essential tool in molecular biology due to its simplicity and high sensitivity (reviewed in [3]).

Several approaches have been developed using *Xenopus* egg extracts to monitor various aspects of chromatin biology, including DNA replication and termination, transcription, DNA damage signaling and repair, and chromatin organization [4–7]. Protocols for different types of *Xenopus* egg extracts have been established, allowing preparation of different protein fractions from different stages of the cell cycle and oocyte development [8]. In this chapter, we describe a modified method of ChIP for analyzing protein

Lata Balakrishnan and Jason A. Stewart (eds.), *DNA Repair: Methods and Protocols*, Methods in Molecular Biology, vol. 1999, https://doi.org/10.1007/978-1-4939-9500-4_10,

interactions with plasmid DNA incubated in *Xenopus* egg extracts. By coupling ChIP with highly synchronized extract reactions, a detailed analysis of protein recruitment, retention, and displacement from DNA can be performed, revealing dynamic protein activity not easily observed in cells [9–12].

2 Materials

Prepare all solutions using analytical grade reagents and ultrapure water (prepared by purifying deionized water to a sensitivity of 18 MΩ-cm at 25 °C).

2.1 Reagents

1. Protein A-Sepharose Fast Flow (PAS-FF) beads. Beads are washed with 1× PBS and stored as a 33% slurry with 0.01% Sodium Azide at 4 °C.
2. 0.5 and 1.5 mL siliconized microfuge tubes.
3. 50 mL conical tubes.
4. Bio-Rad Bio-Spin Columns.
5. Antibodies (commercial or serum) (*see* **Note 1**).
6. Ultrafine pipette tips.
7. 2 mg/mL RNase.
8. 20 mg/mL Pronase.
9. 20 mg/mL glycogen.
10. Oligonucleotide primers (*see* Subheading 3.10, **step 1**).
11. 384-well PCR plates and plate sealing films.

2.2 Buffers and Solutions

1. 1× phosphate-buffered saline (PBS): 10 mM Na_2HPO_4, 1.8 mM KH_2PO_4, 137 mM NaCl, and 2.7 mM KCl in water.
2. 1× Egg Lysis Buffer (ELB): 250 mM sucrose, 2.5 mM $MgCl_2$, 50 mM KCl, 10 mM HEPES pH 7.8. Filter-sterilize and store at 4 °C.
3. ATP Regeneration Mix (ARM): 64.5 mM ATP, 645 mM Phosphocreatine (PC), 2 μM creatine phosphokinase (CPK).
4. Formaldehyde Cross-Linking Buffer: 2% formaldehyde in 1× ELB.
5. Glycine Buffer: 1.25 M glycine in water, pH 7.
6. PMSF solution: 100 mM Phenylmethylsulfonyl (PMSF) in 100% ethanol.
7. Lithium chloride solution: 4 M LiCl in water. Dissolve slowly, as solution becomes very hot.
8. Aprotinin/leupeptin: 5 mg/mL aprotinin, 5 mg/mL leupeptin.

9. 2× Sonic Buffer: 40 mM Tris–HCl (pH 7.5), 300 mM NaCl, 4 mM Ethylenediaminetetraacetic Acid (EDTA) pH 8.0, and 1% NP-40 in water.
10. 1× Sonic Buffer: 50% 2× Sonic Buffer, 2 mM PMSF, 5 μg/mL Aprotinin/Leupeptin (*see* **Note 2**) in water.
11. 1× Sonic Salts Buffer: 50% 2× Sonic Buffer, 2 mM PMSF, 5 μg/mL Aprotinin/Leupeptin (*see* **Note 2**), 650 mM NaCl, 100 mM KCl in water.
12. ChIP Wash Buffer: 10 mM Tris–HCl (pH 7.5), 0.25 M LiCl, 1 mM EDTA pH 8.0, 0.5% NP-40, 0.5% SDS.
13. TE Buffer: 10 mM Tris–HCl (pH 7.5), 1 mM EDTA pH 8.0.
14. ChIP Elution Buffer: 50 mM Tris–HCl (pH 7.5), 10 mM EDTA pH 8.0, 1% SDS.
15. Sodium acetate solution: 3 M NaAc (pH 5.2).
16. 5 M NaCl in water.
17. De-cross-linking mix: 70 μL TE buffer, 20 μL Pronase (20 mg/mL), 10 μL 5 M NaCl in water.
18. 1-to-1 phenol–chloroform solution: prepare by carefully mixing 1.25 volumes of buffered phenol solution with 1 volume of chloroform. Once the two are mixed, the solution should settle into ~2 volumes of phenol–chloroform and 0.25 volumes of aqueous buffer.
19. 100% and 70% ethanol.
20. Applied Biosystems Power-Up SYBR Green master mix or similar (containing SYBR Green dye, DNA Polymerase, and dNTP mix).

2.3 Equipment

1. Diagenode Bioruptor UCD-600 TS or similar sonicator.
2. Rotating wheel or rotisserie.
3. Thermocycler (or heat block).
4. Bio-Rad Real-time PCR machine CFX-384 or similar.
5. Horizontal centrifuge drum rotor (*see* **Note 3**).
6. Swinging platform microfuge to spin down PCR plates.

3 Methods

3.1 Preparation

1. Aliquot 48 μL of Formaldehyde Cross-Linking Buffer into 0.5 mL microfuge tubes for each sample. Cap tubes and store at room temperature (RT).
2. Prepare a Bio-Spin column with provided wash and collection tubes for each sample. According to the manufacturer's instructions: suspend resin by inverting each column several

times. Loosen the cap of the column and then snap off the bottom to allow drainage of column storage buffer. Allow columns to drain into wash tube at 4 °C until needed (*see* Subheading 3.3, **step 3**).

3.2 Replication Reactions

1. Replication of plasmid DNA in High-Speed Supernatant (HSS) and NucleoPlasmic Extract (NPE) can be performed as described (*see* [8]).
2. Thaw and prepare extracts on ice. Supplement HSS with ARM (0.6 μL ARM per 20 μL HSS; final concentration 18 mM PC, 1.8 mM ATP, 0.057 μM CPK) and 10 μM nocodazole. Supplement NPE with ARM (0.6 μL ARM per 20 μL NPE; final concentration 18 mM PC, 1.8 mM ATP, 0.057 μM CPK), 10 μM DTT, and ELB. Add plasmid DNA to HSS at 7.5 ng/μL (*see* **Note 4**) and incubate at RT for 20 min to allow formation of prereplication complexes (pre-RCs).
3. After pre-RC formation, Add 2 volumes of NPE mixture to 1 volume of HSS/plasmid DNA for a final concentration of 2.5 ng/μL. Mix thoroughly by pipetting up and down or gently flicking the tube several times. For most experiments, addition of NPE is considered to be the start of the reaction.

3.3 Formaldehyde Cross-Linking

1. For each condition or time point to be analyzed by ChIP, remove 2 μL of reaction sample and add to a prepared 0.5 mL siliconized microfuge tube containing 48 μL of Formaldehyde Cross-Linking Buffer (*see* **Note 5**) (Fig. 1). Mix thoroughly by pipetting the Formaldehyde Cross-Linking Buffer up and down or gently flicking the tube several times. Incubate samples in Formaldehyde Cross-Linking Buffer at RT for exactly 10 min.

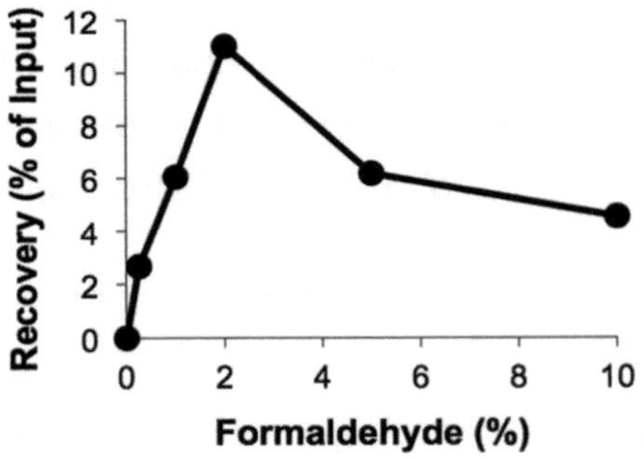

Fig. 1 Cross-linked plasmid DNA was replicated for 60 min in HSS/NPE and then incubated with a range of formaldehyde concentrations in the Formaldehyde Cross-Linking Buffer (*see* Subheading 3.3, **step 1**) prior to immunoprecipitation with anti-RPA antibodies [9]. Percent recovery of DNA near the cross-link is plotted for each formaldehyde concentration tested

2. During the Formaldehyde Cross-Linking Buffer incubation, finish preparing a Bio-Spin column for each sample being cross-linked. Decant flowthrough storage buffer from the wash tube and remove residual storage buffer by spinning the column at 1000 RCF for 2 min. Discard the wash tube and place the dried column into a collection tube (*see* **Note 6**).
3. Immediately after the 10 min Formaldehyde Cross-Linking Buffer incubation, quench the cross-linking reaction with 5 μL of Glycine Buffer (*see* **Note 7**). After glycine addition, mix samples by pipetting up and down twice and apply to the resin of the dried Bio-Spin column.
4. Spin samples through the Bio-Spin column at 1000 RCF for 4 min. Discard the column and add 950 μL of 1× Sonic Buffer (made fresh with PMSF and Aprotinin/Leupeptin) to the eluate, mix gently by inversion, and store on ice until sonication.

3.4 Sonication

1. Transfer 100 μL of each sample to a 0.5 mL siliconized microfuge tube for sonication. The rest of the sample may be aliquoted and snap frozen for storage at −80 °C (*see* **Note 8**).
2. Briefly spin samples to remove bubbles and collect liquid from the sides of the tube.
3. Sonicate samples using the Diagenode Bioruptor or similar sonicator to produce DNA fragments ~300–500 bp in length (Fig. 2). For the Diagenode Bioruptor, prechill the water bath to 4 °C, and sonicate for 25 cycles: 30 s ON, 60 s OFF, at HIGH power. The total run time is about 35 min.

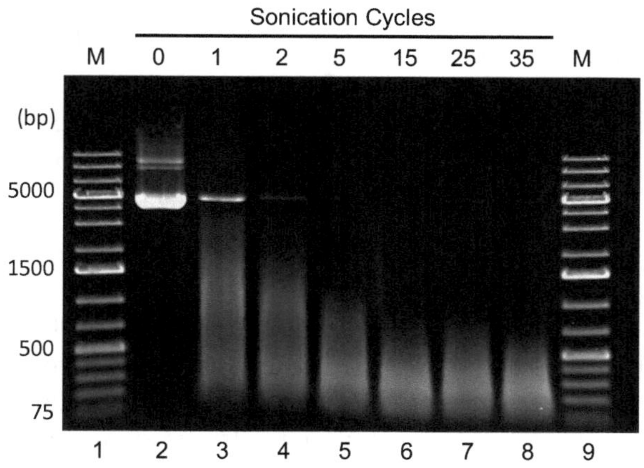

Fig. 2 250 ng of Plasmid DNA (5.2 kb in size) was sonicated using a Diagenode Bioruptor for increasing cycles of: 30 s ON, 60 s OFF, at HIGH power (*see* Subheading 3.4, **step 3**). Samples were then separated by 1% agarose gel electrophoresis and visualized by ethidium bromide staining. M, GeneRuler 1 KB plus DNA ladder

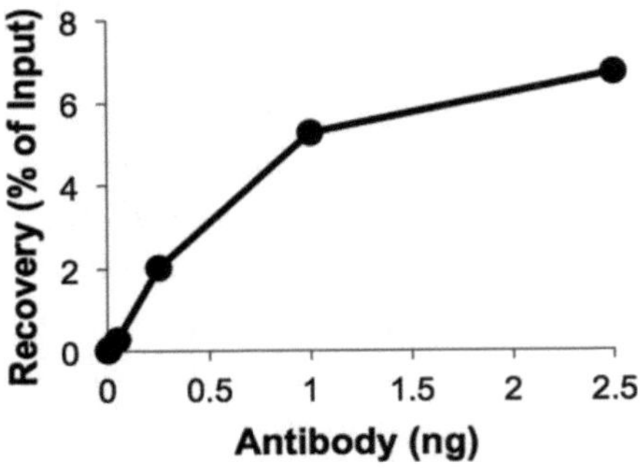

Fig. 3 Cross-linked plasmid DNA was replicated for 60 min in HSS/NPE and then immunoprecipitated with an increasing amount of anti-RPA antibody (*see* Subheading 3.5, **step 2**). Percent recovery of DNA near the cross-link is plotted for each amount of antibody tested

3.5 Antibody Incubation

1. Transfer 10 μL from each sonicated sample into a separate 0.5 mL siliconized microfuge tube and store at 4 °C for "INPUT" samples (*see* Subheading 3.7, **step 4** below).
2. Add 0.5–2 μg of antibody to the remaining 90 μL of each sonicated sample for immunoprecipitation (IP) (*see* **Note 9**) (Fig. 3).
3. Make sure all tubes are capped tightly and place them into a 50 mL conical or other container. A balled up tissue or paper towel can be used to fill extra space and prevent tubes from sliding around and possibly coming open.
4. Incubate IP samples at 4 °C overnight using a rotating wheel or rotisserie.

3.6 Antibody Coupling to Beads

1. Prepare 10 μL of beads for each IP by collectively washing ~1.2× the total amount of beads needed (*see* **Note 10**).
2. Resuspend 33% PAS-FF bead slurry by inverting the tube several times. For each IP, remove ~36 μL of slurry and transfer to a 1.5 mL microfuge tube using a wide-bore or cut pipette tip (e.g., For 6 IPs, remove 216 μL of 33% slurry for 72 μL of beads total).
3. Spin down beads at 1500 RCF for 60 s in a horizontal drum rotor. Remove the supernatant by aspirator or pipette, but do not completely dry beads throughout washes. Adding Sonic buffer to dry beads will create bubbles, making it difficult to aliquot.
4. Wash PAS-FF beads by resuspending with 1 mL 1× Sonic Buffer, spinning at 1500 RCF for 60 s, and removing supernatant. Repeat two times for three total washes.

5. After the final wash, add an equal volume of 1× Sonic Buffer to beads for a 50% bead slurry (e.g., For 6 IPs, add 72 μL of 1× Sonic Buffer to ~72 μL of washed beads).
6. Resuspend beads by pipetting up and down or inverting the tube several times. Using a fresh wide-bore or cut pipette tip to prevent clogging and inconsistent results, transfer 20 μL of 50% bead slurry (10 μL of beads) to each IP sample.
7. Make sure all tubes are capped tightly and place them into a 50 mL conical.
8. Incubate IP samples at RT for 2 h using a rotating wheel or rotisserie (*see* **Note 11**).

3.7 Bead Washing and Elution

1. Spin down beads at 1500 RCF for 60 s in a horizontal drum rotor. Remove the supernatant by aspirator or pipette, but do not completely dry beads throughout washes.
2. For each wash, resuspend PAS-FF beads with 450 μL of the appropriate buffer (*see below*). Cap tubes tightly, place them into a 50 mL conical, and rotate at RT for 5 min (*see* **Note 12**). Spin tubes at 1500 RCF for 60 s and remove supernatant. Wash once with each of the following buffers, totaling four washes:
   (a) 1× Sonic Buffer (made fresh with PMSF and Aprotinin/Leupeptin)
   (b) 1× Sonic Salts Buffer (made fresh with PMSF and Aprotinin/Leupeptin)
   (c) 1× ChIP Wash Buffer
   (d) TE Buffer
3. After the final wash, use an ultrafine tip to remove all traces of buffer from the beads and resuspend beads with 115 μL of ChIP elution buffer.
4. Add 90 μL of ChIP elution buffer to the INPUT samples from Subheading 3.5, **step 1**.
5. Incubate both IP and INPUT samples at 65 °C for 20 min using a heat block or thermocycler.
6. Add 1 μL of RNase (2 mg/mL stock) to INPUT samples and incubate at 37 °C for 30 min (*see* **Note 13**).

3.8 De-Cross-Linking

1. Aliquot 100 μL of de-cross-linking mix into a new 0.5 mL siliconized microfuge tube for each IP sample.
2. For IP samples, spin at 1500 RCF for 1 min to pellet beads. Transfer 100 μL of the supernatant (eluted IP sample) to the prepared 0.5 mL siliconized microfuge tubes containing 100 μL of de-cross-linking mix.

3. For INPUT samples, add 100 μL of de-cross-linking mix directly to the tubes containing 10 μL of sonicated sample and 90 μL of ChIP elution buffer.
4. Incubate IP and INPUT samples at 42 °C for 6 h (protease digestion) and 70 °C for 9 h (de-cross-linking) using a thermocycler or heat block. Samples can be stored at 4 °C for several days until proceeding to DNA extraction.

3.9 DNA Extraction and Precipitation

1. Add 150 μL of 1:1 phenol–chloroform solution to each IP and INPUT sample. Cap tubes tightly, mix by inverting the tube several times, and then spin in a microfuge at 16,000 RCF for 5 min.
2. For each sample, transfer as much of the aqueous (upper) layer as possible to a new 0.5 mL siliconized microfuge tube containing 150 μL of chloroform. Cap tubes tightly, mix by inverting the tube several times, and then spin in a microfuge at 16,000 RCF for 5 min.
3. For each sample, transfer 180 μL of the aqueous (upper) layer to a new 1.5 mL microfuge tube containing 18 μL 3 M NaOAc (pH 5.5) and 2 μL Glycogen (20 mg/mL) (*see* **Note 14**).
4. Add 525 μL of ice-cold 100% EtOH to each tube, mix by inverting the tube several times, and then incubate on ice for 15 min.
5. Spin samples in a microfuge at 16,000 RCF for 30 min at 4 °C.
6. Remove all but ~25 μL of the supernatant by aspirator or pipette.
7. Add 300 μL of ice-cold 70% EtOH to each tube and spin in a microfuge at 16,000 RCF for 10 min.
8. Carefully remove all of the supernatant using a pipette without disturbing the pellet.
9. Air-dry the pellets at RT with tubes uncapped for 10 min to remove residual ethanol. Pellets will appear translucent when dry.
10. Add 40 μL of 10 mM Tris–HCl, pH 7.5 to each pellet. Resuspend pellets by gently flicking the tube several times, and then spin briefly in a microfuge to collect liquid from the sides of the tube. Store DNA samples at −20 °C and avoid multiple freeze/thaw cycles to prevent degradation (*see* **Note 15**).

3.10 Quantitative PCR

1. To analyze recovery of DNA by quantitative real-time PCR, prepare an RT Master Mix for each primer pair to be tested. In general, primers should be 17–25 bases in length, have a melting temperature of ~60 °C, a G/C content of ~50%, and produce an amplification product ~100 bp in size (*see* **Note**

16). For analysis using a 384-well plate, prepare mixes based on the following single reaction size volumes:

(a) 5 μL of 2× SYBR Green Master Mix

(b) 0.5 μL Forward primer (10 μM)

(c) 0.5 μL Reverse primer (10 μM)

(d) 2 μL water.

2. First, aliquot 8 μL of RT Master Mix into each well needed to analyze: IP and INPUT samples, a dilution series of plasmid DNA (standard curve), and a no-template control. Each sample should be loaded in triplicate.
3. Next, add 2 μL of each DNA sample or 10 mM Tris–HCl, pH 7.5 for the no-template control.
4. Seal the plate using optically clear PCR plate sealing film and spin in a swinging platform microfuge at 100 RCF for 1 min to collect liquid from the sides of the well.
5. Load the plate onto a quantitative real-time PCR machine and analyze using SYBR Detection (520 nm) with the following conditions for reaction cycle (stage 1 and 2) and melting curve (stage 3 and 4):

Stage 1	95 °C	10 min (denature)
Stage 2 (×45 cycles)	95 °C	10 s (denature)
	60 °C	15 s (annealing)
	72 °C	30 s (extension, fluorescence reading)
Stage 3	95 °C	15 s (denature)
Stage 4 (×70 cycles)	60 °C	15 s (annealing, fluorescence reading)
	+0.5 °C each cycle	

3.11 Results Analysis

1. Use a standard curve (Fig. 4) to determine the absolute starting quantity of DNA recovered for each IP and INPUT reaction.
2. Determine the median starting quantity for each IP and INPUT sample set run in triplicate.
3. Calculate the recovery of IP samples as a percentage of INPUT: Percent of INPUT = (90 μL) × (median IP quantity)/(10 μL) × (median INPUT quantity) × 100 (Fig. 5a, b).
4. To compare recovery between different IPs, percent of INPUT values can be normalized by setting the peak value as 100% (Fig. 5c).

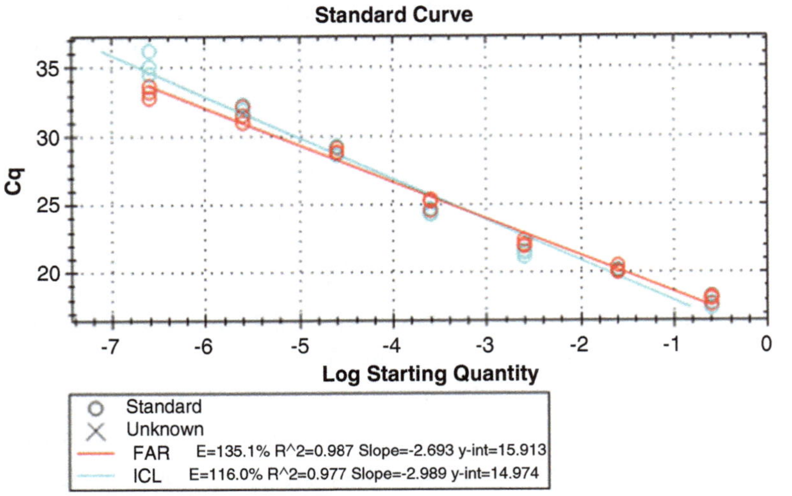

Fig. 4 Representative standard curves are shown using primer pairs adjacent to a DNA interstrand cross-link (ICL) or on the opposite side of the plasmid (FAR). Plasmid DNA concentrations tested range from 2.5×10^{-1} to 2.5×10^{-7} μM. The quantitation cycle (Cq) indicates the cycle number where the curvature of the amplification curve is maximal

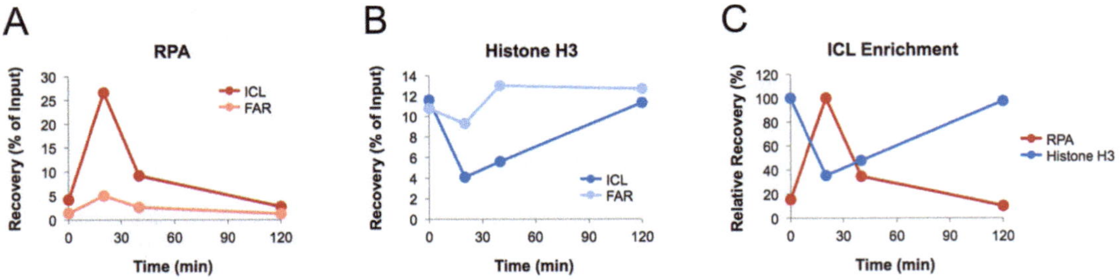

Fig. 5 Cross-linked plasmid DNA was incubated in HSS/NPE and samples were withdrawn at different time points for analysis by ChIP with anti-RPA (**a**) and anti-Histone H3 (**b**) antibodies. Percent recovery of DNA is shown for both the ICL (dark red and blue traces) and FAR (light red and blue traces) regions of DNA. (**c**) Relative recovery of RPA and Histone H3 IPs was normalized based on peak values in each time course and graphed together for comparison

4 Notes

1. Both commercial antibodies and antiserum can be used for immunoprecipitation. To achieve consistent results, total IgG antibodies should be purified from the antiserum.
2. PMSF, aprotinin, and leupeptin are added to buffers "fresh" and solutions should be kept on ice.
3. A standard microfuge can also be used, but does not pellet beads to the bottom of tubes. After spinning in a fixed angle

microfuge, allow beads to settle before removing supernatant during wash steps.

4. Plasmid DNA concentration can be changed depending on the reaction conditions and experimental setup. Plasmids used here were isolated from bacterial cultures and purified by standard miniprep spin columns.
5. Larger reaction samples can be processed by scaling the sample and formaldehyde buffer incubation up to the capacity of the Bio-Spin column or by collecting additional samples for the same condition and processing them separately.
6. If collecting multiple time points in rapid succession, several spin columns can be decanted in advance. However, they should not sit dry for more than ~1 h.
7. Consistency is crucial for formaldehyde incubations. Add glycine to cross-linked samples exactly 10 min after they are added to formaldehyde. For extremely rapid time points, multiple samples can also be treated with Glycine successively and then stored at RT for a short time (up to ~20 min) before being processed together. Incomplete mixing or differences in the duration of formaldehyde incubation will change the efficiency of cross-linking and sample recovery.
8. Samples may degrade with repeated freeze-thaw cycles, which will reduce recovery by immunoprecipitation.
9. Less sample can be used for antibodies that have good recovery efficiency. In general, the amount of antibody and sample required for efficient IP must be tested empirically.
10. 10 μL of PAS-FF beads are used per IP to improve recovery after washes. The binding capacity of 10 μL of PAS-FF beads greatly exceeds what is needed for antibody binding.
11. Incubation times will change the level of background observed, so be sure to incubate each "set" of IPs together.
12. The 5 min bead washing steps can be adjusted anywhere from 3 to 10 min, allowing you to wash two to three batches of samples consecutively. However, bead wash times should be kept consistent between samples.
13. RNA is removed from IP samples during wash steps and does not typically impact qPCR. However, RNAse can be added to all samples if desired.
14. By removing a fixed volume from each extraction sample, results are more precise and reproducible.
15. For studies of interstrand cross-link (ICL) repair, resuspended DNA fragments are digested with BbsI, which cuts immediately adjacent to the cross-link, to allow efficient amplification of DNA near the ICL.

16. Compared to cell lysates, *Xenopus* egg extracts are relatively free of nucleic acids, so designing specific primers is straightforward. However, each primer pair must be tested empirically to ensure specific amplification of the DNA region of interest in prepared samples.

Acknowledgments

This work was supported by NIH grant R35GM119512 to D.T.L. and T32 CA193201 to K.B.

References

1. Gilmour DS, Lis JT (1985) In vivo interactions of RNA polymerase II with genes of *Drosophila melanogaster*. Mol Cell Biol 5(8):2009–2018
2. Gilmour DS, Lis JT (1984) Detecting protein-DNA interactions in vivo: distribution of RNA polymerase on specific bacterial genes. Proc Natl Acad Sci U S A 81(14):4275–4279
3. Collas P (2010) The current state of chromatin immunoprecipitation. Mol Biotechnol 45 (1):87–100
4. Bernis C, Forbes DJ (2014) Analysis of nuclear reconstitution, nuclear envelope assembly, and nuclear pore assembly using Xenopus in vitro assays. Methods Cell Biol 122:165–191
5. Halpin D et al (2011) Mitotic spindle assembly around RCC1-coated beads in Xenopus egg extracts. PLoS Biol 9(12):e1001225
6. Gillespie PJ, Gambus A, Blow JJ (2012) Preparation and use of Xenopus egg extracts to study DNA replication and chromatin associated proteins. Methods 57(2):203–213
7. Eisenhardt N, Schooley A, Antonin W (2014) Xenopus in vitro assays to analyze the function of transmembrane nucleoporins and targeting of inner nuclear membrane proteins. Methods Cell Biol 122:193–218
8. Lebofsky R, Takahashi T, Walter JC (2009) DNA replication in nucleus-free Xenopus egg extracts. Methods Mol Biol 521:229–252
9. Long DT et al (2014) BRCA1 promotes unloading of the CMG helicase from a stalled DNA replication fork. Mol Cell 56 (1):174–185
10. Park J et al (2013) The MCM8-MCM9 complex promotes RAD51 recruitment at DNA damage sites to facilitate homologous recombination. Mol Cell Biol 33(8):1632–1644
11. Long DT et al (2011) Mechanism of RAD51-dependent DNA interstrand cross-link repair. Science 333(6038):84–87
12. Fullbright G et al (2016) p97 promotes a conserved mechanism of helicase unloading during DNA cross-link repair. Mol Cell Biol 36 (23):2983–2994

Chapter 11

Cellular Assays to Study the Functional Importance of Human DNA Repair Helicases

Sanket Awate, Srijita Dhar, Joshua A. Sommers, and Robert M. Brosh Jr.

Abstract

DNA helicases represent a specialized class of enzymes that play crucial roles in the DNA damage response. Using the energy of nucleoside triphosphate binding and hydrolysis, helicases behave as molecular motors capable of efficiently disrupting the many noncovalent hydrogen bonds that stabilize DNA molecules with secondary structure. In addition to their importance in DNA damage sensing and signaling, DNA helicases facilitate specific steps in DNA repair mechanisms that require polynucleotide tract unwinding or resolution. Because they play fundamental roles in the DNA damage response and DNA repair, defects in helicases disrupt cellular homeostasis. Thus, helicase deficiency or inhibition may result in reduced cell proliferation and survival, apoptosis, DNA damage induction, defective localization of repair proteins to sites of genomic DNA damage, chromosomal instability, and defective DNA repair pathways such as homologous recombination of double-strand breaks. In this chapter, we will describe step-by-step protocols to assay the functional importance of human DNA repair helicases in genome stability and cellular homeostasis.

Key words Helicase, DNA repair, Genomic stability, DNA damage, Replication stress, Genetic disease

1 Introduction

DNA helicases represent a large class of enzymes that play integral roles in the maintenance of cellular homeostasis and chromosomal stability. Inherited genetic defects in DNA repair helicases can lead to various genetic disorders including (but not limited to) Werner syndrome, Bloom's syndrome, Rothmund–Thomson syndrome, Baller–Gerold syndrome, RAPADILINO, Fanconi anemia (FA), Trichothiodystrophy, Xeroderma pigmentosum, and Cockayne's syndrome [1–3]. In a nucleoside triphosphate-dependent manner, helicases catalytically unwind double helical nucleic acid molecules (DNA–DNA, DNA–RNA, and/or RNA–RNA) that are stabilized by many noncovalent hydrogen bonds between the bases of the two strands [4–7]. In addition to conventional duplex DNA substrates, certain DNA helicases unwind specialized non-B form DNA structures such as triplex DNA [8, 9], G-quadruplex DNA [10–12], and

Lata Balakrishnan and Jason A. Stewart (eds.), *DNA Repair: Methods and Protocols*, Methods in Molecular Biology, vol. 1999, https://doi.org/10.1007/978-1-4939-9500-4_11,

Z-DNA [13]. Such alternatively arranged DNA structures are believed to interfere with normal cellular DNA transactions, but the mechanisms are still being investigated. Additionally, specialized DNA helicases resolve more complex duplex DNA structures which represent replication and repair intermediates including stalled or regressed replication forks, four-stranded Holliday junctions, and three-stranded displacement (D)-loop or flap structures. A subset of DNA helicases, particularly those of the RecQ family, stimulate annealing of complementary single-stranded DNA [1, 14, 15], but the physiological importance of this biochemical activity has not yet been revealed. Altogether, DNA helicases are crucial enzymes in numerous cellular processes including DNA repair, replication, and transcription.

DNA helicases are upregulated in various cancer cell lines, including breast, colorectal, hypopharyngeal, ovarian, and lung cancers [16, 17]. Increased expression of a given DNA helicase in cancer cells has been associated with increased proliferation and resistance to chemotherapeutic drugs [16]; conversely, inhibition of DNA helicase activity either by its depletion through RNA interference (RNAi) [18, 19] or pharmacological inhibition of helicase-catalyzed DNA unwinding [20–23] has been shown to lead to reduced cell proliferation, accumulation of DNA damage, and increased cellular apoptosis. Thus, DNA repair helicases are proposed to represent a novel class of enzymes which may serve as therapeutic targets for increasing the efficacy of chemotherapeutic agents [20, 24, 25]. In this protocol chapter, we will describe the experimental procedures by which we study the biological importance of human DNA repair helicases in response to endogenous or exogenously induced stress in terms of cellular proliferation and survival, apoptosis, and DNA damage induction as marked by phosphorylation of the histone protein H2AX.

Stalled replication forks can lead to double strand breaks (DSBs) due to broken replication forks, whereas ionizing radiation or certain DNA damaging agents (e.g., bleomycin) can directly introduce DSBs. DSBs introduced by either mechanism require efficient correction by nonhomologous end-joining or homologous recombination (HR) repair. Disruption of these DSB repair pathways can lead to chromosomal instability, age-related diseases, and cancer [26]. Thus, a quantitative method to measure each of these pathways in vivo can provide novel insights into the role of a given DNA repair protein in regulating these pathways. We use a strategy developed originally by the Jasin lab to examine the effect of helicase deficiency on HR repair in vivo [27] (elaborated below). Essentially, U2OS cells with a stably integrated recombination susceptible DNA sequence characterized by a disrupted gene encoding Green Fluorescence Protein (GFP) are scored for HR by restoration of the intact GFP, resulting in cellular immunofluorescence which can be detected by flow cytometry.

Altogether the combined approaches to measure cell proliferation, colony survival, apoptosis, immunofluorescent accumulation of DNA damage markers or DNA repair proteins to sites of damage, cytogenetic analysis for chromosomal stability, and HR repair using an in vivo reporter assay provide a comprehensive battery of techniques to begin to assess the effects of helicase deficiency or helicase inhibition in a biological setting. The techniques described in this book chapter provide a simple list of reagents, disposables, and equipment with step-by-step protocols to perform these assays.

We will begin by describing the laboratory procedures to measure total cell number by Coulter counting and cell proliferation assay by scoring metabolic activity of untreated cells compared to those exposed to the DNA damaging drug (Fig. 1a). Because cell proliferation assays are typically spectrophotometric, they can be easily set up in a high-throughput manner if so desired such as in the case of an RNAi screen. In addition, cell proliferation assays are typically conducted over a shorter period of time (such as 3–4 days), so these are more amenable to screens in which many treatments/conditions are being assessed for their effect on cell viability. For example, the assay can be adapted to use in a 96-well or 384-well plate to conduct large-scale screens to identify novel genetic or chemical interactions. Cell proliferation is measured using a tetrazolium salt such as sodium 5-(2,4-disulfophenyl)-2-(4-iodophenyl)-3-(4-nitrophenyl)-2H-tetrazolium inner salt (WST-1). The WST-1 assay is principally very similar to the MTT assay in which NADH produced by metabolically active cells reduces the positively charged tetrazolium salt 2-(4,5-dimethyl-2-thiazolyl)-3,5-diphenyl-2H-tetrazolium bromide (MTT) to form a purple-colored insoluble formazan dye [28]. Formazan that is produced by the cleavage of the tetrazole ring of the MTT is then solubilized, and the metabolic activity of the cells is measured by colorimetric quantification. Reduction of WST-1 by the NADH produced in proliferating cells leads to the formation of soluble yellow or orange-colored formazan dye, thereby eliminating the need for solubilization of formazan as required in the MTT assay. Thus, the WST-1 assay is a more convenient and time-saving assay for the determination of cellular proliferation compared to the MTT assay.

Often researchers desire to assess the impact of helicase deficiency on cell growth and division, a truer measure of survival than the cell proliferation assay described above, which can be readily assessed by colony formation (Fig. 1b). The appropriate human wild-type and helicase mutant cell lines are treated with a specified compound that induces DNA damage or replication stress. Often, stably transfected cell lines exogenously expressing wild-type or mutant helicase proteins in a genetically null background are characterized for their sensitivity to various DNA damaging or replication stalling drugs. Alternatively, human cells depleted of a specific

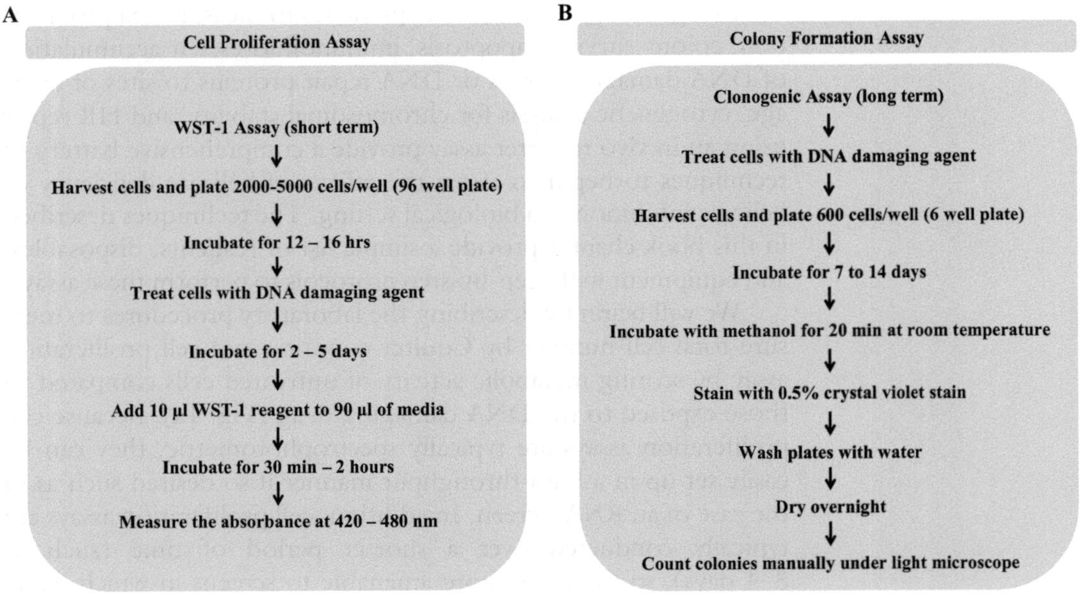

Fig. 1 Flowchart representing assays to measure cell proliferation or colony formation. Panel (**a**) Protocol for determining cell proliferation using an MTT assay variant, WST-1 assay. NADP produced by metabolically active cells can produce a colored formazan dye by the reduction of WST-1. Colorimetric analysis of the soluble formazan dye is used to quantify cellular proliferation. Panel (**b**) Protocol for determining the potential of individual cells to proliferate by using colony formation assay. Change in proliferative ability of the cells before and after treatment with a DNA damaging drug can be studied by allowing individual cell to form colonies (50 cells or more). Further, quantification of colonies of cells with or without a helicase allows us to assess the role of the helicase in overcoming DNA damage as induced by the drug

helicase by RNAi can be assayed for colony formation and compared to those cells that are transfected with a control (nontarget) siRNA. The simplest method to assess the number of cells before and after treatment with drug is by using a cell counting method utilizing a Coulter counter. This method can be used to determine the effect of treatment on a wide range of adherent and suspended cells. We also use a Coulter counter to seed the desired number of cells for subsequent analysis by WST-1, colony formation or apoptosis assay. However, one drawback of the Coulter counter method is that it fails to distinguish between live, dead and metabolically inactive cells. The colony forming assay can be used to measure the ability of human cells to survive and divide after they are treated with various DNA damaging agents. Using this assay, we can determine the capacity of every single cell in the selected sample to continuously divide and form colonies (which typically represent 50 cells or more). The colony formation assay is an accurate method to quantitatively assess the effect of drug treatment on the continued growth of the cells deficient or proficient for a given helicase.

As mentioned earlier, helicases such as FANCJ can resolve specialized non-B DNA structures such as G-quadruplexes thereby maintaining DNA replication and reducing DNA damage. Thus, FANCJ-depleted cells treated with a G-quadruplex stabilizer accumulate DNA damage and in-turn undergo apoptosis [29]. Likewise, pharmacological inhibition of WRN helicase by a small molecule WRN-specific inhibitor in cancer cells leads to accumulation of double-strand DNA breaks and increased apoptosis [22]. To assess the biological effects of such agents, the apoptosis assay is a convenient and useful tool.

DNA fragmentation is one of the hallmarks of cell undergoing apoptosis and can be exploited for an accurate measurement of apoptosis. Various Ca^{2+}/Mg^{2+}-dependent endonucleases are activated during apoptosis, and they cleave the genomic DNA into mononucleosomes and multinucleosomes. These apoptotic nucleases cleave the accessible internucleosomal DNA to produce DNA–histone fragments or nucleosomes [30]. This principle has been utilized by several assays including TUNEL and ELISA to detect cytoplasmic DNA fragments indicative of apoptosis. In this chapter we will discuss ELISA to quantitate cytoplasmic nucleosomes as an indicator of apoptosis. This method relies on the fact that during apoptosis the chromosomal nucleosome units travel to the cytoplasm and can then be extracted from the dying cells and quantitated using a colorimetric method.

In our lab, we have used immunofluorescence confocal microscopy to detect various DNA damage markers in cells mutated for a given helicase, depleted of a target helicase, or inhibited pharmacologically by a helicase-specific small molecule (Table 1). Similarly, the formation of DNA repair and/or replication protein foci in helicase-deficient or helicase-inhibited cells exposed to a DNA damaging agent can be visualized by fluorescence microscopy (Table 2). Recruited proteins can act as DNA damage sensors, amplifiers, or downstream effectors at sites of DNA damage or stalled replication forks. We have also detected the colocalization of DNA repair proteins with DNA helicases in human cells exposed to agents that impose replication stress or DNA damage, suggesting their interactive functions or common pathways of DNA metabolism (Table 3). A representative example of microscopic imaging is shown in Fig. 2. Here, the immunofluorescence data suggest that nuclear FANCJ foci are larger and better colocalize with the single-stranded DNA binding protein RPA after ionizing radiation (IR) or mitomycin C (MMC) exposure to human fibroblasts that express FANCA, an upstream player in the FA pathway of interstrand cross-link repair.

Immunofluorescence imaging of DNA repair proteins has provided novel insights into DNA damage signaling and the repair process [31]. For example, in response to replication stress ATR acts as one of the early sensors and is required for the formation of

Table 1
Immunofluorescent detection of endogenous or exogenously induced DNA damage in helicase-compromised cells

Helicase deficiency/inhibitor	Agent	Marker[a]	DNA structure	Reference
FANCJ-RNAi, U2 OS cells	Telomestatin	γ-H2AX	Double-strand break	[36]
FANCJ mutant DT40 cells	Telomestatin	G4-DNA antibody	G-quadruplex	[37]
FANCJ mutant fibroblasts	Mitomycin C	γ-H2AX	Double-strand break	[38]
RECQL1-RNAi, HeLa cells	Camptothecin	53BP1	Double-strand break	[39]
WRNi NSC 617145[b], HeLa cells	None	γ-H2AX	Double-strand break	[22]
		PCNA	Stalled replication fork	

[a]DNA damage marker or protein foci are elevated upon treatment with agent in helicase-compromised cells
[b]WRNi NSC 617145 is a WRN-specific small molecule helicase inhibitor

Table 2
Immunofluorescent detection of DNA repair proteins in helicase-compromised cells challenged with a DNA damaging agent

Helicase deficiency/ inhibitor	Agent(s)	Repair protein foci	Function	Reference
WRN inhibitor NSC 617145	MMC	DNA-PKcs pS2056	Nonhomologous end-joining	[21]
		ATM	DNA damage sensing and signaling	[21]
		Rad51	Strand exchange, HR repair	[21]
RECQL1-RNAi	Camptothecin	RPA[a]	Single-stranded DNA binding	[39]
		Rad51[a]	Strand exchange, HR repair	[39]
		Mre11	Nucleolytic trimming, HR repair	[39]

[a]Unless stated otherwise, protein foci are elevated upon treatment with agent in helicase-compromised cells. However, in the case of RPA or RAD51, foci are decreased in RECQL1-RNAi depleted cells exposed to camptothecin

RPA and Rad9 foci which act as amplifiers. Formation of RPA and Rad9 foci lead to the activation of Chk1 which acts as the effector molecule that is known to cause cell cycle arrest [32]. Immunofluorescence microscopy is a very powerful technique that helps to identify and monitor nuclear relocalization of these proteins, protein–protein interactions, and histone modifications. Some DNA repair proteins, like ATR, ATRIP, Rad9, WRN, MRE11, BRCA1, and RPA, do not form detectable foci until the fraction of proteins that is not bound to the chromatin is washed away. In situ

Table 3
Immunofluorescent detection of DNA repair proteins colocalizing with human DNA

Helicases	Agent	Colocalizing protein	Function	Reference
WRN	Hydroxyurea	RPA	Single-stranded DNA binding	[40]
WRN	Mitomycin C 4-nitroquinoline-1-oxide Methylmethanesulfonate	FEN-1	Processing of DNA replication and repair intermediates	[41]
BLM	Hydroxyurea	FANCJ	ICL repair, replication stress response	[42]
FANCJ	MMC Ionizing radiation Hydroxyurea	RPA	Single-stranded DNA binding	[43]

fractionation can be used to remove the excess of the nucleoplasmic proteins not bound to the chromatin to enhance the detection of the proteins bound to the chromatin. We have adapted and modified a technique originally described by Mirzoeva and Petrini [33] to enhance image quality, and this is described in the Methods section.

Although immunofluorescent detection of DNA damage by γ-H2AX or 53BP1 foci to visualize DSBs is useful, often an analysis of chromosomal aberrations is warranted. In this chapter, we describe a very simple and useful technique to obtain high-quality chromosome spreads for subsequent microscopic analysis of chromosomal integrity. This technique can be used to identify aneuploidy, insertions, deletions, and genomic rearrangement within the same chromosome or between two different chromosomes.

Disruption or inhibition of a key DNA damage response protein leading to chromosomal instability requires further studies to delineate the precise pathway of DNA repair that is affected. Given that chromosomal instability may result from a defect in DSB repair, and HR is a preferred pathway of DSB repair during S/G2-phases of dividing cells, a quantitative assessment of HR repair is useful for the analysis of cell lines deficient in a DNA helicase, particularly those of the RecQ family but also the Fe-S helicase FANCJ. The DNA recombination substrate originally described by the Jasin lab [27] which we have used consists of a repeat sequence (SceGFP) that is nonfunctional due to the integration of the 18 bp recognition site for the I-SceI endonuclease. Downstream of the SceGFP gene is an internal GFP fragment (iGFP). Transfection of the cells with the plasmid expressing I-SceI endonuclease induces a double stranded break. HR using iGFP as a template restores the full-length functional GFP. GFP-positive cells are then detected by flow cytometry. We have used these reporter cell lines to study the effect of FANCJ helicase deficiency on the frequency of HR repair [34] (Fig. 3).

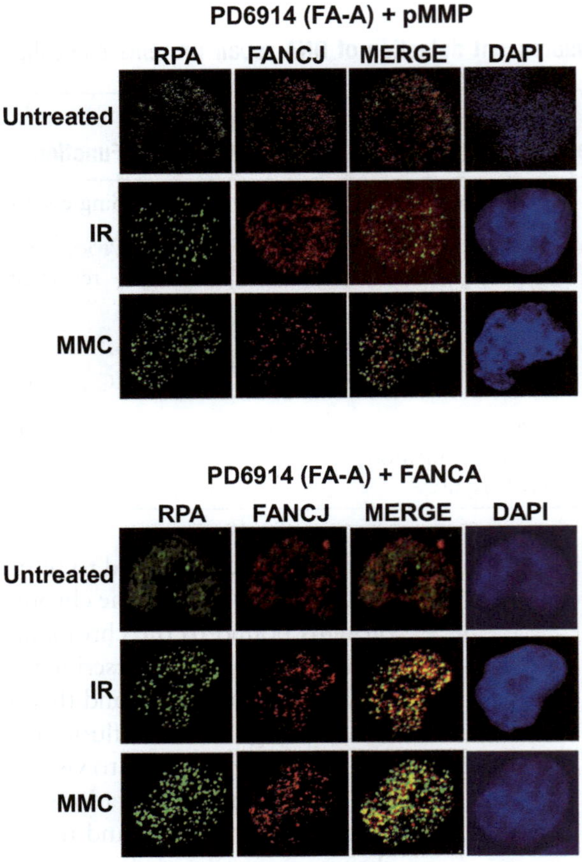

Fig. 2 Colocalization of FANCJ helicase and RPA is dependent on FANCA expression in cells exposed to DNA damaging agents. FA-A fibroblasts (PD6914 + pMMP vector) and corrected fibroblasts (PD6914 + FANCA) [35] were treated with the DNA-damaging agent MMC (500 ng/mL) for 16 h or exposed to 10 Gy IR followed by 6 h recovery. Cells were fixed (3.7% formaldehyde), permeabilized, and incubated with mouse anti-RPA34 Ab-2 (Millipore Sigma) and rabbit anti-FANCJ (Millipore Sigma) antibodies. For immunofluorescent detection, cells were incubated with Alexa Fluor 488 goat anti-mouse IgG and Alexa Fluor 568 goat anti-rabbit IgG (Thermo Fisher Scientific) followed by ProLong Gold mountant with DAPI (Thermo Fisher Scientific). Immunofluorescence was imaged on a Zeiss LSM 510 META inverted Axiovert 200M laser scan microscope (Carl Zeiss, Jena, Germany) with a Plan-Apochromat 63×/1.4 oil DIC objective. After treatment with MMC or IR, RPA (green) localizes in nuclear foci that coincide with FANCJ (red) foci as shown in the overlapped images (yellow) in the presence of FANCA protein. In the absence of FANCA, RPA and FANCJ staining does not overlap. DAPI staining of the nucleus in each cell is indicated by the color blue

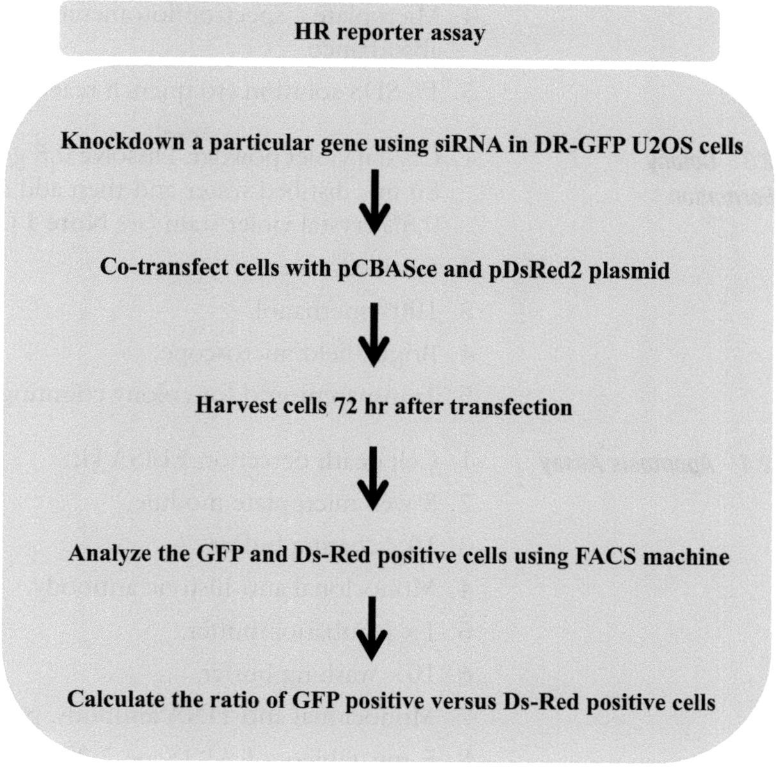

Fig. 3 Flowchart representing homologous recombination repair assay. Cells are transfected with pCBASce and DsRed2 plasmids previously described in ref. [27]. I-SceI generates DSB that at integrated GFP sequence stimulates HR. GFP and Ds-Red signals are assayed after 3 days on an FACSCalibur flow cytometer

2 Materials

2.1 Cell Counting Using Coulter Counter

1. 6-well tissue culture dish.
2. Dulbecco's Modified Eagle's medium (DMEM) supplemented with 10% FBS and 1% Penicillin-Streptomycin.
3. DNA damaging agent.
4. Sterile phosphate buffered saline (PBS).
5. Trypsin–EDTA:0.25% trypsin solution with 1 mM EDTA.
6. Cell counter vials.
7. Coulter counter.
8. Isotonic diluent.

2.2 WST-1 Cell Proliferation Assay

1. 96-well tissue culture plate.
2. Multichannel pipette.
3. Cell proliferation reagent WST-1.

4. Microplate spectrophotometer reader to measure the absorbance.
5. 1% SDS solution (to quench reaction).

2.3 Colony Formation

1. Crystal violet powder. Dissolve 0.5 g of crystal violet powder in 80 mL distilled water and then add 20 mL methanol to make 0.5% crystal violet stain (*see* **Note 1**).
2. Distilled water.
3. 100% methanol.
4. Bright-field microscope.
5. Transparent grid for colony counting.

2.4 Apoptosis Assay

1. Cell death detection ELISA kit.
2. 8-well microplate module.
3. 10× coating buffer.
4. Monoclonal anti-histone antibody.
5. 1× incubation buffer.
6. 10× washing buffer.
7. Monoclonal anti-DNA antibody, peroxidase conjugated.
8. 5 mg tablets of ABTS or 2,2′-azino-bis(3-ethylbenzothiazoline-6-sulfonic acid) substrate.
9. 1× substrate buffer.
10. Adhesive foil to cover the microplate modules for incubation.
11. Microplate spectrophotometer reader to measure the absorbance.

2.5 Cellular Immunofluorescence

1. U2-OS Cells (ATCC).
2. Fetal bovine serum.
3. Dulbecco's Modified Eagle's medium (DMEM) supplemented with 10% FBS and 1% penicillin–streptomycin.
4. Camptothecin.
5. Triton X-100.
6. Tween-20.
7. 1 M HEPES pH 7.4.
8. 1 M $MgCl_2$.
9. 5 M NaCl.
10. Sucrose.
11. Halt EDTA-free protease and phosphatase inhibitors.
12. Chamber slides.
13. PBS, pH 7.4.

14. Preextraction buffer 1: 1% Triton X-100, 10 mM HEPES pH 7.4, 10 mM NaCl, and 3 mM $MgCl_2$ supplemented with 1× Halt EDTA-free protease and phosphatase inhibitors.
15. Preextraction buffer 2: 0.5% Triton X-100, 20 mM HEPES pH 7.4, 50 mM NaCl, 3 mM $MgCl_2$, and 300 mM sucrose supplemented with 1× Halt EDTA-free protease and phosphatase inhibitors.
16. Fixing solution: Add 0.1% Triton X-100 to 4% Paraformaldehyde solution in 1× PBS. 100% ice-cold methanol can also be used for fixation.
17. Blocking Buffer: 5% normal horse serum in 1× PBS containing 0.3% Triton X-100. The normal serum is from the same species as the secondary antibody. SuperBlock Blocking buffer can also be used for blocking.
18. Antibody dilution buffer: 5% BSA in 1× PBS containing 0.1% Tween.
19. Primary antibody: ATR Antibody.
20. Secondary antibody: Fluorescein horse anti-mouse IgG antibody.
21. DAPI.

2.6 Chromosome Spreads

1. U2-OS cells.
2. Fetal bovine serum.
3. Dulbecco's Modified Eagle's medium (DMEM) supplemented with 10% FBS and 1% penicillin–streptomycin.
4. Trypsin.
5. Methanol.
6. Glacial acetic acid.
7. Carnoy's fixative → 3:1 ratio of methanol and glacial acetic acid.
8. KCl.
9. Giemsa stain.
10. Colcemid.
11. Mounting media.

2.7 Double-Strand Break HR Reporter Assay

1. U2-OS cells containing a stably integrated DR-GFP reporter made by the Jasin lab [27].
2. Fetal bovine serum.
3. Dulbecco's Modified Eagle's medium (DMEM) supplemented with 10% FBS and 1% penicillin–streptomycin.
4. Amaxa® Cell Line Nucleofector® Kit V.
5. Nucleofector™ 2b Device.

6. 2 μg of I-SceI expressing plasmid DNA (pCBASce) and 100 ng of ds-Red expressing plasmid DNA (pDsRed2-N1).
7. BD FACSCalibur Flow Cytometer Cell Analyzer.

3 Methods

3.1 Coulter Counter Assessment of Cell Number

1. For adherent cancer cell lines such as HeLa or U2-OS, plate ~15×10^4 cells are seeded per well in 1.5 mL of DMEM supplemented with 10% fetal bovine serum (FBS), 100 μg/mL streptomycin, and 100 U/mL penicillin (*see* **Note 2**).
2. Treat the cells with a predetermined concentration of DNA damaging agents and incubate for the appropriate duration (*see* **Notes 3** and **4**). This treatment can be done in duplicates to get more accurate and reliable data.
3. Remove media and wash the cells with PBS (*see* **Note 5**).
4. Add trypsin–EDTA solution to the well (cover the entire surface area of the well).
5. Incubate at 37 °C for 2–3 min to detach the cells.
6. Collect the cells in a tube by adding medium.
7. Centrifuge the tubes at $1000 \times g$ for 5 min.
8. Aspirate out the medium and resuspend the pellet in appropriate volume of fresh medium.
9. Dilute 500 μL of cell suspension in 10 mL of the isotonic diluent in the cell counter vials. Resuspend well.
10. Place the cell counter vials on the Coulter counter lever. Make sure the tip of the Coulter counter with the aperture is completely immersed in the cell suspension (*see* **Note 6**).
11. Count the total number of cells using a dilution factor of 21 (*see* **Note** 7). To get accurate reading this process can be repeated several times and the average value can be used to determine cell number.

3.2 WST-1 Cell Proliferation Assay

1. The isogenic pair of wild-type and helicase mutant cell lines, or the target helicase-shRNA and conrol-shRNA treated cells are seeded in a 96-well plate. For cancer cell lines (including U2-OS and HeLa), plate 700–1000 cells per well in 100 μL in DMEM media (as mentioned above, *see* **Note 2**).
2. Treat the cells with appropriate concentration of DNA damaging agent (*see* **step 2** in Subheading 3.1) 12–16 h after seeding, when the cells have attached (*see* **Note 3**). For control wells, treat the cells with the solvent (e.g., DMSO) that the drug is dissolved in (*see* **Note 8**). All the treatments and controls should be done in triplicates to obtain more accurate readings.

3. Incubate the cells for 2–5 days at 37 °C and 5% CO_2 (*see* **Note 9**).
4. Equilibrate the temperature of WST-1 to room temperature (*see* **Note 10**). Add 10 μL WST-1 reagent to each well (*see* **Note 11**). Make sure the ratio of media to WST-1 is 9:1. For blank wells add 10 μL of WST-1 to 90 μL of culture medium (*see* **Note 12**). Tilt plate to mix or put it in a shaker for 30 s (*see* **Note 13**).
5. Incubate cells at 37 °C for 30 min to 2 h. To stop the assay 10 μL of 1% SDS may be added.
6. Measure the absorbance of the plate using a microplate reader or spectrophotometer at OD between 420 and 480 nm (*see* **Note 14**).

3.3 Colony Forming Assay

1. Treat the adherent cells with DNA damaging agents and harvest the cells after 24 h using trypsinization. For detailed procedure refer to the above protocol (*see* **steps 1–7** in Subheading 3.1).
2. Resuspend the cells in fresh medium to the appropriate dilution, based on the kind of cell and treatment.
3. Count the cells using a Coulter counter (*see* **steps 9–11** in Subheading 3.1) and plate around 500 cells in each well of a 6-well plate (*see* **Note 15**).
4. Incubate the plates for 7–14 days till the colonies are formed.
5. Carefully aspirate out the medium from the dish and gently add 100% methanol so as to not disturb the colonies. Make sure the entire plate is covered with methanol (*see* **Note 16**).
6. Incubate the dish at room temperature for 20 min. Keep the dish covered.
7. Rinse the cells with water and add 2 mL of 0.5% crystal violet stain. Incubate at room temperature 5–10 min.
8. Remove the crystal violet stain and wash the cells with water. Place the plates in a water bath or water filled sink to remove any excess dye (*see* **Note 17**).
9. The plates can be left upright to air dry in room temperature alternatively they can be left inverted on tissue paper. Leave the plates overnight to ensure proper drying.
10. Count the colonies manually using a bright-field microscope and a transparent grid (*see* **Note 18**).

3.4 Apoptosis Assay

1. Seed the appropriate number of cells in a 6-well plate and treat with or without the DNA damaging agent for up to 72 h (*see* **Note 19**). Harvest the adherent cells using trypsinization method and resuspend in fresh medium.

2. Using the Coulter counter method, count the cells and transfer 5×10^4 cells into an eppendorf tube (*see* **steps 9–11** in Subheading 3.1).
3. Pellet the cells by centrifuging the tubes at 1000–1500 × *g* for 5 min and resuspend the cell pellet well in 500 μL of incubation buffer.
4. Lyse the cells by incubating them for 30 min at room temperature.
5. Centrifuge the tube at 20,000 × *g* to separate the cell cytoplasmic fraction from the cell nuclear fraction.
6. Carefully aspirate out the supernatant into a fresh Eppendorf tube. The samples can now stored at −20 °C overnight.
7. Incubate the 8-well microplate module with 100 μL of the coating buffer with anti-histone antibody (*see* **Notes 20** and **21**) for 1 h at room temperature or overnight at 4 °C. Make sure that the microplate module is covered properly.
8. After removing the coating solution, add around 200 μL of incubation buffer to the microplate module.
9. Cover the microplate module carefully and incubate at room temperature for 30 min. This blocks the nonspecific binding sites on the surface of the coated microplate module.
10. Wash the microplate wells thrice using 300 μL of washing buffer (*see* **Notes 22** and **23**).
11. Dilute one part of the samples obtained in **step 6** with nine parts of the incubation buffer. Add 100 μL of the diluted sample to the microplate module. Cover the microplate and incubate for 90 min at room temperature. To get "blank" absorbance values, add only incubation buffer without any sample lysate.
12. Repeat **step 10**.
13. Add 100 μL of freshly prepared incubation buffer with the anti-DNA-peroxidase conjugated antibody (9:1) to the microplate modules. Do not add this anti-DNA antibody for blank wells.
14. Cover the microplate module tightly and incubate for 90 min at room temperature.
15. Repeat **step 10**.
16. Add 100 μL of the ABTS substrate (*see* **Note 24**) to the microplate module.
17. Plate the microplate modules on a shaker incubator for 10–20 min at 250 rpm. The incubation time should be long enough for the development of color to perform the colorimetric quantification.

18. The contents of the microplate should be mixed thoroughly by tapping the sides of the microplate module and the absorbance can be read at 405 nm using a microplate spectrophotometer.

3.5 Cellular Immunofluorescence Detection

1. Maintain U2-OS cells in 5% CO_2 at 37 °C in DMEM media. Split cells after every 3 days. Count cells using the cell counter and seed the cells onto the chamber slides. Seed 60,000 cells per well for four chamber slides or 30,000 cells per well for eight chamber slides.
2. Incubate cells in the 5% CO_2 incubator and grow cells until they are 50–60% confluent.
3. Add 2 μM Camptothecin directly to the chamber slides for 2 h (*see* **Note 25**).
4. Add the preextraction buffer 1 (*see* **Notes 26** and **27**). Keep the slides at 4 °C for 5 min.
5. Wash with 1× PBS.
6. Add preextraction buffer 2 (*see* **Notes 26** and **27**). Keep the slides at 4 °C for 5 min.
7. Wash with 1× PBS.
8. Add 4% paraformaldehyde with 0.1% Triton X-100 at room temperature and make sure the cells are covered to a depth of ~3 mm. Cells can also be fixed with 100% methanol for 10 min at −20 °C.
9. After fixing the cells in paraformaldehyde for 15 min, wash the cells with 1× PBS three times.
10. Incubate cells in the Blocking buffer.
11. Dilute primary antibody (1:250) in PBS/T. Add the primary antibody on top of the cells in the chamber slides and incubate overnight at 4 °C.
12. Wash chamber slides three times for 5 min with PBS/T (*see* **Note 28**).
13. Dilute fluorescence labeled secondary antibody (1:500) in PBS/T. Add the secondary antibody on top of the cells and incubate for 3 h at room temperature in the dark.
14. Wash the chamber slides three times for 5 min with PBS/T in low lighting (*see* **Note 28**).
15. Aspirate off excess PBS and remove the chambers. Air dry the slides for 3 min in the dark.
16. Add a small drop of DAPI to each chamber and cover with glass coverslips. Gently tap the coverslips to remove the air bubbles and then seal the coverslips with a transparent nail polish. Store the slides at 4 °C in a dark box.

17. Images of the nuclei can be obtained by using a fluorescence microscope at 63× or 100× magnification. If Z-sections are used, then all the sections should be merged so that all the foci are on a single visible plane.
18. Count at least 50 nuclei from three independent experiments and determine average number of foci per nucleus (*see* **Note 29**).

3.6 Cytogenetic Analysis of Chromosome Spreads

1. Maintain U2-OS cells in 5% CO_2 atmosphere at 37 °C in DMEM media. Grow cells in a 10 cm dish and split cells after every 3 days.
2. When cells reach ~80% confluency, add colcemid to a final concentration of 200 ng/mL and incubate cells in the 37 °C incubator for 3 h (*see* **Note 30**).
3. Gently wash the cells with 1× PBS and aspirate off the PBS.
4. Add 1 mL of trypsin, swirl and keep the flask in 37 °C incubator for 3 min.
5. Tap the flask gently to detach the cells. Once the cells are detached, add 5 mL of media and transfer the cell suspension to a 15 mL conical tube.
6. Centrifuge at 180 × *g* for 5 min. Remove supernatant and resuspend pellet in 10 mL of 0.075 M KCl solution. Make sure that the KCl solution is prewarmed at 37 °C. Vortex the tube to mix the cells properly in the KCl solution (*see* **Note 31**).
7. Incubate cells in the 37 °C water bath for 10 min. Centrifuge at 180 × *g* for 5 min. Remove supernatant and resuspend pellet in 6 mL of fresh Carnoy's Fixative (*see* **Note 32**).
8. Centrifuge at 180 × *g* for 5 min. Remove supernatant and resuspend pellet in 500 μL of Carnoy's Fixative. Cells can be stored up to several months at 4 °C at this point.
9. Remove the top of the p1000 tip box and place the slide against one side of the box so that the slide is placed at an angle.
10. Hold a glass pasture pipet approximately 6 in. above the slide and release three drops of cell suspension on the glass slides at three different areas of the slide (*see* **Note 33**).
11. Place the slides in the chemical hood and allow it to dry (*see* **Note 34**).
12. Once the slide is dry, place the slides on a staining rack over the sink. Add the Giemsa staining solution on in the staining rack on top of the slides and stain for 5 min.
13. Wash the slides with distilled water, drain and allow the slides to air dry.
14. Add mounting media to the cover slips and place it on top of the slides. Put four drops of a clear nail polish on the corners of

the coverslips and wait for 2 min to dry. Then add more nail polish on to the edges to seal the coverslip to the slides. Slides can be stored at −80 °C in slide box for up to 4 months.

15. Take pictures with a phase contrast microscope using 100× magnification. To obtain enough analyzable metaphase spread images, prepare at least eight glass slides of spreads for each data. Count at least 20 best metaphase spreads. Good spreads are the ones where chromosome spread is from a single cell. Calculate the individual chromosomes and ignore the ones where spreads from two chromosomes overlap. An individual chromosome is the one that does not overlap upon itself or onto the neighboring chromosomes. Both numerical and structural aberrations can be detected by this method.

3.7 Double-Strand Break HR Reporter Assay

1. Maintain U2-OS cells containing the stably integrated DR-GFP reporter in 5% CO_2 atmosphere at 37 °C in a humidifier incubator in DMEM media. Grow cells in a T-25 flask and split cells after every 3 days.
2. When cells are around 70–80% confluent, transfect them using Amaxa Nucleofector per manufacturer's instructions. In short, first trypsinize the cells and resuspend them in 100 μL of 1× transfection reaction solution. For each transfection add 18 μL of the supplement to 82 μL of Nucleofector Solution to make 100 μL of 1× transfection reaction solution. To the 1× transfection reaction add 2 μg of I-SceI expressing plasmid and 100 ng of ds-Red plasmid (*see* **Note 35**).
3. Transfer cell suspension into certified cuvette. All the sample must cover the bottom of the cuvette. Make sure there are no air bubbles. Close the cuvette with the cap.
4. Insert the cuvette with cell suspension into the Nucleofector cuvette holder and start the program number specific for U2-OS cells (X-001).
5. Take the cuvette out of the holder once the program is finished. Immediately add ~500 μL of warmed culture medium to the cuvette and gently transfer the sample into the 6-well plate containing 2 mL of DMEM media.
6. After 18–24 h, aspirate the media, wash cells with 1× PBS, and add fresh media. 72 h after transfection harvest the I-SceI and ds-Red cotransfected cells and the control cells. Control cells must include (1) cells transfected without I-SceI and ds-Red plasmids, (2) cells transfected with only I-SceI plasmid, and (3) cells transfected with only ds-Red plasmid.
7. Resuspend the cells in 300–500 μL of 1× PBS and transfer the mixture in the FACS tubes. Analyze the cells immediately after harvesting. Calibrate FACS by analyzing the cells transfected

with only I-SceI plasmid, only ds-Red plasmid, and without I-SceI/ds-Red plasmids (*see* **Note 36**).

8. The percentage of GFP-positive cells represents the population of cells with accurate DSB repair and the ds-Red positive cells indicate the transfection efficiency. The relative efficiency of DSB repair is calculated as the ratio of the GFP positive cells to ds-Red positive cells (*see* **Note 37**).

4 Notes

1. Freshly made crystal violet stain solution can be stored in the dark at room temperature and used for up to 2 months.
2. The culture conditions for each cell line should be optimized as to provide the cells with sufficient surface area to grow without being confluent thereby avoiding contact mediated inhibition of growth.
3. To study the effect of DNA damaging agents on cellular proliferation, the culture conditions, including the concentration of the DNA damaging agents and duration of treatment, should be optimized according to cell type.
4. It is recommended to have "control" wells where the cells are only treated with the solvent that the DNA damaging drug is dissolved in. The number of cells in the "control" wells can be utilized to get the fold decrease or percentage decrease as a function of drug concentration.
5. Gentle washing of the cells with PBS is recommended. If cells are detached during washing, the PBS supernatant should be saved and centrifuged to recollect any detached cells.
6. The aperture should be free of any debris. To ensure this, the aperture should be washed with distilled water, "flushed" with the isotonic solution, and cleaned gently with a soft bristle brush before and in between measurements.
7. Since 500 μL of cell suspension was diluted with 10 mL of the isotonic diluent, a dilution factor of 21 is used. However, the cells can be further diluted and the dilution factor changed accordingly within the total volume ranging from 10 to 15 mL.
8. The absorbance value from the control cells should be used to quantify cell proliferation with drug treatment, as a fold change or percentage of the control value.
9. The duration of the incubation can be adjusted based on the seeding density, type of cell and kind of treatment.
10. Extended exposure of WST-1 to light should be avoided as it may cause increased reduction in WST-1 and increase the background absorbance readings.

11. The volume of media usually decreases around 10% over an incubation period of 3–5 days in a 96-well plate. Adjust the volume of WST-1 added according to the volume of medium in well.
12. The "blank" value is the background absorbance OD value and it generally increases with the incubation time. The blank value should be deducted from all readings while quantifying cell proliferation data.
13. Avoid forming bubbles while mixing the media. Introduction of air bubbles to the sample will lead to incorrect absorbance values.
14. Avoid exceeding the sensitivity of the microplate spectrophotometer reader (typically an optical density (OD) value range of 0–3). If the OD value exceeds the upper limit of the spectrophotometer, the number of cells seeded can be reduced and/or the incubation period minimized.
15. The number of cells can be adjusted according to the size of the plate and kind of treatment. Pilot experiments should be performed using serially diluted cells to determine the correct number of cells that allows the counting separate colonies without any overlap.
16. The washing and adding of methanol must be done carefully as colonies can detach from the plates. Tilting the tissue culture plate while washing and adding methanol can minimize the risk of dislodging the cells from the plate.
17. The colonies are now fixed and gently washing them with water will not dislodge the colonies.
18. Plates can be stored at room temperature for up to 50 weeks.
19. A cell culture control is recommended to measure the level of apoptosis in untreated cells that are seeded in a manner similar to those same cells exposed to a DNA damaging drug treatment. This control value will serve as a measurement of spontaneous apoptosis under the culture conditions used without treatment. Further, the "control" absorbance value can be used to normalize the absorbance values from cells treated with different concentrations of the drug.
20. Coating solution is prepared by diluting 10× Coating buffer to 1× with ddH_2O and adding 1 mL of reconstituted anti-histone antibody to 9 mL of the 1× coating buffer. This should be prepared freshly before use.
21. The anti-histone antibody is biotin conjugated that in turn binds to the streptavidin coated microplate module.
22. Reconstitute the washing buffer from 10× to 1× by diluting with ddH_2O at room temperature.

23. The wells should be washed carefully but thoroughly to avoid disturbing the coated surface of the microplate modules.
24. The ABTS substrate is prepared by dissolving 1× 5 mg tablet in 5 mL of the substrate buffer at room temperature.
25. To obtain an optimal number of foci, we recommend standardizing the dose of DNA damaging agent and the treatment conditions by doing a pilot study prior to an actual experiment.
26. Some of the cell lines detach easily. We recommend adding the extraction and the washing buffers very gently and to check under microscope that the cells are not detaching.
27. Washes with the extraction buffer (**steps 4** and **6**) are required to obtain distinct ATR, ATRIP, Rad9, and RPA foci, and reduce background staining. This step is optional for some of the DNA Damage response proteins like γH2AX.
28. Longer duration of washes after the addition of the primary and secondary antibodies may help to reduce nonspecific signals.
29. This protocol can also be slightly modified to detect single stranded DNA by BrdU incorporation. Treat cells with 20 μM BrdU for 24 h. Repeat the steps exactly as mentioned above. Add anti-BrdU primary antibody (1:100, BD Pharmingen™) and goat anti-mouse secondary (1:1000 Invitrogen).
30. Incubation time with colcemid is crucial. Longer exposure time can result in shorter and thicker chromosomes and shorter exposure can result in fewer metaphase spreads.
31. Molarity of hypotonic solution is very important. It must be sufficient to swell the cells without lysing them.
32. Add the Carnoy's Fixative solution to the pellet and vortex.
33. Drop the cells in the fixative agent onto the slides from a height of 6 in. and make sure that the slide is at a 45° angle.
34. Place the slides in 50% humidity and at 25 °C for quicker drying.
35. The quality of plasmid may affect the transfection efficiency. To obtain accuracy and consistency try to use same plasmid mix for one whole experiment.
36. Count at least 30,000 cells for each treatment group.
37. This protocol can be slightly modified to examine the effect of a gene product on HR repair. Cells can be treated with inhibitors/siRNA against a particular protein/gene before transfection with I-SceI and ds-Red plasmids.

Acknowledgments

This work is supported by the National Institutes of Health, National Institute on Aging.

References

1. Estep KN, Brosh RM Jr (2018) RecQ and Fe-S helicases have unique roles in DNA metabolism dictated by their unwinding directionality, substrate specificity, and protein interactions. Biochem Soc Trans 46:77–95. https://doi.org/10.1042/bst20170044
2. Suhasini AN, Brosh RM Jr (2013) Disease-causing missense mutations in human DNA helicase disorders. Mutat Res 752(2):138–152. https://doi.org/10.1016/j.mrrev.2012.12.004
3. van Brabant AJ, Stan R, Ellis NA (2000) DNA helicases, genomic instability, and human genetic disease. Annu Rev Genomics Hum Genet 1:409–459. https://doi.org/10.1146/annurev.genom.1.1.409
4. Byrd AK, Raney KD (2012) Superfamily 2 helicases. Front Biosci (Landmark Ed) 17:2070–2088
5. Gilman B, Tijerina P, Russell R (2017) Distinct RNA-unwinding mechanisms of DEAD-box and DEAH-box RNA helicase proteins in remodeling structured RNAs and RNPs. Biochem Soc Trans 45(6):1313–1321. https://doi.org/10.1042/bst20170095
6. Raney KD, Byrd AK, Aarattuthodiyil S (2013) Structure and mechanisms of SF1 DNA helicases. Adv Exp Med Biol 767:17–46. https://doi.org/10.1007/978-1-4614-5037-5_2
7. Trakselis MA (2016) Structural mechanisms of hexameric helicase loading, assembly, and unwinding. F1000Res 5. F1000 Faculty Rev-111. https://doi.org/10.12688/f1000research.7509.1
8. Brosh RM Jr, Majumdar A, Desai S, Hickson ID, Bohr VA, Seidman MM (2001) Unwinding of a DNA triple helix by the Werner and Bloom syndrome helicases. J Biol Chem 276(5):3024–3030. https://doi.org/10.1074/jbc.M006784200
9. Guo M, Hundseth K, Ding H, Vidhyasagar V, Inoue A, Nguyen CH, Zain R, Lee JS, Wu Y (2015) A distinct triplex DNA unwinding activity of ChlR1 helicase. J Biol Chem 290(8):5174–5189. https://doi.org/10.1074/jbc.M114.634923
10. Mohaghegh P, Karow JK, Brosh RM Jr, Bohr VA, Hickson ID (2001) The Bloom's and Werner's syndrome proteins are DNA structure-specific helicases. Nucleic Acids Res 29(13):2843–2849
11. Sun H, Karow JK, Hickson ID, Maizels N (1998) The Bloom's syndrome helicase unwinds G4 DNA. J Biol Chem 273(42):27587–27592
12. Vaughn JP, Creacy SD, Routh ED, Joyner-Butt C, Jenkins GS, Pauli S, Nagamine Y, Akman SA (2005) The DEXH protein product of the DHX36 gene is the major source of tetramolecular quadruplex G4-DNA resolving activity in HeLa cell lysates. J Biol Chem 280(46):38117–38120. https://doi.org/10.1074/jbc.C500348200
13. Bacolla A, Wang G, Jain A, Chuzhanova NA, Cer RZ, Collins JR, Cooper DN, Bohr VA, Vasquez KM (2011) Non-B DNA-forming sequences and WRN deficiency independently increase the frequency of base substitution in human cells. J Biol Chem 286(12):10017–10026. https://doi.org/10.1074/jbc.M110.176636
14. Sharma S, Doherty Kevin M, Brosh Robert M (2006) Mechanisms of RecQ helicases in pathways of DNA metabolism and maintenance of genomic stability. Biochem J 398(Pt 3):319–337. https://doi.org/10.1042/BJ20060450
15. Wu Y (2012) Unwinding and rewinding: double faces of helicase? J Nucl Acids 2012:140601. https://doi.org/10.1155/2012/140601
16. Brosh RM (2013) DNA helicases involved in DNA repair and their roles in cancer. Nat Rev Cancer 13(8):542–558. https://doi.org/10.1038/nrc3560
17. Sharma S (2014) An appraisal of RECQ1 expression in cancer progression. Front Genet 5:426. https://doi.org/10.3389/fgene.2014.00426
18. Arai A, Chano T, Futami K, Furuichi Y, Ikebuchi K, Inui T, Tameno H, Ochi Y, Shimada T, Hisa Y, Okabe H (2011) RECQL1 and WRN proteins are potential therapeutic targets in head and neck squamous cell carcinoma. Cancer Res 71(13):4598–4607. https://doi.org/10.1158/0008-5472.can-11-0320

19. Sharma S, Brosh RM Jr (2007) Human RECQ1 is a DNA damage responsive protein required for genotoxic stress resistance and suppression of sister chromatid exchanges. PLoS One 2(12):e1297. https://doi.org/10.1371/journal.pone.0001297
20. Aggarwal M, Banerjee T, Sommers JA, Brosh RM Jr (2013) Targeting an Achilles' heel of cancer with a WRN helicase inhibitor. Cell Cycle 12(20):3329–3335. https://doi.org/10.4161/cc.26320
21. Aggarwal M, Banerjee T, Sommers JA, Iannascoli C, Pichierri P, Shoemaker RH, Brosh RM Jr (2013) Werner syndrome helicase has a critical role in DNA damage responses in the absence of a functional Fanconi anemia pathway. Cancer Res 73(17):5497–5507. https://doi.org/10.1158/0008-5472.can-12-2975
22. Aggarwal M, Sommers JA, Shoemaker RH, Brosh RM Jr (2011) Inhibition of helicase activity by a small molecule impairs Werner syndrome helicase (WRN) function in the cellular response to DNA damage or replication stress. Proc Natl Acad Sci U S A 108 (4):1525–1530. https://doi.org/10.1073/pnas.1006423108
23. Nguyen GH, Dexheimer TS, Rosenthal AS, Chu WK, Singh DK, Mosedale G, Bachrati CZ, Schultz L, Sakurai M, Savitsky P, Abu M, McHugh PJ, Bohr VA, Harris CC, Jadhav A, Gileadi O, Maloney DJ, Simeonov A, Hickson ID (2013) A small molecule inhibitor of the BLM helicase modulates chromosome stability in human cells. Chem Biol 20(1):55–62. https://doi.org/10.1016/j.chembiol.2012.10.016
24. Estep KN, Butler TJ, Ding J, Brosh RM Jr (2017) G4-interacting DNA helicases and polymerases: potential therapeutic targets. Curr Med Chem. https://doi.org/10.2174/0929867324666171116123345
25. Hengel SR, Spies MA, Spies M (2017) Small-molecule inhibitors targeting DNA repair and DNA repair deficiency in research and cancer therapy. Cell Chem Biol 24(9):1101–1119. https://doi.org/10.1016/j.chembiol.2017.08.027
26. Brosh RM Jr, Bohr VA (2007) Human premature aging, DNA repair and RecQ helicases. Nucleic Acids Res 35(22):7527–7544. https://doi.org/10.1093/nar/gkm1008
27. Pierce AJ, Johnson RD, Thompson LH, Jasin M (1999) XRCC3 promotes homology-directed repair of DNA damage in mammalian cells. Genes Dev 13(20):2633–2638
28. Berridge MV, Herst PM, Tan AS (2005) Tetrazolium dyes as tools in cell biology: new insights into their cellular reduction. Biotechnol Annu Rev 11:127–152. https://doi.org/10.1016/s1387-2656(05)11004-7
29. Wu Y, Shin-ya K, Brosh RM Jr (2008) FANCJ helicase defective in Fanconia anemia and breast cancer unwinds G-quadruplex DNA to defend genomic stability. Mol Cell Biol 28 (12):4116–4128. https://doi.org/10.1128/mcb.02210-07
30. Khodarev NN, Sokolova IA, Vaughan AT (1998) Mechanisms of induction of apoptotic DNA fragmentation. Int J Radiat Biol 73 (5):455–467
31. Bennett BT, Bewersdorf J, Knight KL (2009) Immunofluorescence imaging of DNA damage response proteins: optimizing protocols for super-resolution microscopy. Methods 48 (1):63–71. https://doi.org/10.1016/j.ymeth.2009.02.009
32. Awasthi P, Foiani M, Kumar A (2015) ATM and ATR signaling at a glance. J Cell Sci 128 (23):4255–4262. https://doi.org/10.1242/jcs.169730
33. Mirzoeva OK, Petrini JH (2001) DNA damage-dependent nuclear dynamics of the Mre11 complex. Mol Cell Biol 21 (1):281–288. https://doi.org/10.1128/MCB.21.1.281-288.2001
34. Suhasini AN, Sommers JA, Muniandy PA, Coulombe Y, Cantor SB, Masson JY, Seidman MM, Brosh RM Jr (2013) Fanconi anemia group J helicase and MRE11 nuclease interact to facilitate the DNA damage response. Mol Cell Biol 33(11):2212–2227. https://doi.org/10.1128/mcb.01256-12
35. Litman R, Peng M, Jin Z, Zhang F, Zhang J, Powell S, Andreassen PR, Cantor SB (2005) BACH1 is critical for homologous recombination and appears to be the Fanconi anemia gene product FANCJ. Cancer Cell 8(3):255–265. https://doi.org/10.1016/j.ccr.2005.08.004
36. Bharti SK, Sommers JA, George F, Kuper J, Hamon F, Shin-ya K, Teulade-Fichou MP, Kisker C, Brosh RM Jr (2013) Specialization among iron-sulfur cluster helicases to resolve G-quadruplex DNA structures that threaten genomic stability. J Biol Chem 288:28217–28229
37. Henderson A, Wu Y, Huang YC, Chavez EA, Platt J, Johnson FB, Brosh RM Jr, Sen D, Lansdorp PM (2014) Detection of G-quadruplex DNA in mammalian cells. Nucleic Acids Res 42:860–869
38. Wu Y, Sommers JA, Suhasini AN, Leonard T, Deakyne JS, Mazin AV, Shin-Ya K, Kitao H, Brosh RM Jr (2010) Fanconi anemia group J mutation abolishes its DNA repair function by

uncoupling DNA translocation from helicase activity or disruption of protein-DNA complexes. Blood 116:3780–3791

39. Banerjee T, Sommers JA, Huang J, Seidman MM, Brosh RM Jr (2015) Catalytic strand separation by RECQ1 is required for RPA-mediated response to replication stress. Curr Biol 25:2830–2838
40. Constantinou A, Tarsounas M, Karow JK, Brosh RM, Bohr VA, Hickson ID, West SC (2000) Werner's syndrome protein (WRN) migrates Holliday junctions and co-localizes with RPA upon replication arrest. EMBO Rep 1:80–84
41. Sharma S, Otterlei M, Sommers JA, Driscoll HC, Dianov GL, Kao HI, Bambara RA, Brosh RM Jr (2004) WRN helicase and FEN-1 form a complex upon replication arrest and together process branchmigrating DNA structures associated with the replication fork. Mol Biol Cell 15:734–750
42. Suhasini AN, Rawtani NA, Wu Y, Sommers JA, Sharma S, Mosedale G, North PS, Cantor SB, Hickson ID, Brosh RM Jr (2011) Interaction between the helicases genetically linked to Fanconi anemia group J and Bloom's syndrome. EMBO J 30:692–705
43. Gupta R, Sharma S, Sommers JA, Kenny MK, Cantor SB, Brosh RM Jr (2007) FANCJ (BACH1) helicase forms DNA damage inducible foci with replication protein A and interacts physically and functionally with the single-stranded DNA-binding protein. Blood 110:2390–2398

Chapter 12

A Mammalian Genetic Complementation Assay for Assessing Cellular Resistance to Genotoxic Compounds

Nicole M. Reilly and Douglas L. Pittman

Abstract

A complementation assay was developed to determine whether alleles of DNA repair genes are necessary for repairing specific types of damage. The assay was established by measuring the resistance capacity of *Rad51d*-deficient mouse embryonic fibroblasts (MEFs) transfected with mammalian expression constructs. Here, we describe the methods used to assess colony survival following the treatment of transfected cells with genotoxic compounds. This approach provides a time-efficient and stringent strategy to screen genetic alleles for identifying regions or specific amino acid residues critical for function or regulation of DNA repair pathways.

Key words Complementation assay, Transfection, Antibiotic resistance, Interstrand cross-links, Mitomycin C, RAD51D, Homologous recombination

1 Introduction

Traditionally, stable expression of constructs in mammalian cells has been used to demonstrate cellular resistance to genotoxic compounds [1]. Although a useful technique, establishing stable cell lines that express a construct of interest or genome editing using CRISPR/Cas9 approaches can be time-consuming and tedious. With the method described herein, it is possible to assess cellular resistance to compounds in the short-term using transient transfection methodology [2]. The MTT (3-(4,5-dimethylthiazolyl-2)-2,5-diphenyltetrazolium bromide) tetrazolium reduction assay is an alternative short-term cytotoxicity assay that assesses cell viability by measuring cellular metabolic activity [3]. While useful, this assay does not directly measure proliferation and results can vary depending on the cell line and cytotoxic agent used, thus making it difficult to measure differences depending upon the forms of DNA damage [4]. Quantitation of the MTT assay results also requires the use of a spectrophotometer. In contrast, the complementation assay allows for the measurement of cell proliferation through the direct

Lata Balakrishnan and Jason A. Stewart (eds.), *DNA Repair: Methods and Protocols*, Methods in Molecular Biology, vol. 1999, https://doi.org/10.1007/978-1-4939-9500-4_12,

visualization of colony formation. It provides a stringent functional test of individual alleles and may also be useful for testing multiple allelic combinations.

2 Materials

2.1 Plastic Materials

1. Sterile 6-well plates.
2. Sterile 150 mm dishes (two dishes per plasmid).

2.2 Mammalian Cells

1. Cells should be deficient, preferably a null-allele, for the specific DNA repair gene of interest (*see* **Note 1**).

2.3 Transfection Reagents

1. Plasmid DNA (*see* **Note 2**). An "empty vector" plasmid should be included as a negative control. A plasmid encoding a fully functional protein of interest should be included as a positive control. All plasmids must contain a resistance gene specific to an antibiotic that can be used with mammalian cells (*see* **Note 3**).
2. Lipofectamine LTX with Plus transfection reagent (Thermo Fisher Scientific) or optimized transfection protocol for the cell line and gene expression systems being used [5, 6].
3. Opti-MEM I reduced serum medium (Thermo Fisher Scientific) (*see* **Note 4**).

2.4 Cell Culture Solutions

1. Phosphate buffered saline (PBS; potassium phosphate monobasic, sodium chloride, sodium phosphate dibasic).
2. 0.05% trypsin–EDTA.
3. Cell culture medium with serum and antibiotics.

2.5 Treatment Solutions

1. "Hygromycin B Only": Transfer the total volume of appropriate cell culture medium needed to a clean glass bottle (*see* **Note 5**). Add Hygromycin B to a final concentration of 200 μg/mL. Prepare solution fresh prior for each treatment.
2. "Hygromycin B + MMC": Transfer total volume of medium needed to a clean glass bottle. Add Hygromycin B to a final concentration of 200 μg/mL. Add mitomycin C (MMC) to a final concentration of 4 ng/mL (*see* **Note 6**). Prepare solution fresh prior to each treatment.

2.6 Colony Staining

1. 100% ice-cold methanol.
2. Giemsa: Using a 25-mL serological pipette, transfer 20 mL Giemsa to a clean 1 L glass bottle. Add distilled water to a final volume of 1 L. Store at room temperature.

3 Methods

This method is designed for adherent cell lines, and an overview is provided in Fig. 1. Subheadings 3.1–3.5 should be performed under a laminar flow hood certified for mammalian cell culture, and Subheading 3.6 should be performed under a chemical fume hood. The transfection protocol (Subheading 3.2) described is specific to the Lipofectamine LTX with Plus reagent and can be adjusted based on the transfection procedure used.

3.1 Plate Cells in Preparation for Transfection (Day−3)

1. Plate cells in a 6-well plate (one well per plasmid) in appropriate medium containing serum. The density should be sufficient so that the cells are 80% confluent after 24 h. Incubate at 37 °C and 5% CO_2 for 24 h.

3.2 Plasmid Transfection (Day−2)

1. Prepare the DNA + Plus solution: Add 125 μL (microliters) Opti-MEM reduced serum medium to a clean 1.5 mL centrifuge tube (one tube per plasmid). Add 2 μg (micrograms) of the appropriate plasmid DNA to each tube. Add 2 μL Plus reagent to each tube. Incubate at room temperature for 15 min.
2. Dilute the Lipofectamine LTX in Opti-MEM reduced serum medium: Add 125 μL per "number of wells + 1" to a clean 1.5 mL tube (*see* **Note** 7). Add 6 μL Lipofectamine LTX per "number of plasmids + 1" to the tube containing the medium. Incubate at room temperature for 15 min.
3. After 15 min, add 125 μL of "Lipofectamine LTX in Opti-MEM" solution to each tube containing the plasmid DNA. Incubate for 15 min at room temperature.

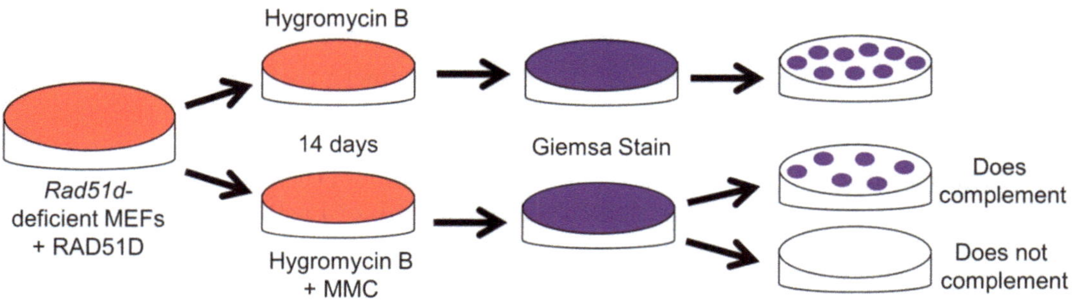

Fig. 1 Workflow of the complementation assay. A plasmid expressing wild-type RAD51D is transfected into *Rad51d*-deficient mouse embryonic fibroblasts (MEFs). The cells are maintained for 14 days under selection with Hygromycin B alone or Hygromycin B with the genotoxic agent mitomycin C (MMC). At the end of the treatment period, cells are fixed and stained with Giemsa. Colony formation following Hygromycin B + MMC indicates that the expressed construct complements *Rad51d* deficiency, while the absence of colonies indicates the allele is necessary to repair MMC-induced DNA damage (no complementation)

4. After 15 min, add 250 μL of the "DNA + Plus + Lipofectamine LTX" solution to the appropriate well ***dropwise***. Add the drops evenly across the well to ensure that all of the cells are exposed to the transfection solution and swirl the plates to ensure even distribution of the transfection reagent.
5. Incubate at 37 °C and 5% CO_2 for 24 h.

3.3 Splitting Cells into 150 mm Dishes (Day−1)

1. Prepare 150 mm dishes by adding 14 mL cell culture medium to each dish (two dishes per plasmid).
2. Wash transfected cells twice with 1× PBS.
3. Add 500 μL 0.01% trypsin–EDTA to each well and incubate at 37 °C and 5% CO_2 until all cells have detached from the plate (typically 3–5 min).
4. Add 1.5 mL cell culture medium to each well and resuspend cells by pipetting.
5. Transfer 1 mL cell suspension to two 150 mm dishes. Gently swirl the plate to ensure even distribution of the cells. Label one plate "Set A" and label the other plate "Set B".
6. Incubate at 37 °C and 5% CO_2 for 24 h.

3.4 Initiation of Drug Treatment (Day 0)

1. Prepare the "Hygromycin B Only" and "Hygromycin B + MMC" solutions.
2. For the "Set A" plate, remove the medium and replace with 15 mL "Hygromycin B Only" medium.

   For the "Set B" plate, remove the medium and replace with 15 mL "Hygromycin B + MMC" medium.
3. Incubate at 37 °C and 5% CO_2 for 3 days (approximately 72 h).

3.5 Maintaining Selection and Drug Treatment (Days 3, 7, 11)

1. Every 3 days (approximately 72 h), remove medium on the plates in Set A and replace with "Hygromycin B Only" medium. Remove the medium on the plates in Set B and replace with "Hygromycin B + MMC" medium (*see* **Note 8**).
2. The selection and drug treatment are maintained for 14 days (*see* **Note 9**).

3.6 Giemsa Staining

1. At the end of the treatment time (Day 14), remove the medium and wash the cells with 10 mL of 1× PBS.
2. Add 10 mL of *ice-cold (100%)* methanol slowly to each dish to fix the cells (*see* **Note 10**). Make sure that the methanol covers the bottom of the dish. Let sit for 20 min at room temperature (*see* **Note 11**).
3. Remove the methanol and place the dish upside down on a paper towel to air dry.

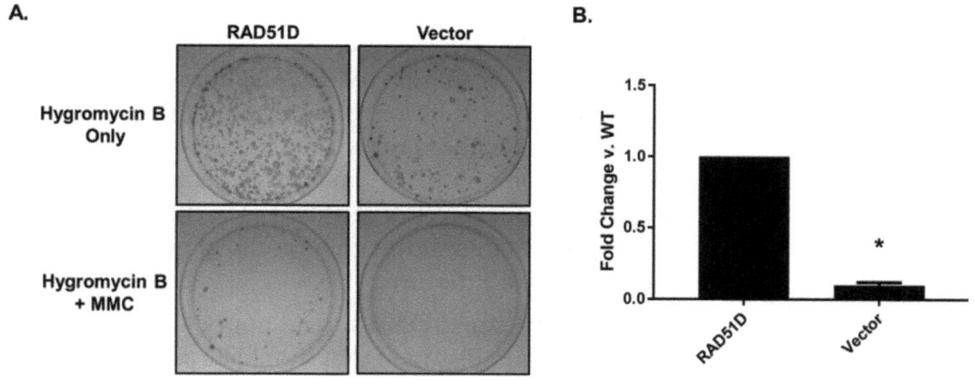

Fig. 2 Complementation screening in *Rad51d-deficient* mouse embryonic fibroblasts. (**a**) Representative Giemsa stained plates. Equally divided subpopulations of *Rad51d*-deficient fibroblasts transfected with either a plasmid encoding *RAD51D* or a vector (pcDNA3.1/Hygro) plasmid were selected with 200 μg/mL Hygromycin B either in the presence (bottom row) or absence (top row) of 4 ng/mL mitomycin C (MMC). (**b**) The fold change compared to wild-type of MMC resistant colonies is displayed. The data presented are from three independent experiments. Error bars represent standard error of the mean and * indicates $p < 0.05$ calculated using a one-way ANOVA

4. Add 10 mL Giemsa stain solution to each dish. Make sure that the solution covers the bottom of the dish. Let sit for at least 30 min at room temperature (*see* **Note 12**).
5. Remove the Giemsa stain. Wash each dish twice with 10–15 mL single distilled water.
6. Place the dish upside down on a paper towel and allow to air dry overnight.

3.7 Colony Counting

1. Most colonies can be counted by eye or, if necessary, using a stereo dissecting microscope. One colony is greater than or equal to 50 cells. An example of expected results is shown in Fig. 2.

3.8 Calculating Percent Resistance

1. Calculate the percent resistance using the following equation:

$$\text{Percent resistance} = \frac{\text{number of colonies on Hygromycin B} + \text{MMC plate}}{\text{number of colonies on Hygromycin B Only plate}} \times 100$$

4 Notes

1. All materials are sterile and opened only under a BSL2+ certified laminar flow hood approved for mammalian cell culture work [7] being careful to follow proper regulations when disposing of waste materials. For transfections, DNA prepared

using midi or maxi prep kits is optimal (i.e., Qiagen Plasmid Midi Kit). The genotoxic compound (mitomycin C) at the indicated dose is specific to *Rad51d*-deficient mouse embryonic fibroblast cells. Prior to performing this assay, it is necessary to establish the sensitivity of the desired cell line to the genotoxic compound of interest by colony forming units [8]. The optimal level of sensitivity for the complementation assay is the minimum dose that results in near 100% cell death after 14 days of exposure.

2. Low-quality plasmid DNA should not be used for mammalian cell transfection. Plasmid DNA should be prepared using a midi or maxi prep kit that reduces endotoxins and is appropriate for transfection.
3. Hygromycin B and puromycin are examples of antibiotics that can be used on mammalian cells. Here, we describe the use of a plasmid with Hygromycin B resistance.
4. The serum can potentially affect the formation of the DNA/reagent complex; therefore, reduced serum medium is recommended for this step.
5. It is recommended that the total medium volume for 150 mm dishes be equivalent to 15 mL × (number of dishes + 1).
6. For this assay, the concentration of drug used should be the lowest possible to achieve 100% cell death in the absence of the gene of interest.
7. Always add an extra volume of reagent. For example, if six plasmids are being transfected, the volume of Opti-MEM medium prepared should be 125 μL × (6 plasmids + 1) = 875 μL medium.
8. It is necessary to make the "Hygromycin B Only" and "Hygromycin B + MMC" solutions immediately before adding the medium to the appropriate plates.
9. If no colonies are detected on the negative control (empty vector) plate in "Hygromycin B + MMC" by Day 11 of treatment, the medium on all plates can be replaced with regular medium (no antibiotic or drug). This will allow any cell colonies that have formed to grow and be detectable on Day 14.
10. All subsequent steps should be performed in a certified chemical fume hood, and not in a cell culture hood.
11. The fixation step can be performed overnight at room temperature, if needed.
12. The Giemsa staining step can be performed overnight at room temperature, if needed.

Acknowledgments

This work was supported in part by a grant from the American Cancer Society (RSG-03-158-01-GMC), a Helen and Harold McMaster Endowment, and the National Institute of General Medical Sciences of the National Institutes of Health Award Number R15 GM110615. The content is solely the responsibility of the authors and does not necessarily represent the official views of the National Institutes of Health. Additional support was provided to NMR from a SPARC Graduate Research Grant from the Office of the Vice President for Research at the University of South Carolina.

References

1. Condreay JP, Witherspoon SM, Clay WC, Kost TA (1999) Transient and stable gene expression in mammalian cells transduced with a recombinant baculovirus vector. Proc Natl Acad Sci U S A 96(1):127–132
2. Gruver AM, Miller KA, Rajesh C, Smiraldo PG, Kaliyaperumal S, Balder R, Stiles KM, Albala JS, Pittman DL (2005) The ATPase motif in RAD51D is required for resistance to DNA interstrand crosslinking agents and interaction with RAD51C. Mutagenesis 20(6):433–440
3. Mosmann T (1983) Rapid colorimetric assay for cellular growth and survival: application to proliferation and cytotoxicity assays. J Immunol Methods 65(1–2):55–63
4. Stepanenko AA, Dmitrenko VV (2015) Pitfalls of the MTT assay: direct and off-target effects of inhibitors can result in over/underestimation of cell viability. Gene 574(2):193–203
5. Longo PA, Kavran JM, Kim MS, Leahy DJ (2013) Transient mammalian cell transfection with polyethylenimine (PEI). Methods Enzymol 529:227–240
6. Dyson MR (2016) Fundamentals of expression in mammalian cells. Adv Exp Med Biol 896:217–224
7. Phelan K, May KM (2017) Mammalian cell tissue culture. Curr Protoc Hum Genet 94: A.3g.1–A.3g.22
8. Pochampally R (2008) Colony forming unit assays for MSCs. Methods Mol Biol 449:83–91

Chapter 13

Small-Molecule Inhibitor Screen for DNA Repair Proteins

John J. Turchi and Pamela S. VanderVere-Carozza

Abstract

With the recent interest in targeting the DNA damage response (DDR) and DNA repair, new screening methodologies are needed to broaden the scope of targetable proteins beyond kinases and traditional enzymes. Many of the proteins involved in the DDR and repair impart their activity by making specific contacts with DNA. These protein–nucleic acid interactions represent a tractable target for perturbation with small molecules. We describe a high throughput, solution-based equilibrium binding fluorescence polarization assay that can be applied to a wide array of protein–nucleic acid interactions. The assay is sensitive, stable, and able to identify small molecules capable of blocking DNA–protein interactions.

Key words DNA damage response, DNA repair, High-throughput screening, Fluorescence polarization, Nucleic acid enzymology

1 Introduction

Numerous DNA repair pathways and the DNA damage response pathway are tractable targets in the therapeutic intervention in the treatment of cancer [1]. The development of poly-ADP-ribose polymerase (PARP) inhibitors and their approval for the treatment of ovarian and breast cancer has restimulated interest in targeting pathways for detecting, signaling and repairing DNA damage [2]. These pathways can be targeted in the context of synthetic lethal interactions to provide a therapeutic window and to increase the efficacy of traditional chemotherapeutics whose mechanism of action involves the formation of DNA damage [3]. The majority of DDR and DNA repair targets to date have revolved around serine–-threonine kinases including ATM, ATR, DNA-PK, WEE1, and CHK1/2 [4]. Advances in the development of receptor tyrosine kinases and other growth signaling kinases have supported these efforts [5]. Only a small percentage of the proteins involved in the DDR and repair are kinases whereas many proteins impart their activity via binding to DNA. The difficulties in targeting a protein–DNA interaction lie in the relatively large areas that can be in

Lata Balakrishnan and Jason A. Stewart (eds.), *DNA Repair: Methods and Protocols*, Methods in Molecular Biology, vol. 1999, https://doi.org/10.1007/978-1-4939-9500-4_13,

contact, the potential for multiple independent interactions, and the hydrophobic nature of the interactions such that compounds identified are often poorly soluble. Advances in medicinal chemistry and the expanding array of modular precursors, pharmacophores, and bioisosteres can alleviate many of these issues. False positives remain a consistent challenge with all high throughput assays. A number of assays have been established to identify inhibitors of DNA repair, replication or recombination of which we were the first [6–11]. One advantage of a biochemical assay lies in the limited number of variables and potential interaction that could result in a signal but not necessarily by the intended mechanism. In a HTS for a DNA--protein interaction the biggest concern is the identification of compounds that inhibit the protein–DNA interaction by binding to the DNA and not the target protein. This issue is addressed by either prescreening the library or screening hits for a direct DNA interaction. An alternative is to screen the selected library against two individual protein targets. Any compound that scores positive in both screens would be deemed nonspecific, potentially via an interaction with DNA. Fluorescence polarization or anisotropy assays are based on the differential rotation of a fluorescent molecule as a function of Stokes radius [12]. When excited with plane polarized light, a static fluorescent molecule will emit plane polarized light while a highly mobile, rapidly rotating molecule will emit depolarized light. The binding of a large molecule to the fluorescent molecule will slow the rotation and movement and reduce the depolarization. As such, the FP assay represents a solution-based equilibrium-binding assay. The advantage is also that the assay can be set up as a mix and read assay minimizing the processing steps that can decrease robustness and increase the costs of the assay [12].

2 Materials

All reagents were of analytic or molecular biology grade purchased from standard suppliers.

2.1 Buffers

1. 10× Reaction Buffer: 30 mM HEPES pH 7.0, 0.2% NP40. Diluted tenfold prior to use.
2. HI Buffer: 30 mM HEPES, 0.25% myoinositol, 1 mM DTT, 0.25 mM EDTA, 0.02% NP-40.

2.2 DNA Substrates

1. Single stranded DNA oligonucleotides were purchased from IDT. The F indicates a 5′-fluorescein modification.

   JT/ILH 2.0 5′ F-$(T)_{12}$.

   JT/ILH 2.1 5′ CCCCTATCCTTTCC**GCG**TCCTTACTT CCCC 3′.

   JT/ILH 2.2 5′ F-GGGGAAGTAAGGACGCGGAAAGGAT AGGGG 3′.

2.3 Target Proteins

1. Recombinant human RPA heterotrimer was purified as previously described [13] and diluted in HI prior to use.
2. Recombinant human XPA was purified as previously described [14] and diluted in HI prior to use.
3. Compound library: Mother plates: 10 mM in DMSO stored at −80 °C in 384 well plates.

3 Methods

3.1 DNA Substrates

The DNA substrate to assess XPA binding is a duplex 30-mer containing a single cisplatin 1,3-d(GpCpG) adduct. The substrate is prepared as previously described [13]. Briefly, the JT/ILH2.1 substrate is incubated with cisplatin at a 2:1 molar ratio of Pt to DNA oligo for 24 h in the dark in 1 mM $NaHPO_4$, pH 7.0. The modified oligonucleotide, purified from the unreacted cisplatin by ethanol precipitation, is annealed to the fluorescein labeled JT/ILH 2.2 oligonucleotide at a 1.5:1 ratio. The annealing reaction is heated to 95 °C for 5 min and then brought to RT over a 3–4 h period. The resulting duplex DNA can be assessed by native polyacrylamide electrophoresis to ensure complete annealing and Pt modification (*see* **Note 1**). The single-stranded fluorescent DNA substrate for RPA binding was used without modification.

3.2 Compound Dilution

384 well Mother plates contain compounds at 10 mM in 100% DMSO and are stored at −80 °C. Working plates are prepared by dilution to 1.25 mM in 100% DMSO and can be stored at −20 °C. Preassay plates contain compound diluted 0.05 mM in Reaction buffer and are prepared the day of the assay. 10 μL of compound will be dispensed into assay plates to achieve a compound concentration of 10 μM for screening (*see* **Note 2**).

3.3 Assay Plate

1. Master reaction mixes are prepared containing the target protein diluted in reaction buffer and the appropriate DNA substrate. The final concentration of RPA in reactions was 30 nM and XPA was 60 nM. DNA was added to a final concentration 5 nM. Protein and DNA mixes were prepared at 1.25× of the target screening concentrations such that dispensing of 40 μL would results in the final target concentration in 50 μL binding reactions.
2. Dispense 40 μL of the master reaction mix is into 352 wells of the 384 wells.
3. Controls: no compounds will be included in columns 23 and 24. Positive control (100 ng heparin in 10 μL) is included and the negative control have only the vehicle DMSO added (*see* **Note 3**).

4. Dispense 10 μL of the compounds from the preassay plate to the Assay plate in columns 1–22.
5. Incubate plates for 30 min at RT. Equilibrium is achieved within 30 min and the signal is stable for up to 90 min.
6. Measure fluorescence emission at 535 nm following excitation of 485 nm using vertical and horizontal polarizers (*see* **Note 4**).
7. Polarization is calculated as described [15] and the percent inhibition determined relative to the heparin control.

4 Notes

1. If there is substantial unannealed ssDNA or nonplatinated DNA, the duplex DNA can be purified via preparative native gel electrophoresis on a 12% polyacrylamide gel. The duplex DNA can be visualized via UV shadowing on a TLC plate and band excised, eluted and purified.
2. Preparation of daughter plates and preassay plates can be achieved using multichannel pipettors but is most efficiently carried out using automated liquid handling systems.
3. Many DNA binding proteins bind very effectively to heparin sulfate and this can be used as a positive control. Alternatively, an excess nonfluorescent DNA can be used as a control.
4. Determination of fluorescence polarization is most efficiently measured in an automated plate reader with automated plate stacking. The stability of the reaction is likely protein dependent and should be assessed for each protein preparation used in the screening process.

References

1. Gavande NS, VanderVere-Carozza PS, Hinshaw HD, Jalal SI, Sears CR, Pawelczak KS, Turchi JJ (2016) DNA repair targeted therapy: the past or future of cancer treatment? Pharmacol Ther 160:65–83
2. Lord CJ, Tutt AN, Ashworth A (2015) Synthetic lethality and cancer therapy: lessons learned from the development of PARP inhibitors. Annu Rev Med 66:455–470
3. Brown JS, O'Carrigan B, Jackson SP, Yap TA (2017) Targeting DNA repair in cancer: beyond PARP inhibitors. Cancer Discov 7:20–37
4. Murai J (2017) Targeting DNA repair and replication stress in the treatment of ovarian cancer. Int J Clin Oncol 22(4):619–628
5. Bhullar KS, Lagaron NO, McGowan EM, Parmar I, Jha A, Hubbard BP, Rupasinghe HPV (2018) Kinase-targeted cancer therapies: progress, challenges and future directions. Mol Cancer 17:48
6. Arora S, Heyza J, Zhang H, Kalman-Maltese V, Tillison K, Floyd AM, Chalfin EM, Bepler G, Patrick SM (2016) Identification of small molecule inhibitors of ERCC1-XPF that inhibit DNA repair and potentiate cisplatin efficacy in cancer cells. Oncotarget 7:75104–75117
7. Chapman TM, Wallace C, Gillen KJ, Bakrania P, Khurana P, Coombs PJ, Fox S, Bureau EA, Brownlees J, Melton DW, Saxty B (2015) N-Hydroxyimides and hydroxypyrimidinones as inhibitors of the DNA repair complex ERCC1-XPF. Bioorg Med Chem Lett 25:4104–4108

8. Goglia AG, Delsite R, Luz AN, Shahbazian D, Salem AF, Sundaram RK, Chiaravalli J, Hendrikx PJ, Wilshire JA, Jasin M, Kluger HM, Glickman JF, Powell SN, Bindra RS (2015) Identification of novel radiosensitizers in a high-throughput, cell-based screen for DSB repair inhibitors. Mol Cancer Ther 14:326–342
9. Klitgaard RN, Lobner-Olesen A (2018) A novel fluorescence based screen for inhibitors of the initiation of DNA replication in bacteria. Curr Drug Discov Technol. https://doi.org/10.2174/1570163815666180423115514
10. Liu W, Zhou M, Li Z, Li H, Polaczek P, Dai H, Wu Q, Liu C, Karanja KK, Popuri V, Shan SO, Schlacher K, Zheng L, Campbell JL, Shen B (2016) A selective small molecule DNA2 inhibitor for sensitization of human cancer cells to chemotherapy. EBioMedicine 6:73–86
11. Mohammed MZ, Vyjayanti VN, Laughton CA, Dekker LV, Fischer PM, Wilson DM III, Abbotts R, Shah S, Patel PM, Hickson ID, Madhusudan S (2011) Development and evaluation of human AP endonuclease inhibitors in melanoma and glioma cell lines. Br J Cancer 104:653–663
12. Huang X, Aulabaugh A (2016) Application of fluorescence polarization in HTS assays. Methods Mol Biol 1439:115–130
13. Patrick SM, Turchi JJ (1999) Replication protein a (RPA) binding to duplex cisplatin-damaged DNA is mediated through the generation of single-stranded DNA. J Biol Chem 274:14972–14978
14. Hermanson IL, Turchi JJ (2000) Overexpression and purification of human XPA using a Baculovirus expression system. Protein Expr Purif 19:1–11
15. Andrews BJ, Turchi JJ (2004) Development of a high-throughput screen for inhibitors of replication protein A and its role in nucleotide excision repair. Mol Cancer Ther 3:385–391

Part IV

In Vitro Biochemical and Biophysical Methods to Study DNA Repair

Chapter 14

Assembling the Human Resectosome on DNA Curtains

Michael M. Soniat, Logan R. Myler, and Ilya J. Finkelstein

Abstract

DNA double-strand breaks (DSBs) are a potentially lethal DNA lesions that disrupt both the physical and genetic continuity of the DNA duplex. Homologous recombination (HR) is a universally conserved genome maintenance pathway that initiates via nucleolytic processing of the broken DNA ends (resection). Eukaryotic DNA resection is catalyzed by the resectosome—a multicomponent molecular machine consisting of the nucleases DNA2 or Exonuclease 1 (EXO1), Bloom's helicase (BLM), the MRE11-RAD50-NBS1 (MRN) complex, and additional regulatory factors. Here, we describe methods for purification and single-molecule imaging and analysis of EXO1, DNA2, and BLM. We also describe how to adapt resection assays to the high-throughput single-molecule DNA curtain assay. By organizing hundreds of individual molecules on the surface of a microfluidic flowcell, DNA curtains visualize protein complexes with the required spatial and temporal resolution to resolve the molecular choreography during critical DNA-processing reactions.

Key words DNA curtains, Homologous recombination, Bloom's syndrome helicase (BLM), DNA nuclease (DNA2), Exonuclease 1 (EXO1)

1 Introduction

Each of our cells is confronted with up to 50 DNA double-strand breaks (DSBs) per replication cycle [1, 2]. DSBs arise as a result of cellular metabolism, but are extremely hazardous because they disrupt both the physical and genetic continuity of the genome. Unrepaired DSBs are primary causes of gross chromosomal rearrangements, loss of heterozygosity, and genomic instability. Thus, all organisms have developed multiple pathways to rapidly identify and repair DSBs.

Homologous recombination (HR) is a universally conserved double-strand break repair pathway that uses the intact sister chromatid to promote error-free repair (Fig. 1) [3–6]. HR is initiated when the Mre11-Rad50-Nbs1 (MRN) complex identifies the DNA ends [7–10]. MRN is a multifunctional enzyme. The Mre11 subunit contains a 3′ to 5′ exonuclease and an endonuclease activity which is important for processing DNA ends [11]. The Rad50

Lata Balakrishnan and Jason A. Stewart (eds.), *DNA Repair: Methods and Protocols*, Methods in Molecular Biology, vol. 1999, https://doi.org/10.1007/978-1-4939-9500-4_14,

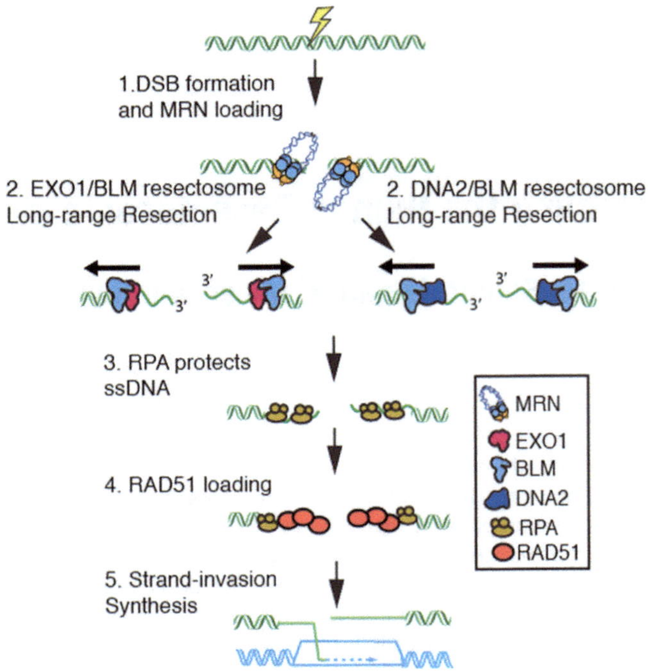

Fig. 1 Overview of double-strand break resection. (*1*) The MRE11-RAD50-NBS1 (MRN) complex rapidly identifies genomic double-strand breaks (DSBs). (*2*) MRN loads Bloom's syndrome helicase (BLM) and Exonuclease 1 (EXO1) complex or the BLM/DNA2 complex at DNA ends. EXO1 or DNA2 then nucleolytically processes (resect) the free DNA ends to produce long 3′-ssDNA ends. (*3*) The resulting ssDNA is rapidly bound by Replication Protein A (RPA). (*4*) RPA is then replaced by the recombinase RAD51. 5. RAD51 catalyzes the search for a homologous DNA sequence in a sister chromatid. Finally, the missing genetic information is resynthesized to repair the genome

subunit is an ATPase that interacts with Mre11 to promote the assembly of the MR complex. Rad50 also contains nonspecific DNA binding activity and extended ~50 nm coiled-coiled arms that interact via an apical zinc hook motif. The zinc hook is essential for cellular viability and the activation of the DNA damage response (DDR) [12]. The Nbs1 subunit does not contain any enzymatic activity, but mediates nuclear import of the MRN complex [13, 14]. Nbs1 also promotes protein interactions with ATM kinase and other DDR signaling proteins [8, 12, 15]. MRN initiates HR by processing the free DNA ends via a two-step reaction. First, MRN's endonucleolytic activity creates a nick tens of nucleotides upstream of the free DNA end [16–18]. This activity is not strictly required at DSBs that produce "clean" DNA ends that are free of modified bases or protein-DNA adducts. MRN, along with CtIP, are critical for initiating processing of DSBs that contain protein–DNA adducts or other lesions [16, 18–20]. MRN's exonucleolytic activity can process in a 3′ to 5′ direction toward the free

DNA end [21]. Next, MRN recruits helicases and nucleases that coordinate long-range DNA resection for downstream homologous recombination.

Long-range resection is carried out by two partially redundant multiprotein molecular complexes [4]. The first pathway couples exonuclease 1 (EXO1) and the Bloom's Syndrome helicase (BLM) [3, 20, 22–27]. While the EXO1/BLM resection machinery appears to be the major pathway in human cells, an alternative pathway uses the DNA2 helicase/nuclease along with BLM [28]. Together, these DNA resection complexes nucleolytically process the genome to generate kilobase-length stretches of single-stranded DNA (ssDNA) [4]. The functional significance of how these resection machineries are selected and whether this modulates repair outcomes remains an outstanding question in the field. Furthermore, many of the biophysical properties (e.g., velocity, processivity, and regulation by interacting partners) remain active areas of research. To address these outstanding questions, our lab has reconstituted the first key steps of human HR in vitro and with single-molecule resolution. Here, we provide methods for purifying, labeling, and single-molecule imaging of these resectosome components on the DNA curtains assay. The DNA curtains assay enables high-throughput single-molecule imaging of individual protein molecules on organized arrays of DNA molecules within a microfluidic flowcell. Compared to other ensemble and single-molecule fluorescence methods, DNA curtains offer three key advantages: [1] long DNA substrates (~48 kb) permit direct observation of kilobase-length DNA resection and repair, [2] by using multilaser illumination, multiple protein components can be monitored simultaneously with millisecond temporal resolution, and [3] the construction of multichannel, microfluidic DNA curtains facilitates high-throughput data acquisition. Bottom-up assembly of DNA resection with single-molecule sensitivity is shedding new light on the functions and regulation of these critical guardians of genome integrity. DNA curtains is a high-throughput single-molecule assay that facilitates direct imaging of DNA resection proteins using fluorescence microscopy (Fig. 2) [30–33].

2 Materials

2.1 Purification

1. BLM Lysis Buffer: 50 mM KH_2PO_4, 500 mM KCl, 10% (vol/vol) glycerol, 2.5 mM imidazole, 400 μL 50% Tween-20, 400 μL of phenylmethylsulfonyl fluoride (PMSF: 17 mg/mL), 20 mM β-mercaptoethanol.
2. EXO1/DNA2 Lysis Buffer: 25 mM Tris–HCl pH 8.0, 100 mM NaCl, 10% (vol/vol) glycerol, 400 μL of PMSF (17 mg/mL), 20 mM β-mercaptoethanol.

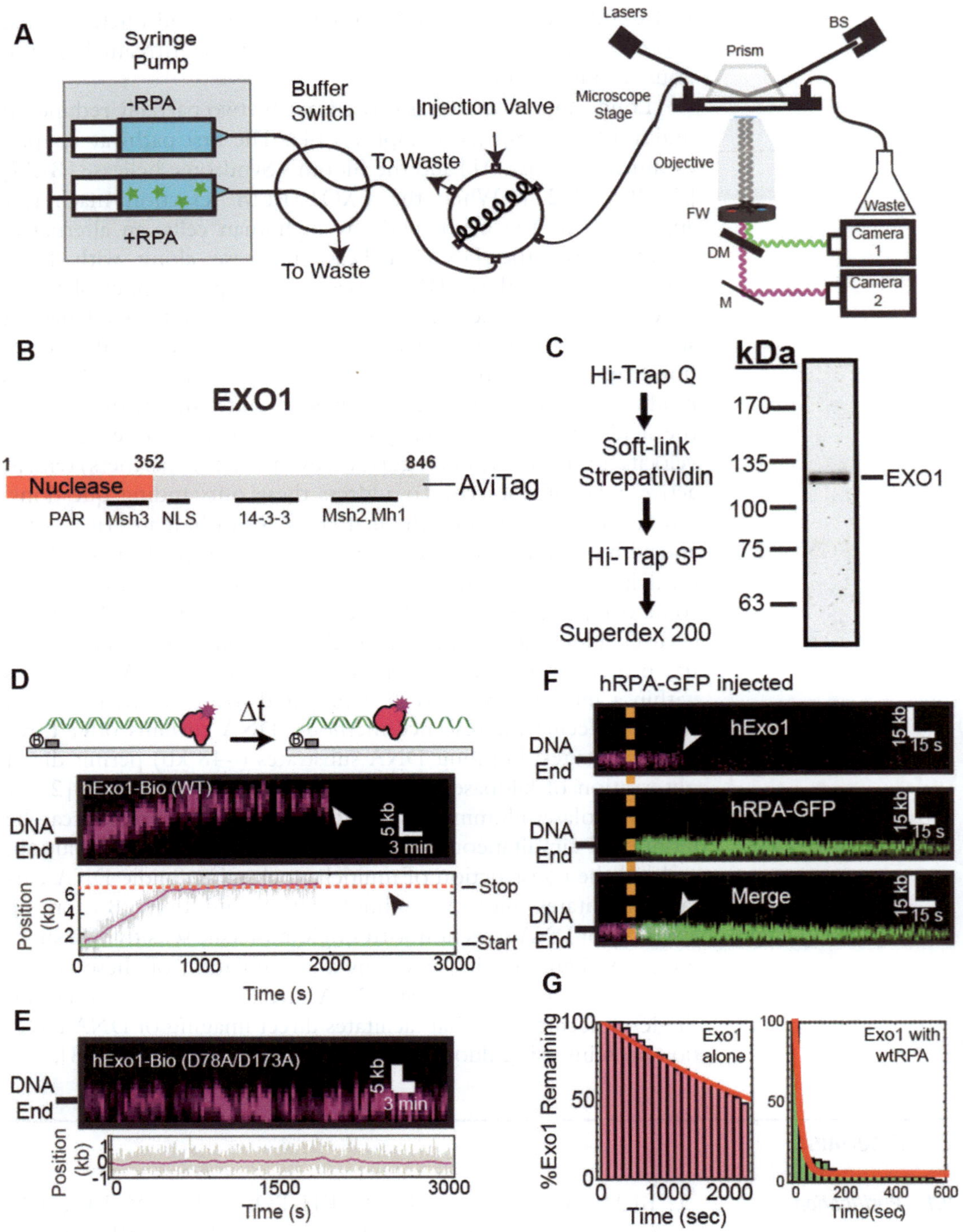

Fig. 2 Single-Molecule Imaging of EXO1. (**a**) Overview of the custom-built total internal reflection fluorescence (TIRF) microscope for DNA curtain experiments. A 488 nm laser beam passes through a computer-controlled shutter and neutral density filter (not shown). The laser is directed through a prism at a total internal reflection angle. The beam terminates in a beam stop (BS). This generates an evanescent excitation wave that illuminates the surface-bound DNA and protein molecules. The resulting fluorescent signals are collected via a water immersion high numerical aperture objective, passed through two excitation cleanup filters

3. Ni^{2+} B Buffer: 50 mM KH_2PO_4, 500 mM KCl, 10% (vol/vol) glycerol, 250 mM imidazole, 20 mM β-mercaptoethanol.
4. A Buffer: 25 mM Tris–HCl pH 8.0, 100 mM NaCl, 10% (vol/vol) glycerol, 1 mM DTT.
5. B Buffer: 25 mM Tris–HCl pH 8.0, 1 M NaCl, 10% (vol/vol) glycerol, 1 mM DTT.
6. PBS: 137 mM NaCl, 2.7 mM KCl, 4.3 mM Na_2HPO_4, 1.47 mM KH_2PO_4.
7. RPA Ni^{2+} A Buffer: 40 mM Tris–HCl pH 7.5, 1 M NaCl, 10 mM imidazole, pH 8.0, 20% glycerol, 4 mM β-mercaptoethanol, 1 mM PMSF.
8. RPA Ni^{2+} B Buffer: 20 mM Tris–HCl pH 7.5, 500 mM NaCl, 500 mM imidazole pH 8.0, 10% glycerol, 2 mM β-mercaptoethanol.
9. RPA Heparin Buffer A: 20 mM Tris–HCl pH 7.5, 50 mM KCl, 10% glycerol, 0.5 mM EDTA, 1 mM DTT.
10. RPA Heparin Buffer B: 20 mM Tris–HCl pH 7.5, 500 mM KCl, 10% glycerol, 0.5 mM EDTA, 1 mM DTT.
11. RPA Storage Buffer: 10 mM Tris pH 7.5, 100 mM KCl, 0.1 mM EDTA, 50% glycerol, 1 mM DTT.
12. Ni^{2+}-NTA resin.
13. Anti-FLAG M2 resin.
14. Streptavidin resin.
15. Dounce homogenizer.
16. Disposable columns.
17. 3× FLAG peptide.
18. Hi-Trap SP column (GE Healthcare).
19. Hi-Trap Q column (GE Healthcare).
20. Superdex 200 Increase 10/300 GL column (GE Healthcare).
21. HiLoad 16/600 Superdex 200 pg column (GE Healthcare).

Fig. 2 (continued) (FW) and dispersed through a dichromic mirror (DM) onto two electron-multiplied charge coupled device (EMCCD) cameras. A computer-controlled dual-syringe pump and two digitally actuated injections valves permit rapid buffer switching or the injection of different protein complexes. (**b**) Domain map of EXO1 showing key known interacting partner domains (**c**) Purification scheme of EXO1-AviTag (**d**) Kymograph of a single EXO1 resecting a DNA substrate. White arrow indicates when the molecule dissociates from the DNA. Below is a time-dependent single-particle trajectory of the same EXO1 molecule. Black arrow indicates protein dissociation. (**e**) Kymograph of nuclease-dead (D78A/D173A) EXO1 molecule. Gray line indicates the raw particle-tracking trajectory and the magenta line is a 15-s sliding average. (**f**) Kymograph of EXO1 (magenta, top) as it is displaced by RPA-GFP (green, middle). Merged images: bottom. The orange line indicates when RPA-GFP enters the flowcell, and white arrowheads indicate EXO1 dissociation. (**g**) Lifetime analysis of DNA end-bound EXO1 in the absence and in the presence of 1 nM RPA-GFP. Images in panel (**c–g**) are reprinted with permission from PNAS [29]

2.2 Microscopy

1. Lipids Buffer: 40 mM Tris–HCl pH 8.0, Bovine Serum Albumin (BSA; 0.2 mg/mL in H_2O).
2. Imaging Buffer: 40 mM Tris–HCl pH 8.0, 60 mM NaCl, 1 mM $MgCl_2$, 2 mM DTT, bovine serum albumin (BSA; 0.2 mg/mL in H_2O).
3. Biotinylated anti-FLAG M2 antibody (Sigma-Aldrich).
4. Streptavidin-conjugated quantum dots (QDs) 705 nm (ThermoFisher).

2.3 Equipment

1. Sonicator.
2. Ultracentrifuge.
3. High-speed centrifuge.
4. AKTA FPLC (GE Healthcare).
5. 488 nm laser (Sapphire 488–200 CW CDRH; Coherent).
6. Eclipse TI-E Inverted microscope (Nikon).
7. 60× water-immersion objective (1.2NA) (Nikon).
8. 500 nm long-pass filter (Chroma).
9. 638 nm dichroic beam splitter (Chroma).
10. iXon X3 DU897 EMCCD cameras (Andor).
11. Placeholder TextSyringe pumps.

3 Methods

3.1 Purification and Single-Molecule Imaging of EXO1

Since its discovery in 1992, the 5′ to 3′ exonuclease 1 (EXO1) has been identified as a key player in DNA double-strand break repair, mismatch repair, telomere maintenance, and replication fork restart [5, 34–37]. EXO1 translocates on DNA without hydrolyzing ATP by acting as a Brownian ratchet, stabilizing the transient opening of the DNA to allow for phosphodiester bond cleavage [38, 39]. In the cell, the ssDNA that is generated by EXO1 is rapidly bound by replication protein A (RPA), a ubiquitous eukaryotic ssDNA-binding protein. However, the molecular details of EXO1 processivity and its regulation by RPA remained controversial, largely because ensemble-biochemical methods cannot distinguish the precise choreography of both proteins at the ssDNA-dsDNA junction [23, 40, 41].

We recently established a single-molecule assay that visualizes EXO1 on high-throughput DNA curtains [29]. In these assays, EXO1 is purified with a C-terminal epitope tag that can be used for conjugation with fluorescent antibodies or quantum dots (QDs) (Fig. 2b, c). QDs are relatively small (~10 nm radius) fluorophores that have a high quantum yield. Moreover, QDs do not photobleach over several hours of illumination (*see* **Note 1**) [42]. We have evaluated several C-terminal epitopes for fluorescent EXO1 labeling. These include GFP, FLAG, and the 15-amino acid-long AviTag (GLNDIFEAQKIEWHE). AviTag is biotinylated on the lysine residue in insect cells that are coinfected with both the EXO1 and *E. coli* biotin ligase (BirA) viruses [43]. Coinfection of EXO1-AviTag with BirA resulted in ~50% biotinylation efficiency, as determined by streptavidin band-shift on an SDS-PAGE gel [44]. A streptavidin column was used in the protocol described below to further enrich for biotinylated protein. The fully biotinylated EXO1 concentration was ~300–400 nM, which is sufficient for single-molecule studies. Direct labeling of the EXO1-biotin using streptavidin-conjugated QDs resulted in excellent fluorescent imaging on DNA curtains (Fig. 2d, e). The following protocols detail the purification and fluorescent imaging of EXO1-biotin via single-molecule resection assays.

3.1.1 Purification of EXO1

1. Grow Sf21 insect cells with baculoviruses harboring EXO1-biotin and BirA (biotin ligase) following manufacturer-suggested protocols (Thermo). Briefly, the FastBac plasmid containing human EXO1, BirA, Tn7 transposon segments are transformed into DH10bac cells. The baculovirus is then produced by transfecting the bacmid into a small culture of insect cells and then amplifying the titer. To amplify the baculovirus, incubate the virus in a 15-cm plate of insect cells for 72 h (first amplification). Next, incubated the first amplification in the same manner in several 15-cm plates of cells to create the second amplification. Following amplification, add 600 μL of virus to 15×10^6 cells in sixty 15-cm dish containing 25 mL of Sf-900 II Serum Free Media containing penicillin–streptomycin. Incubate for 72 h.
2. Harvest Sf21 cells 72 h after infection. Centrifuge at 4000 × *g* to pellet the cells, snap-freeze in liquid nitrogen, and store at −80 °C until purification.
3. Thaw the pellet quickly at room temperature.
4. Homogenize pellet in 40 mL of EXO1 Lysis Buffer in a 40 mL Dounce homogenizer with 50 strokes of a B pestle.
5. Sonicate cells on ice three times for 30 s each time with a 30 s recovery on ice.
6. Centrifuge the mixture at 100,000 × *g* for 1 h at 4 °C to obtain soluble extract. Keep a small aliquot for analysis.
7. Run supernatant through a 5-mL Hi-Trap Q using A Buffer and B Buffer on an AKTA FPLC. Perform a 100% B Buffer elution to retain as much protein as possible. Collect 1 mL fractions.

8. Incubate elution with 1 mL of equilibrated streptavidin resin for 1 h at 4 °C with gentle rocking.
9. Centrifuge the sample at 500 × *g* for 3 min. Pack the resin into a column to put onto an AKTA FPLC.
10. Wash the resin with A Buffer until the UV peak stabilizes (~10 column volumes).
11. Elute with 5 mM biotin dissolved in A Buffer in 1 mL fractions.
12. Run sample through a Hi-Trap SP using A Buffer and B Buffer on an AKTA FPLC. Perform a 100% Buffer B elution to retain as much protein as possible. Collect 1 mL fractions and concentrate the sample before gel filtration.
13. The most concentrated fraction from the SP column is then developed on a Superdex 200 column equilibrated in A Buffer on an AKTA FPLC. Collect 1 mL fractions and pool highest concentration fractions.
14. Determine proteins concentration using Bradford assay.
15. Aliquot sample, snap-freeze with liquid nitrogen, and store at −80 °C.
16. The typical yield from 2 L of Sf21 cells is ~80–100 μg of ~300–400 nM EXO1-biotin.
17. To measure EXO1 biotinylation efficiency: incubate purified EXO1-bio with a large excess of streptavidin (~2 μM) for 10 min on ice. Mix with loading dye and run on an SDS-PAGE gel (do not boil sample to preserve the EXO1-Streptavidin interaction). Measure biotinylation efficiency by measuring the percentage of EXO1 that shift above the molecular weight of EXO1 on the gel.

3.1.2 Purification of RPA-GFP

1. Grow hRPA-GFP-His6 plasmid in Rosetta(DE3)/pLysS cells.
2. Inoculate single colony into 50 mL of LB in a 500 mL flask with 50 μg/mL of carbenicillin and 34 μg/mL chloramphenicol.
3. Grow overnight at 37 °C.
4. Next day, inoculate 2 L of LB with 15 mL of overnight per liter.
5. Grow until an O.D. 600 nm of 0.6–1.0 then induce cultures at 16 °C for 16–18 h with 1 mM IPTG.
6. Harvest cells for 15 min at 4000 × *g*.
7. Resuspend pellet in 20 mL of 1× PBS and respin for 15 min at 4000 × *g*.
8. Cells can be flash frozen in liquid nitrogen and store in −80 °C until ready to use.
9. Resuspend pellet in 25 mL of RPA Ni^{2+} A Buffer.

10. Sonicate cells on ice for a total of 90 s (75 amplitude; 15 s bursts with 90 s rests in between).
11. Centrifuge the mixture at 100,000 × *g* for 35 min at 4 °C to obtain soluble extract. Keep a small aliquot for analysis.
12. Run supernatant through a 5 mL HisTrap column using RPA Ni^{2+} A Buffer and RPA Ni^{2+} B Buffer. Perform a linear gradient over 10 CVs to elute protein. Collect 1 mL fractions. (RPA elutes at about 250 mM imidazole).
13. To remove NaCl, dialyze for 4 h-overnight with RPA Heparin Buffer A.
14. Following dialysis, if there are aggregates centrifuge at 4000 × *g* for 15 min.
15. Run sample through a 1 mL Heparin column using RPA Heparin Buffer A and RPA Heparin Buffer B. Perform a linear gradient over 10 CVs to elute protein. Collect 1 mL fractions. Monitor the fractions with two wavelengths. 280 nm for protein absorption and 488 nm for GFP fluorescence. Analyze samples on a 10–12% SDS-PAGE gel.
16. Pool fractions and run sample through a HiLoad 16/600 Superdex 200 pg column using RPA Heparin Buffer A. Collect 1 mL fractions. RPA should elute about 60 mL. Analyze samples on a 10–12% SDS-PAGE gel.
17. To concentrate sample, pool fractions and run sample through a 1 mL HiTrap Q column using RPA Heparin Buffer A and RPA Heparin Buffer B. Perform a linear gradient over 10 CVs to elute protein. Collect 1 mL fractions. RPA should elute at about 250 mM KCl. Analyze samples on a 10–12% SDS-PAGE gel.
18. Pool fractions and dialyze overnight in 1 L RPA Storage Buffer.
19. Determine proteins concentration using Bradford assay.
20. Aliquot sample and snap-freeze with liquid nitrogen and store at −80 °C.
21. The typical yield from 2 L of *E. coli* cells is ~0.5–1 mg of ~1–10 μM protein.

3.1.3 Single-Molecule Analysis of EXO1 Nuclease Activity

A homemade prism-type total internal reflection fluorescence (TIRF) microscope is used to image DNA curtains, as previously described (Fig. 2a) [45]. Briefly, quartz wafers are coated with UV-sensitive photoresist, exposed to UV through a high-resolution photomask, and then developed. Following development, ~13-nm layer of Cr is deposited on the wafer surface and are diced into six 50 mm × 22 mm quartz slides and further drilled. The flowcell is then created by combining one of the quartz slides with a glass coverslip using double-sided tape. Connectors are then added to drilled holes using a glue gun. Flowcells are then pre-equilibrated in Lipids Buffer and then coated with a with a lipid bilayer composed of a mixture of DOPC (97.7 mol%), DOPE-biotin (0.3 mol%), and DOPE-mPEG2K (2 mol%; Avanti Lipids). Flowcells are next incubated with Imaging Buffer for 10 min. Next, streptavidin is injected in to the flowcell followed by biotinylated λ-DNA. Flowcells are then mounted on a custom-machined microscope stage. For these experiments, DNA molecules are tethered to the lipid bilayer at one end [30, 46]. The second DNA end remaining free for EXO1 binding and DNA resection. DNA curtains are first preassembled in the flowcell. Next, EXO1 is injected through a 100 μL injection loop at a flow rate of 200 μL/min. Fluorescence emitted by fluorescent EXO1 and/or resectosome cofactors is split by dichroic optics that allows for simultaneous multicolor observation on two electron-multiplied charge coupled devices (EM-CCD cameras). Using this approach, we recently dissected how RPA rapidly displaces EXO1 from DNA ends (Fig. 2f, g) [29]. For this assay, a dual syringe setup allowed rapid injection of RPA (or RPA-GFP) into the flowcell. One syringe contained just Imaging Buffer, whereas the second syringe had Imaging Buffer with 1 nM RPA or RPA-GFP. The detailed protocol describing this experiment is presented below.

1. Assemble flowcells and single-tethered DNA curtains as described previously [45]. Using DNA with a long 3′ ssDNA overhang (78 nt) facilitates the loading of EXO1 onto DNA ends (*see* **Note 2**).
2. Add 800 fmol (2 μL of 400 nM) of EXO1 to 1 nmol (1 μL of 1 μM streptavidin QDs) in 7 μL of Imaging Buffer. Incubate for 10 min on ice.
3. Dilute to 200 μL with Imaging Buffer and add 5 μL saturating free biotin (~10 mM) dissolved in PBS.
4. Inject EXO1 at a rate of 200 μL/min in a 100 μL loop (*see* **Note 3**).
5. Increase the flow rate to 400 μL/min after EXO1 is bound to the DNA. The DNA substrate is extended to ~90% of its B-form crystallographic lengths at this flow rate.
6. Experiments are performed with the flowcell and microscope objective both equilibrated to 37 °C. Images are collected using Nikon Elements (Nikon). Data is acquired with a 488 nm laser (~100 mW), 200 ms frame rate collected every second, 10 MHz camera readout mode, 300× EM gain, and 5× conversion.
7. For analysis of the effect of RPA on resection, set up RPA in a second syringe at 1 nM concentration in Imaging Buffer.
8. After loading EXO1, use a digitally actuated valve to switch to the RPA-containing buffer.

3.1.4 Analyzing EXO1 Resection Data

1. Collect imaging data and export data into a TIFF (tagged-image file format) stack.
2. Adjust for imaging drift by picking out a stationary particle on the flowcell surface and track its x and y positions by fitting the point-spread function to a 2D-Gaussian. Shift every frame of the original movie using the tracked data (FIJI script available upon request).
3. The precise location of moving EXO1 molecules in each frame is fit to a 2D Gaussian using a custom-written FIJI plugin, (code available upon request).
4. Tracked data can be used to calculate the velocity, processivity, and DNA-binding lifetimes of individual molecules.

3.2 Imaging DNA2 Resection by Tracking RPA-GFP

DNA2 was first discovered as a key enzyme in Okazaki flap maturation, but is also implicated in other nucleolytic transactions in DSB repair, mitochondrial DNA replication/repair, and telomere maintenance [4]. DNA2 encodes an ATP-dependent helicase and ATP-independent nuclease domains in a single polypeptide (Fig. 3a). How these two activities are coupled, and how DNA2 processes long DNA substrates is not entirely clear. Recently it has been shown that the helicase activity of DNA2 accelerates DNA resection in the presence of RPA. This is further stimulated by BLM helicase. Based on these biochemical reconstitutions, an emerging model posits that BLM and DNA2 form a bidirectional motor where BLM is the lead helicase and the helicase activity of DNA2 acts as a ssDNA translocase to promote DNA resection [31, 47]. The nuclease activity of DNA2 is critical to all of DNA2 functions; however, little is known about the role of the DNA2's helicase activity and whether this activity is required for efficient DSB resection in vivo [31–33, 47]. A recent single-molecule study found that the nuclease-dead DNA2 exhibits processive helicase activity, suggesting that the helicase activity is autoregulated by the nuclease activity [32]. Furthermore, structural studies have also defined the basis for DNA2 interaction with RPA and its preference for DNA ends [48]. RPA directs the 5′ to 3′ nuclease activity of DNA while inhibiting the 3′ to 5′ nuclease activity [49]. Though the helicase and nuclease activity of DNA2's role in DNA repair has been shown, the long-range resection activity of DNA2 protein is still undefined. Here, we describe the purification of recombinant human DNA2-FLAG in Sf21 cells using the Bac-to-Bac (Life Tech.) expression system (*see* **Note 4**) (Fig. 3b). We also describe single-molecule analysis of DNA2 translocation on DNA curtains using RPA-GFP as a readout of DNA resection (*see* **Note 5**) (Fig. 3c, d):

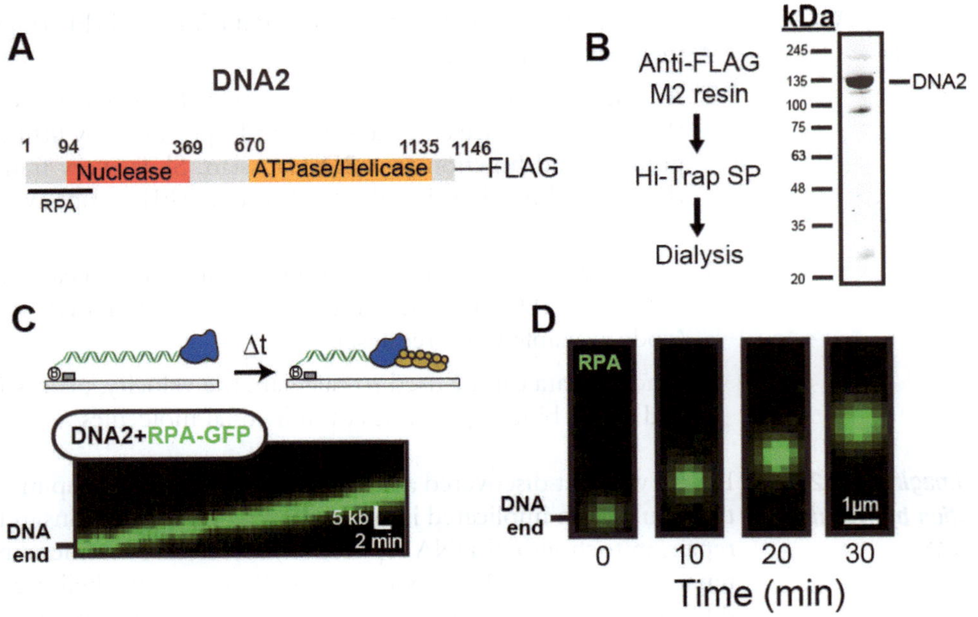

Fig. 3 Indirect Imaging of DNA2 resection by RPA-GFP tracking. (**a**) Domain map of DNA2 used for single-molecule assays. Key interacting partners are highlighted in the map. (**b**) DNA2-FLAG purification scheme and SDS-PAGE gel of the purified protein. (**c**) Kymograph of DNA2-mediated resection by monitoring RPA-GFP signal. (**d**) DNA2 processively generates ssDNA, as indicated by snapshots of RPA-GFP accumulation

3.2.1 Purification of DNA2

1. Grow Sf21 insect cells in 60 15-cm dishes with a baculovirus harboring DNA2-FLAG, as recommended by the manufacturer. See Subheading 3.1.1 for details.
2. Harvest 2 L of Sf21 cells 72 h after infection. Centrifuge at 4000 × *g* to pellet the cells, snap-freeze in liquid nitrogen, and store at −80 °C until purification.
3. Thaw the pellet quickly at room temperature.
4. Homogenize pellet obtained from in 40 mL of DNA2 Lysis Buffer in a 40 mL Dounce homogenizer with 50 strokes of a B pestle.
5. Sonicate cells on ice three times for 30 s each time with a 30 s recovery on ice.
6. Centrifuge the mixture at 100,000 × *g* for 1 h at 4 °C to obtain soluble extract. Keep a small aliquot for analysis.
7. Incubate supernatant with 0.8 mL of equilibrated anti-FLAG M2 Affinity Gel and incubate for 1 h at 4 °C with gentle rocking.
8. Transfer FLAG resin to a disposable column.
9. Wash the resin with 15 mL of A Buffer three times.
10. Add 5 mL of A Buffer supplemented with 100 μL of 3×FLAG peptide (stock: 5 mg/mL) and incubate for 30 min at 4 °C.

11. Collect the 5 mL of sample from FLAG resin. Keep a small aliquot for analysis.
12. Run sample through a 1 mL Hi-Trap SP column using A Buffer and B Buffer on an AKTA FPLC. Perform a linear gradient over 15 CVs to elute protein. Collect 1 mL fractions.
13. Pool peak fractions from the SP column and dialyze overnight in A Buffer.
14. Determine proteins concentration using Bradford assay.
15. Aliquot sample and snap-freeze with liquid nitrogen and store at −80 °C.
16. The typical yield from 2 L of SF21 cells is ~100–125 μg of purified DNA2 at ~400–600 nM.

3.2.2 Single-Molecule Analysis of DNA2 Resection

1. Follow flowcell assembly protocol for single-tethered DNA curtains as previously reported using DNA with a 3′ or 5′ ssDNA overhang to facilitate DNA2 loading on DNA ends [45].
2. Dilute 800 fmol (2 μL of 400 nM) of DNA2 to 200 μL with Imaging Buffer.
3. Inject DNA2 at 200 μL per min in a 100 μL loop.
4. Increase the flow rate to 400 μL per minute after loading for full DNA extension.
5. Switch to Imaging Buffer containing 1 nM RPA-GFP.
6. Experiments are performed at 37 °C. Images are collected using Nikon Elements (Nikon). Data is acquired with a 488 nm laser (~100 mW), 200 ms frame rate collected every 2 s, 10 MHz camera readout mode, 300× EM gain, and 5× conversion.

3.2.3 Analyzing DNA2 Resection Data

1. Collect imaging data and export data into a TIFF (tagged-image file format) stack.
2. Adjust for imaging drift by picking out a stationary particle on the flowcell, and track its x and y positions by fitting the point-spread function to a 2D-Gaussian. Shift every frame of the original movie using the tracked data (FIJI script available upon request).
3. Annotate the moving RPA-GFP molecules and fit a 2D Gaussian to the frames using a custom-written FIJI plugin (code available upon request).
4. Tracked data can be used to calculate the velocity and processivity of individual molecules.
5. Calculate RPA intensity as readout of DNA resection by summing the total pixel intensity over a defined area over every frame in ImageJ.

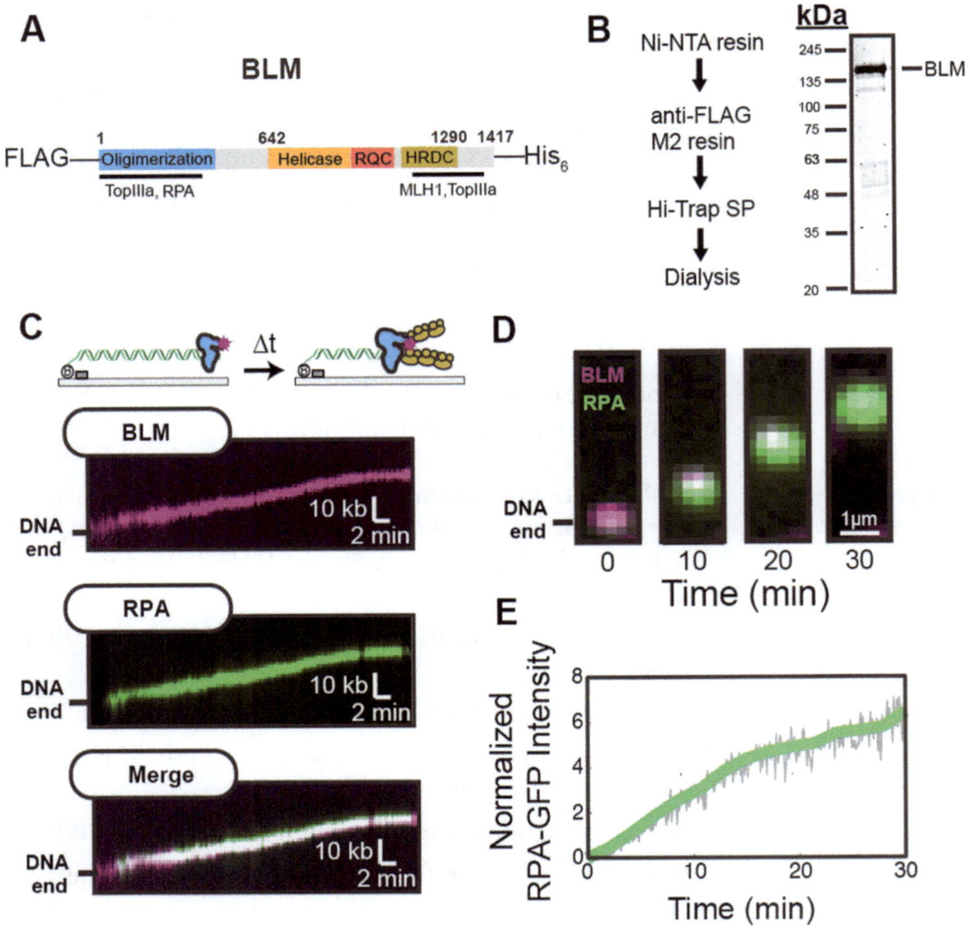

Fig. 4 Direct Imaging of BLM helicase activity. (**a**) Domain map of BLM used for single-molecule assays. (**b**) Purification scheme and SDS-PAGE gel of FLAG-BLM-His$_6$. (**c**) Kymograph of BLM (top) during helicase activity in the presence of RPA-GFP (green). Bottom: merged images. (**d**) BLM helicase generates two strands of ssDNA, which appears as RPA-GFP accumulation at the BLM position. (**e**) Quantification of RPA-GFP intensity. Solid lines represent a twenty-frame moving average filter of the raw particle tracking intensities

3.3 Imaging of BLM Helicase Activity

BLM is a 3′ to 5′ ATP-dependent helicase and one of five helicases found in humans with structural similarity to *E. coli* RecQ [50, 51]. BLM is a key player in DNA double-strand break repair, DNA recombination, DNA replication, and telomere maintenance. The enzyme is comprised of an N-terminal oligomerization domain followed by a conserved core RecQ helicase domain containing the helicase domain, the RecQ C-terminal (RQC) domain and the helicase-and-ribonuclease D-C terminal (HRDC) domains (Fig. 4a). The N-terminus of BLM induces oligomerization of the full-length enzyme, but the role of these structures in promoting BLM's myriad functions is still unknown [52]. The RQC domain directs binding to of BLM to ssDNA–dsDNA junctions, and aids in coupling ATP hydrolysis to DNA unwinding (*see* **Note 6**)

[53, 54]. The HRDC domain has weak affinity for ssDNA and is required in BLM-catalyzed dissolution of double Holliday junctions [54–57]. BLM unwinds a variety of DNA substrates during DNA replication and repair (14). BLM is also critical for DSB resection because it stimulates the DNA resection activities of both EXO1 and DNA2 nucleases [22–25, 40]. Along with EXO1 and DNA2, RPA also physically interacts with BLM and stimulates BLM's helicase activity [58–60]. However, the mechanism of this stimulation is unknown, nor is the functional overlap with the DNA2 helicase activity. One gap in the field is that most biochemical and biophysical studies have focused on an *E. coli*-expressed BLM truncation that retains just the core RecQ helicase domain [61–67]. This truncated BLM likely recapitulates the key features of the motor core but lacks the oligomerization and additional regulatory peptides that facilitate interactions with other repair proteins. Here, we describe the purification of full-length BLM with an N-terminal FLAG and a C-terminal His_6 epitope from Sf21 insect cells using the Bac-to-Bac (Life Tech.) expression system (Fig. 4b). We also describe single-molecule analysis of BLM helicase activity on DNA curtains by tracking both fluorescently tagged BLM and RPA-GFP (Fig. 4c, d).

3.3.1 Purification of BLM

1. Grow Sf21 insect cells in 60 15-cm dishes with a baculovirus harboring FLAG-BLM-His_6, as recommended by the manufacturer. See Subheading 3.1.1 for details.
2. Harvest 2 L of Sf21 cells 72 h after infection. Centrifuge at 4000 × *g* to pellet the cells, snap-freeze in liquid nitrogen, and store at −80 °C until purification.
3. Thaw the pellet quickly at room temperature.
4. Homogenize pellet in 40 mL of BLM Lysis Buffer in a 40 mL Dounce homogenizer with 50 strokes of a B pestle.
5. Sonicate cells on ice three times for 30 s each time with a 30 s recovery on ice.
6. Centrifuge the mixture at 100,000 × *g* for 1 h at 4 °C to obtain soluble extract. Keep a small aliquot for analysis.
7. Incubate supernatant with 8–10 mL of equilibrated Ni-NTA resin for 1 h at 4 °C with gentle rocking.
8. Centrifuge the sample at 500 × *g* for 2 min. Keep a small aliquot for analysis.
9. Transfer Ni-NTA resin to a disposable column.
10. Wash the resin with 50 mL of BLM Lysis Buffer three times.
11. Add 15 mL of Ni^{2+} B Buffer: to resin and incubate for 30 min at 4 °C.

12. Collect the 15 mL of sample from the Ni-NTA resin. Keep a small aliquot for analysis.
13. Add sample to 0.8 mL of equilibrated anti-FLAG M2 affinity resin and incubate for 30 min at 4 °C.
14. Transfer FLAG resin to a disposable column.
15. Wash the resin with 15 mL of A Buffer three times.
16. Add 5 mL of A Buffer +100 μL of 3× FLAG peptide (5 mg/mL) and incubate for 30 min at 4 °C.
17. Collect the 5 mL of sample from FLAG resin. Keep a small aliquot for analysis.
18. Run sample through a 1 mL Hi-Trap SP using A Buffer and B Buffer on an AKTA FPLC. Perform a linear gradient over 10 CVs to elute protein. Collect 1 mL fractions.
19. Pool aliquots containing BLM from the SP column and dialyze overnight in A Buffer.
20. Determine proteins concentration using Bradford assay.
21. Aliquot sample and snap-freeze with liquid nitrogen and store at −80 °C.
22. The typical yield from 2 L of Sf21 cells is ~75–100 μg of ~300–400 nM protein.

3.3.2 Single-Molecule Analysis of BLM Helicase Activity with RPA

1. Follow flowcell assembly protocol for single-tethered DNA curtains as previously described for EXO1 (*see* **Note 7**) (*see* Subheading 3.1.3 above).
2. Dilute a biotinylated anti-FLAG M2 antibody 1:100 in Lipids Buffer.
3. Incubate an 8:1 ratio of diluted antibody to streptavidin-conjugated QDs for 10 min on ice.
4. Add 40 nM FLAG-BLM to the antibody-QD mixture and incubate for another 10 min on ice (*see* **Note 8**).
5. Dilute mixture to 200 μL Imaging Buffer and add 2 μL of saturating biotin (~10 μM).
6. Inject BLM at 200 μL per minute in a 100 μL loop.
7. Increase the flow rate to 400 μL per minute after loading and switch to Imaging buffer containing 1 nM RPA-GFP.
8. Experiments are performed at 37 °C. Images are collected using Nikon Elements (Nikon). Data is acquired with a 488 nm laser (~100 mW), 200 ms frame rate collected every 2 s, 10 MHz camera readout mode, 300× EM gain, and 5× conversion.

3.3.3 Analyzing BLM and RPA Helicase Activity Data

1. Collect imaging data and export data into a TIFF (tagged-image file format) stack.
2. Adjust for imaging drift by picking out a stationary particle on the flowcell and track its *x* and *y* positions by fitting the point-spread function to a 2D-Gaussian. Shift every frame of the original movie using the tracked data (FIJI script available upon request).
3. Select moving BLM molecules and fit a 2D Gaussian to the frames using a custom-written FIJI plugin (code available upon request).
4. Tracked data can be used to calculate the velocity, processivity, and DNA-binding lifetimes of individual molecules.
5. Calculate RPA intensity as a readout of helicase activity by summing the total pixel intensity over a defined area over every frame in ImageJ (Fig. 4d).

4 Notes

1. Quantum dots (QDs), while having a high quantum yield, also have batch-to-batch variation in the percent of "dark" QDs. One estimate indicates that 25–75% of QDs in a particular batch may not be fluorescent [68, 69].
2. EXO1 will load onto nicks and blunt ends with roughly equal affinity. Avoid centrifugation and pipette gently with wide-bore tips to avoid accumulating unwanted nicks in the 48.5 kb-long DNA substrate. Even with the gentlest handling, we routinely observe ~3–5 nicks in freshly prepared DNA substrates. We find that addition of the 78-nt long 3′ ssDNA overhang stimulates loading of ~60% of the EXO1 molecules at the DNA ends as opposed to internal nicks.
3. EXO1 loading on DNA ends is salt sensitive. We did not observe efficient protein binding to the DNA above ~80 mM NaCl (total ionic strength: 103 mM). Preloading the EXO1 and then switching to a buffer at 120 mM NaCl (total ionic strength: 143 mM) retained most of the EXO1 molecules. We typically inject 4 nM of EXO1 in Imaging Buffer supplemented with 60 mM NaCl, resulting in ~1 EXO1 molecule per DNA substrate.
4. Movement of the FLAG tag to the N-terminus of DNA2 results in low protein expression (personal communication).
5. DNA2 activity is inhibited when an anti-Flag antibody-conjugated QD is appended to the C-terminus. This suggests that both termini may be important for regulating helicase/nuclease activity.

6. BLM binding to DNA is nucleotide-dependent. We observe that BLM loads onto DNA curtains nonspecifically in the apo (no nucleotide) or ADP-bound states. To promote specific DNA end binding, 1 mM ATP is added to the reaction buffer.
7. Unlike EXO1, BLM can also load onto DNA substrates containing a blunt end or a 12 nt 5′-overhang.
8. BLM helicase activity and fluorescent labeling efficiency are highest when it is preincubated with the fluorescent antibodies prior to injection onto DNA curtains.

Acknowledgments

We are indebted to Drs. Mauro Modesti and Tanya Paull for plasmids, cell pellets, and other reagents. This work was supported by the National Institutes of Health (GM120554 and CA092584) and the Welch Foundation (F-l808 to I.J.F.). M.M.S. is supported by a Postdoctoral Fellowship, PF-17-169-01-DMC, from the American Cancer Society. L.R.M. is supported by the National Cancer Institute (CA212452).

References

1. Vilenchik MM, Knudson AG (2003) Endogenous DNA double-strand breaks: production, fidelity of repair, and induction of cancer. Proc Natl Acad Sci U S A 100:12871–12876
2. Vilenchik MM, Knudson AG (2006) Radiation dose-rate effects, endogenous DNA damage, and signaling resonance. Proc Natl Acad Sci U S A 103:17874–17879
3. Ciccia A, Elledge SJ (2010) The DNA damage response: making it safe to play with knives. Mol Cell 40:179–204
4. Symington LS (2016) Mechanism and regulation of DNA end resection in eukaryotes. Crit Rev Biochem Mol Biol 51:195–212
5. Symington LS, Gautier J (2011) Double-strand break end resection and repair pathway choice. Annu Rev Genet 45:247–271
6. Jasin M, Rothstein R (2013) Repair of strand breaks by homologous recombination. Cold Spring Harb Perspect Biol 5:a012740
7. Cannavo E, Cejka P (2014) Sae2 promotes dsDNA endonuclease activity within Mre11-Rad50-Xrs2 to resect DNA breaks. Nature 514:122–125
8. Paull TT, Gellert M (1998) The 3′ to 5′ exonuclease activity of Mre11 facilitates repair of DNA double-strand breaks. Mol Cell 1:969–979
9. Shibata A, Moiani D, Arvai AS et al (2014) DNA double-strand break repair pathway choice is directed by distinct MRE11 nuclease activities. Mol Cell 53:7–18
10. Lukas C, Melander F, Stucki M et al (2004) Mdc1 couples DNA double-strand break recognition by Nbs1 with its H2AX-dependent chromatin retention. EMBO J 23:2674–2683
11. Stracker TH, Petrini JHJ (2011) The MRE11 complex: starting from the ends. Nat Rev Mol Cell Biol 12:90–103
12. Lee J-H, Mand MR, Deshpande RA et al (2013) Ataxia telangiectasia-mutated (ATM) kinase activity is regulated by ATP-driven conformational changes in the Mre11/Rad50/Nbs1 (MRN) complex. J Biol Chem 288:12840–12851
13. Tauchi H, Kobayashi J, Morishima K et al (2002) Nbs1 is essential for DNA repair by homologous recombination in higher vertebrate cells. Nature 420:93–98
14. Desai-Mehta A, Cerosaletti KM, Concannon P (2001) Distinct functional domains of nibrin mediate Mre11 binding, focus formation, and nuclear localization. Mol Cell Biol 21:2184–2191
15. Williams RS, Dodson GE, Limbo O et al (2009) Nbs1 flexibly tethers Ctp1 and Mre11-Rad50 to coordinate DNA double-

strand break processing and repair. Cell 139:87–99
16. Deshpande RA, Lee J-H, Arora S et al (2016) Nbs1 converts the human Mre11/Rad50 nuclease complex into an endo/exonuclease machine specific for protein-DNA adducts. Mol Cell 64:593–606
17. Anand R, Ranjha L, Cannavo E et al (2016) Phosphorylated CtIP functions as a co-factor of the MRE11-RAD50-NBS1 endonuclease in DNA end resection. Mol Cell 64:940–950
18. Myler LR, Gallardo IF, Soniat MM et al (2017) Single-molecule imaging reveals how Mre11-Rad50-Nbs1 initiates DNA break repair. Mol Cell 67:891–898. e4
19. Mimitou EP, Symington LS (2008) Sae2, Exo1 and Sgs1 collaborate in DNA double-strand break processing. Nature 455:770–774
20. Zhu Z, Chung W-H, Shim EY et al (2008) Sgs1 helicase and two nucleases Dna2 and Exo1 resect DNA double-strand break ends. Cell 134:981–994
21. Garcia V, Phelps SE, Gray S et al (2011) Bidirectional resection of DNA double-strand breaks by Mre11 and Exo1. Nature 479:241–244
22. Cejka P, Cannavo E, Polaczek P et al (2010) DNA end resection by Dna2-Sgs1-RPA and its stimulation by Top3-Rmi1 and Mre11-Rad50-Xrs2. Nature 467:112–116
23. Nimonkar AV, Genschel J, Kinoshita E et al (2011) BLM-DNA2-RPA-MRN and EXO1-BLM-RPA-MRN constitute two DNA end resection machineries for human DNA break repair. Genes Dev 25:350–362
24. Nimonkar AV, Ozsoy AZ, Genschel J et al (2008) Human exonuclease 1 and BLM helicase interact to resect DNA and initiate DNA repair. Proc Natl Acad Sci U S A 105:16906–16911
25. Niu H, Chung W-H, Zhu Z et al (2010) Mechanism of the ATP-dependent DNA end-resection machinery from *Saccharomyces cerevisiae*. Nature 467:108–111
26. Gravel S, Chapman JR, Magill C et al (2008) DNA helicases Sgs1 and BLM promote DNA double-strand break resection. Genes Dev 22:2767–2772
27. Mimitou EP, Symington LS (2011) DNA end resection—Unraveling the tail. DNA Repair 10:344–348
28. Tomimatsu N, Mukherjee B, Deland K et al (2012) Exo1 plays a major role in DNA end resection in humans and influences double-strand break repair and damage signaling decisions. DNA Repair 11:441–448
29. Myler LR, Gallardo IF, Zhou Y et al (2016) Single-molecule imaging reveals the mechanism of Exo1 regulation by single-stranded DNA binding proteins. Proc Natl Acad Sci U S A 113:e1170–e1179
30. Gallardo IF, Pasupathy P, Brown M et al (2015) High-throughput universal DNA curtain arrays for single-molecule fluorescence imaging. Langmuir 31:10310–10317
31. Levikova M, Pinto C, Cejka P (2017) The motor activity of DNA2 functions as an ssDNA translocase to promote DNA end resection. Genes Dev 31:493–502
32. Levikova M, Klaue D, Seidel R et al (2013) Nuclease activity of *Saccharomyces cerevisiae* Dna2 inhibits its potent DNA helicase activity. Proc Natl Acad Sci U S A 110:E1992–E2001
33. Pinto C, Kasaciunaite K, Seidel R et al (2016) Human DNA2 possesses a cryptic DNA unwinding activity that functionally integrates with BLM or WRN helicases. elife 5:e18574
34. Szankasi P, Smith GR (1992) A DNA exonuclease induced during meiosis of Schizosaccharomyces pombe. J Biol Chem 267:3014–3023
35. Wu P, Takai H, de LT (2012) Telomeric 3′ overhangs derive from resection by Exo1 and Apollo and fill-in by POT1b-associated CST. Cell 150:39–52
36. Modrich P (2006) Mechanisms in eukaryotic mismatch repair. J Biol Chem 281:30305–30309
37. Pasero P, Vindigni A (2017) Nucleases acting at stalled forks: how to reboot the replication program with a few shortcuts. Annu Rev Genet 51:477–499
38. Orans J, McSweeney EA, Iyer RR et al (2011) Structures of human exonuclease 1 DNA complexes suggest a unified mechanism for nuclease family. Cell 145:212–223
39. Shi Y, Hellinga HW, Beese LS (2017) Interplay of catalysis, fidelity, threading, and processivity in the exo- and endonucleolytic reactions of human exonuclease I. Proc Natl Acad Sci U S A 114(23):6010–6015
40. Yang S-H, Zhou R, Campbell J et al (2013) The SOSS1 single-stranded DNA binding complex promotes DNA end resection in concert with Exo1. EMBO J 32:126–139
41. Genschel J, Modrich P (2003) Mechanism of 5′-directed excision in human mismatch repair. Mol Cell 12:1077–1086
42. Finkelstein IJ, Visnapuu M-L, Greene EC (2010) Single-molecule imaging reveals mechanisms of protein disruption by a DNA translocase. Nature 468:983–987

43. Duffy S, Tsao KL, Waugh DS (1998) Site-specific, enzymatic biotinylation of recombinant proteins in *Spodoptera frugiperda* cells using biotin acceptor peptides. Anal Biochem 262:122–128
44. Sorenson AE, Askin SP, Schaeffer PM (2015) In-gel detection of biotin–protein conjugates with a green fluorescent streptavidin probe. Anal Methods 7:2087–2092
45. Soniat MM, Myler LR, Schaub JM et al (2017) Next-Generation DNA curtains for single-molecule studies of homologous recombination. In: Eichman BF (ed) Methods in enzymology. Academic Press, Cambridge, MA, pp 259–281
46. Finkelstein IJ, Greene EC (2011) Supported lipid bilayers and DNA curtains for high-throughput single-molecule studies. Methods Mol Biol 745:447–461
47. Miller AS, Daley JM, Pham NT et al (2017) A novel role of the Dna2 translocase function in DNA break resection. Genes Dev 31:503–510
48. Zhou C, Pourmal S, Pavletich NP (2015) Dna2 nuclease-helicase structure, mechanism and regulation by Rpa. elife 4:e09832
49. Masuda-Sasa T, Imamura O, Campbell JL (2006) Biochemical analysis of human Dna2. Nucleic Acids Res 34:1865–1875
50. Croteau DL, Popuri V, Opresko PL et al (2014) Human RecQ helicases in DNA repair, recombination, and replication. Annu Rev Biochem 83:519–552
51. Bernstein KA, Gangloff S, Rothstein R (2010) The RecQ DNA helicases in DNA repair. Annu Rev Genet 44:393–417
52. Beresten SF, Stan R, van Brabant AJ et al (1999) Purification of overexpressed hexahistidine-tagged BLM N431 as oligomeric complexes. Protein Expr Purif 17:239–248
53. Janscak P, Garcia PL, Hamburger F et al (2003) Characterization and mutational analysis of the RecQ core of the bloom syndrome protein. J Mol Biol 330:29–42
54. Kitano K (2014) Structural mechanisms of human RecQ helicases WRN and BLM. Front Genet 5:366
55. Bernstein DA, Keck JL (2003) Domain mapping of *Escherichia coli* RecQ defines the roles of conserved N- and C-terminal regions in the RecQ family. Nucleic Acids Res 31:2778–2785
56. Kim YM, Choi B-S (2010) Structure and function of the regulatory HRDC domain from human Bloom syndrome protein. Nucleic Acids Res 38:7764–7777
57. Liu Z, Macias MJ, Bottomley MJ et al (1999) The three-dimensional structure of the HRDC domain and implications for the Werner and Bloom syndrome proteins. Structure 7:1557–1566
58. Brosh RM, Li J-L, Kenny MK et al (2000) Replication protein A physically interacts with the Bloom's syndrome protein and stimulates its helicase activity. J Biol Chem 275:23500–23508
59. Doherty KM, Sommers JA, Gray MD et al (2005) Physical and functional mapping of the replication protein A interaction domain of the Werner and Bloom syndrome helicases. J Biol Chem 280:29494–29505
60. Kang D, Lee S, Ryu K-S et al (2018) Interaction of replication protein A with two acidic peptides from human Bloom syndrome protein. FEBS Lett 592:547–558
61. Guo R-B, Rigolet P, Ren H et al (2007) Structural and functional analyses of disease-causing missense mutations in Bloom syndrome protein. Nucleic Acids Res 35:6297–6310
62. Chatterjee S, Zagelbaum J, Savitsky P et al (2014) Mechanistic insight into the interaction of BLM helicase with intra-strand G-quadruplex structures. Nat Commun 5:ncomms6556
63. Yodh JG, Stevens BC, Kanagaraj R et al (2009) BLM helicase measures DNA unwound before switching strands and hRPA promotes unwinding reinitiation. EMBO J 28:405–416
64. Newman JA, Savitsky P, Allerston CK et al (2015) Crystal structure of the Bloom's syndrome helicase indicates a role for the HRDC domain in conformational changes. Nucleic Acids Res 43:5221–5235
65. Swan MK, Legris V, Tanner A et al (2014) Structure of human Bloom's syndrome helicase in complex with ADP and duplex DNA. Acta Crystallogr D Biol Crystallogr 70:1465–1475
66. Nguyen GH, Dexheimer TS, Rosenthal AS et al (2013) A small molecule inhibitor of the BLM helicase modulates chromosome stability in human cells. Chem Biol 20:55–62
67. Karow JK, Chakraverty RK, Hickson ID (1997) The Bloom's syndrome gene product is a 3′–5′ DNA helicase. J Biol Chem 272:30611–30614
68. Yao J, Larson DR, Vishwasrao HD et al (2005) Blinking and nonradiant dark fraction of water-soluble quantum dots in aqueous solution. Proc Natl Acad Sci U S A 102:14284–14289
69. Ebenstein Y, Mokari T, Banin U (2002) Fluorescence quantum yield of CdSe/ZnS nanocrystals investigated by correlated atomic-force and single-particle fluorescence microscopy. Appl Phys Lett 80:4033–4035

Chapter 15

Thin-Layer Chromatography and Real-Time Coupled Assays to Measure ATP Hydrolysis

Christopher W. Sausen, Cody M. Rogers, and Matthew L. Bochman

Abstract

Many chemical reactions in the cell are thermodynamically unfavorable. To overcome this barrier, the energy released from the hydrolysis of adenosine triphosphate (ATP) is coupled to these reactions via ATP hydrolyzing enzymes known as ATPases. These enzymes are ubiquitous in nature and frequently act as molecular motors in processes ranging from DNA replication to protein degradation. Assays that characterize ATPase activity in vitro are important tools to gain insight into their functions in vivo. Here, we describe a direct and flexible thin-layer chromatography method for detecting ATPase activity using radiolabeled ATP. Additionally, we describe a high-throughput coupled reaction assay pairing ATP hydrolysis with nicotinamide adenine dinucleotide (NADH) oxidation to monitor ATP hydrolysis in real time.

Key words ATPase, NADH, Thin-layer chromatography (TLC), Coupled assay

1 Introduction

ATPases are enzymes that catalyze the hydrolysis of adenosine triphosphate (ATP) into adenosine diphosphate (ADP) and inorganic phosphate [1]. This reaction releases energy that is coupled by the ATPase to drive chemical reactions or conformational changes in the protein itself or other molecules. The hydrolysis of ATP by these enzymes is a conserved biochemical activity found in all life forms and in a diverse set of reactions, including DNA replication and transcription, cellular metabolism, ion pumping, and protein assembly and trafficking [2]. Characterizing the ATPase activity of an enzyme is a critical step in determining how it functions both in vitro and in vivo. Many techniques have been described to measure the ATPase activity of various proteins, each with their own merits and limitations. Many methodologies detect the release of inorganic phosphate using radioisotopes [3] or colorimetry [4]. These methods are useful for high-throughput screening but can suffer from a poor signal-to-noise ratio and/or dynamic detection range. The direct detection of ATP hydrolysis can be

Lata Balakrishnan and Jason A. Stewart (eds.), *DNA Repair: Methods and Protocols*, Methods in Molecular Biology, vol. 1999, https://doi.org/10.1007/978-1-4939-9500-4_15,

achieved using radiolabeled ATP (e.g., α^{32}P-ATP) and thin-layer chromatography (TLC) [5]. This technique is sensitive and simple to perform but is not amenable to high-throughput assays. An alternate technique couples multiple enzymatic reactions together, detecting ATP hydrolysis in an indirect manner [6]. This coupled assay is simple to conduct, high throughput, and boasts automated data collection. Relative to the direct assay, large quantities of enzyme are needed for this method however, which can be a limiting factor. Here, we describe both the direct TLC assay and the NADH-coupled 96-well plate assay for the measurement of ATPase activity. The assay best suited for a particular ATPase depends on protein availability, equipment access, optimal reaction temperature, and your specific ATPase. Conducting both assays may be necessary for a thorough characterization of an enzyme's ATPase activity. However, one or the other will be suitable for most situations.

The direct TLC ATPase assay utilizes a positively charged polyethyleneimine (PEI) cellulose membrane as a stationary phase, and a formic acid and lithium chloride solution as the mobile phase. The less polar ADP migrates further in the mobile phase than ATP, thus producing separation on the membrane (Fig. 1). Detection of ATP and ADP is achieved by the use of radiolabeled ATP in the assay and subsequent phosphorimaging. The coupled ATPase assay uses two additional enzymes, pyruvate kinase and lactate dehydrogenase, to regenerate ATP and couple ATP hydrolysis to NADH oxidation (Fig. 2). ATP and ADP absorb light at similar wavelengths and, thus, cannot be easily distinguished. NADH and NAD^+, however, have distinct absorption spectra, and can easily be distinguished using spectrophotometry. Using the reactions in the coupled assay, ATP hydrolysis is equimolar to NADH oxidation, as monitored by absorbance at 340 nm. The TLC ATPase assay benefits from the small quantity of protein needed, the ability to easily assay a wide range of temperatures, and the lack of any necessary specialized equipment [7]. The NADH-coupled assay benefits from its high throughput nature, automated data collection, ease of preparation, and ATP regeneration, which maintains a constant concentration of ATP available for the ATPase [8].

Below, we describe detailed protocols for both the direct TLC assay and NADH-coupled ATPase assay. The TLC assay involves the preparation of the chromatography chamber and plate, taking time points during the assay itself, and development of the TLC plate. The NADH-coupled assay involves the preparation of the ATPase reaction, loading of a 96-well plate, and data collection via a spectrophotometer or plate reader. Both assays are capable of characterizing the ATPase activity of an enzyme, providing valuable insight into its in vitro activity and in vivo function.

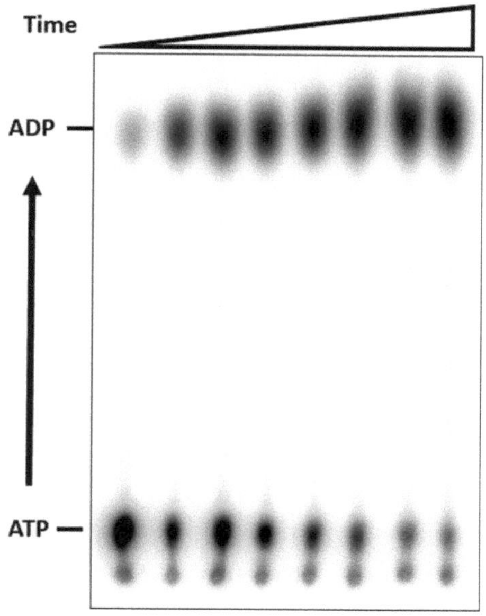

Fig. 1 Representative data of a thin layer chromatography ATPase assay. ADP migrates with the mobile phase up the TLC plate, while ATP migration is minimal. The arrow shows the direction of mobile phase movement. Over the course of the ATPase reaction, ATP is hydrolyzed into ADP and inorganic phosphate. This can be visualized by the presence of more ADP in the later time points of the assay

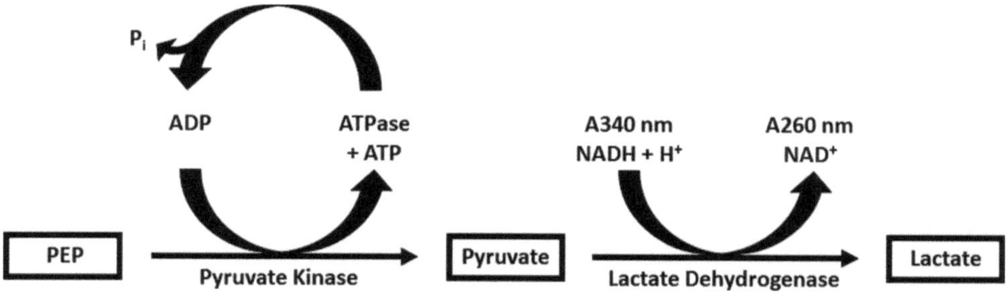

Fig. 2 The reactions taking place in the NADH-coupled ATPase assay. The ATPase converts ATP into ADP and Pi. Pyruvate kinase converts ADP back into ATP, while converting PEP into pyruvate. Pyruvate is then converted into lactate by lactate dehydrogenase. During this reaction, lactate dehydrogenase oxidizes NADH into NAD^+, changing its absorbance peak from 340 to 260 nm. This change in absorbance can be detected by a spectrophotometer or appropriate plate reader in real time

2 Materials

All buffers and other reagents should be made with ultrapure MilliQ water (resistivity of 18.2 MΩ·cm at 25 °C).

2.1 TLC Assay

1. Running buffer: 1.0 M formic acid, 400 mM LiCl. Add 1.70 g of LiCl and 4.24 mL formic acid (23.6 M) to a 100 mL

graduated cylinder. Bring the volume to 100 mL with water, and store in a glass bottle (*see* **Note 1**).

2. 2× ATPase buffer: 50 mM Na-HEPES, pH 8, 10% glycerol, 100 mM NaOAc, pH 7.5, 300 μM NaCl, 0.02% NP-40.
3. Reaction stop buffer: 2% SDS, 20 mM EDTA.
4. α[^{32}P]-ATP (*see* **Note 2**),
5. TLC PEI cellulose sheets 20 × 20 cm.
6. Glass TLC tank (*see* **Note 3**).
7. Vacuum grease (*see* **Note 4**).

2.2 NADH-Coupled Assay

1. 2× ATPase buffer: 50 mM Na-HEPES, pH 8.0, 10% glycerol, 100 mM NaOAc, pH 7.5, 300 μM NaCl, 0.02% NP-40. Store at −20 °C.
2. 100 mM ATP, pH 7.0: 100 mM ATP in water. Adjust the pH to 7.0 with NaOH. Store at −80 °C.
3. 10 mM phosphoenolpyruvic acid (PEP): Store at −80 °C
4. 10 mM NADH, pH 8.0: Dissolve 71 mg β-Nicotinamide adenine dinucleotide in 10 mL 25 mM Na-HEPES, pH 8. Aliquot 1 mL portions into 1.7 mL tubes and store at −80 °C.
5. Rabbit pyruvate kinase.
6. 100 mM $MgCl_2$.
7. Lactate dehydrogenase.
8. 10 μM DNA.
9. Clear, flat-bottom 96-well plate.
10. BioTek Synergy H1 plate reader (*see* **Note 5**).

3 Methods

3.1 TLC Assay

1. To delineate lanes for your reactions, draw a line with a pencil across a thin-layer chromatography sheet 1 cm from the bottom edge (*see* **Note 6**). Starting 0.5 cm from the edge, make a mark every 1 cm along the line, making 20 marks total.
2. Add approximately 50 mL of running buffer to the TLC tank (*see* **Note 7**). Use vacuum grease to line the top edge of the tank (ensuring a tight seal), and cover with the lid. Let the tank equilibrate with the buffer for ~30 min.
3. Add 2 μL of reaction stop buffer to seven tubes for each reaction, on ice.
4. Add 3.75 μL 2× ATPase buffer, 3 μL of 5 μM DNA oligonucleotide, 7.5 μL of 2× concentrated protein in 1× ATPase buffer, and 3.75 μL water to a 0.7 mL tube on ice (*see* **Note 8**). Mix gently with a pipette, and repeat for the number of reactions

desired (*see* **Note 9**). Assemble an additional reaction without protein, adding 7.5 μL 1× buffer instead to reach a volume of 15 μL. This will serve as a control for background ATP hydrolysis.

5. Add 0.6 μL α[^{32}P]-ATP (equal to 6 μCi) into each tube.
6. Keeping the reactions on ice, aliquot 2 μL of each reaction into a stop tube on ice. This is the zero time point (*see* **Note 10**).
7. Move the reactions to the desired temperature(s) (*see* **Note 11**). At 1, 2, 5, 10, 15, and 20 min, aliquot 2 μL of each reaction into a separate stop tube on ice (*see* **Note 12**).
8. Spot 1 μL of a stopped reaction onto a mark on the TLC plate. Repeat for each stop tube. Allow the spots to dry until no longer visible (*see* **Note 13**).
9. Place the dried TLC plate into the tank, submerging the bottom edge in running buffer. Ensure the tank lid is properly sealed around the edges of the tank. Develop the plate for 1 h, allowing the solvent front to reach 1 cm from the top of the TLC plate (*see* **Note 14**).
10. Remove the TLC plate from the tank and allow it to dry, cellulose side up (*see* **Note 15**).
11. Wrap the TLC plate in plastic wrap and expose to a phosphor plate for 4 h (*see* **Note 16**).
12. After exposure, scan the plate using a Typhoon imager. The hydrolyzed ADP will have migrated toward the top of the plate, while unhydrolyzed ATP will remain near the original spot placement (Fig. 2).
13. Quantitate the percent ATP hydrolyzed in each reaction by densitometry using the following equation: ((ADP)/(ADP + ATP)) × 100. The amount of ATP hydrolyzed in the no protein control reactions should be subtracted from each reaction.

3.2 NADH-Coupled Assay

1. Preheat the plate reader to 37 °C.
2. Assemble 60 μL of the 2× reaction cocktail for each reaction condition on ice: 10 mM ATP, pH 7.0, 10 mM $MgCl_2$, 1 mM PEP, 0.8 mM NADH, pH 8.0, 10 U/mL pyruvate kinase, 16 U/mL lactate dehydrogenase, 1× ATPase buffer.
3. Dilute the ATPase to 2× the final concentration in 1× ATPase buffer along with 2 μM DNA (*see* **Note 17**). Make an additional mixture without ATPase for a control for background NADH oxidation.
4. Add 60 μL of the 2× reaction cocktail and 60 μL of 2× ATPase in a tube on ice, mixing with a pipette.
5. Add 100 μL of each reaction mixture into the well of a 96-well plate.

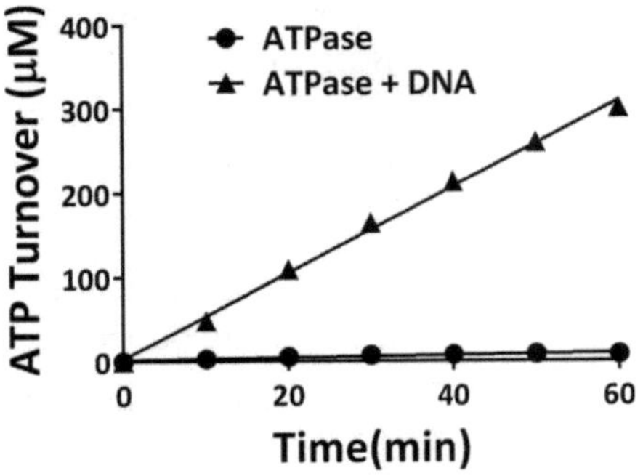

Fig. 3 Representative data of an NADH-coupled ATPase assay. A steady rate of ATP hydrolysis can be quantitated, along with total ATP turnover. Many ATPases involved in DNA metabolism, like the one assayed here, are stimulated by the presence of DNA

6. Insert the 96-well plate into the plate reader. Monitor the absorbance at 340 nm at 37 °C every 60 s for 1 h without shaking (*see* **Note 18**).
7. Once data collection is complete, export the data to a spreadsheet software (e.g., Microsoft Excel).
8. To calculate the NADH concentration and ultimately the ATP hydrolysis rate for each reaction, an NADH standard curve must be generated (*see* **Note 19**). Substitute each absorbance read into the $y = mx + b$ equation from the standard curve, where y = absorbance at 340 nm and x = NADH concentration. Subtract the calculated NADH concentrations from the starting NADH concentration. Assuming 1 µM NADH oxidized is equivalent to 1 µM ATP hydrolyzed, these values are the amount of ADP present at each time point. Thus, the data can be presented as ADP generated or ATP hydrolyzed (Fig. 3) (*see* **Note 20**).

4 Notes

1. TLC buffer should be made fresh for each experiment for optimal resolution.
2. Although this protocol is designed for the measurement of ATPase activity, it can also be used to measure the hydrolysis of other nucleotide triphosphates.
3. In place of a chromatography tank, inexpensive glass tanks can be used instead. A fish tank with a glass lid cut to size works well.
4. Any highly viscous grease (e.g., Vaseline) can be used here.

5. Alternatively, a spectrophotometer can be used to monitor a single reaction.
6. Pencil should always be used instead of ink, which contains pigments that will also separate on the TLC plate. Use light pressure to avoid puncturing the TLC sheet. A blunt, nonmechanical pencil works well.
7. Buffer amount will vary depending on tank size. The buffer should cover the bottom of the tank, but should not rise above 1 cm in depth. Buffer levels higher than the ATP spots will result in diffusion and poor resolution.
8. PCR tubes can also be used. These allow for incubation in a thermocycler but require greater care in ensuring the temperature of the reactions does not rise above ice temperature during reaction assembly.
9. Alternatively, a master mix of components can be assembled, depending on the number of conditions and replicates being assayed. Several master mixes can be assembled for multiple reactions of several conditions (e.g., +/– DNA).
10. This aliquot should be taken as soon as possible after the addition of ATP. Hydrolysis will usually begin to occur even when the samples are on ice.
11. Reactions can be incubated in a gradient of temperatures using an appropriate thermocycler.
12. Time points may need to be taken earlier than 1 min or later than 20 min, depending on the ATPase and reaction conditions.
13. Spots can be aliquoted onto the TLC plate as reactions are stopped or all at once after the reaction has finished. Aliquots are transferred easily if the drop is carefully expelled from the pipette tip, and then gently tapped onto the TLC plate. Spots take approximately 5 min to dry.
14. Depending on the dimensions of your TLC tank, two or more TLC plates can be developed at once in the same tank. The plates should be set up so they are not touching each other, and not overly adhering to the sides of the tank, to ensure proper development.
15. The TLC plate can be placed on top of a few layers of paper towels to facilitate drying.
16. The necessary exposure time is dependent on the activity date of the $\alpha[^{32}P]$-ATP. Fresh $\alpha[^{32}P]$-ATP only requires a few hours of exposure, while older ATP may require overnight exposure.
17. Reagents in the 2× reaction cocktail are doubled to provide 1× concentration in the final reaction. Various conditions to assay include ATPase concentration, DNA oligonucleotide sequence

and length, additional cofactors, or other ATPase-specific condition variants. To eliminate any secondary structure the DNA oligonucleotides may form, the DNA can be boiled for 5 minutes, and snap-cooled on ice prior to addition to the cocktail. A reaction should also be run in the absence of DNA, to determine its magnitude of ATPase stimulation, if any.

18. The temperature of the reaction, as well as the frequency and duration of absorption measurements can be adjusted for the ATPase being assayed. Depending on the hydrolysis rate of the ATPase, reactions may need to be equilibrated at room temperature for 3–5 min prior to absorbance measurement on a plate reader. If the reactions go directly from ice to the reaction temperature (e.g., 37 °C), the first few reads will not be linear as the ATPase rate changes along with the temperature of the reactions.
19. To generate an NADH standard curve, measure the absorbance of known concentrations of NADH within a linear range (we use 0.3 to 800 μM). Plot the concentration of NADH on the x-axis, and the absorbance values on the *y*-axis. Fit the data with a line of best fit. The resultant $y = \mathrm{m}x + b$ line equation yields the slope (m) and *y*-intercept (b) needed to correlate the absorbance values of an ATPase assay with NADH concentration, as in Subheading 3.2, **step 8**.
20. While ATP is regenerated in this assay, NADH is not. Over the course of the reaction, the absorbance reads will decrease as NADH is oxidized. If all of the NADH is oxidized, the absorbance reads will flat-line, suggesting ATP hydrolysis has ceased. However, these data points should be disregarded as they no longer reflect ATPase activity. For extremely fast ATPases, NADH can be depleted in the time it takes to get the 96-well plate on the plate reader. In this scenario, the concentration of NADH in the reactions should be increased.

Acknowledgments

This work was supported by the College of Arts and Sciences, Indiana University, and Indiana University Collaborative Research Grant fund of the Office of the Vice President for Research, a Collaboration in Translational Research Pilot Grant from the Indiana Clinical and Translational Sciences Institute, and the American Cancer Society [RSG-16-180-01-DMC].

References

1. Heppel LA, Hilmoe RJ (1953) Mechanism of enzymatic hydrolysis of adenosinetriphosphate. J Biol Chem 202(1):217–226
2. Ogura T, Wilkinson AJ (2001) AAA(+) superfamily ATPases: common structure-diverse function. Genes Cells 6(7):575–597. https://doi.org/10.1046/j.1365-2443.2001.00447.x
3. King SM (1995) Chapter 21: Measurement of ATPase activity using [γ-32P]ATP. In: Dentler W, Witman G (eds) Methods in cell biology, vol 47. Academic Press, New York, pp 141–145. https://doi.org/10.1016/S0091-679X(08)60802-X
4. Kagami O, Kamiya R (1995) Chapter 22: Non-radioactive method for ATPase assays. In: Dentler W, Witman G (eds) Methods in cell biology, vol 47. Academic Press, New York, pp 147–150. https://doi.org/10.1016/S0091-679X(08)60803-1
5. Rajagopal V, Lorsch JR (2013) Chapter 25: ATP and GTP hydrolysis assays (TLC). In: Lorsch J (ed) Methods in enzymology, vol 533. Academic Press, New York, pp 325–334. https://doi.org/10.1016/B978-0-12-420067-8.00025-8
6. Kiianitsa K, Solinger JA, Heyer WD (2003) NADH-coupled microplate photometric assay for kinetic studies of ATP-hydrolyzing enzymes with low and high specific activities. Anal Biochem 321(2):266–271
7. Andis NM, Sausen CW, Alladin A, Bochman ML (2018) The WYL domain of the PIF1 helicase from the thermophilic bacterium thermotoga elfii is an accessory single-stranded DNA binding module. Biochemistry 57(7):1108–1118. https://doi.org/10.1021/acs.biochem.7b01233
8. Rogers CM, Wang JC, Noguchi H, Imasaki T, Takagi Y, Bochman ML (2017) Yeast Hrq1 shares structural and functional homology with the disease-linked human RecQ4 helicase. Nucleic Acids Res 45(9):5217–5230. https://doi.org/10.1093/nar/gkx151

Chapter 16

Gel-Based Assays for Measuring DNA Unwinding, Annealing, and Strand Exchange

Cody M. Rogers, Christopher W. Sausen, and Matthew L. Bochman

Abstract

Efficient replication and repair of the genome requires a multitude of protein–DNA transactions. These interactions can result in a variety of consequences for DNA such as the unwinding of double-stranded DNA (dsDNA) into single-stranded DNA (ssDNA), the annealing of complementary ssDNAs, or the exchange of ssDNA with one strand of a dsDNA duplex. Some DNA helicases possess all three activities, but many DNA-interacting proteins can also catalyze one or more of these reactions. Assays that quantify these activities are an important first step in characterizing these protein–DNA interactions in vitro. Here, we describe methods for the formation of dsDNA substrates and the assays that can be used to biochemically characterize proteins that can unwind, anneal, and/or exchange DNA strands.

Key words Radiolabeled DNA, DNA helicase, DNA unwinding, DNA annealing, DNA strand exchange

1 Introduction

The melting of double-stranded DNA (dsDNA) is an essential step in multiple aspects of DNA replication, recombination, and repair [1]. In the case of DNA helicases, identifying the types of substrates preferentially unwound by the protein is an informative first step in understanding how the enzyme functions. There are a plethora of techniques to measure DNA unwinding, but the complexity and output from these methods can vary considerably. Recent advances in single-molecule microscopy have allowed for analysis of DNA melting at a molecular level. Some of these modern techniques are summarized in ref. 2, and the use of magnetic tweezers to measure helicase activity is reviewed in ref. 3. However, gel-based methods for measuring DNA unwinding remain a common and highly accessible technique to assess helicase activity and can be performed with common laboratory equipment. Gel-based methods are advantageous in that they are simple to set up and flexible in the types of experiments that can be performed (e.g., time

Lata Balakrishnan and Jason A. Stewart (eds.), *DNA Repair: Methods and Protocols*, Methods in Molecular Biology, vol. 1999, https://doi.org/10.1007/978-1-4939-9500-4_16,

courses vs. titrations) [4, 5]. Indeed, valuable data can be extracted from a thorough analysis of the substrates that are preferentially unwound by a helicase.

In addition to DNA unwinding, many helicases and other DNA binding proteins are capable of annealing complementary single-stranded DNA (ssDNA) strands [6]. Some helicases anneal ssDNAs so readily that DNA unwinding is masked [7]. Annealing can be observed using a method similar to measuring DNA unwinding with just a few simple modifications. Here, we describe gel-based methods to monitor DNA unwinding and annealing that we use for characterizing DNA helicases and describe helpful techniques to measure DNA unwinding by helicases that strongly anneal ssDNA [5]. Additionally, proteins involved in unwinding or annealing may also participate in DNA strand exchange [8]. Briefly, this involves swapping a strand from dsDNA with complementary ssDNA, a process which is typified by strand exchange in homologous recombination [9]. These experiments are described such that the equipment and reagents required to run each assay are nearly identical.

To begin, we describe a protocol for 5′ labeling of DNA with ^{32}P and suggest alternative methods for DNA radiolabeling. Once DNA is labeled, many gel-based techniques can be used to measure protein–DNA interactions. We describe, in detail, a method for preparing reactions to observe DNA unwinding, annealing, and strand exchange. Finally, we describe how to prepare and run these gels, as well as how they can be used to quantify the desired protein–DNA transaction. Taken together, these assays are an accessible first step to characterize many protein–DNA interactions, especially in the case of DNA helicases.

2 Materials

All buffers and other reagents should be made with ultrapure MilliQ water (resistivity of 18.2 MΩ·cm at 25 °C).

2.1 DNA Labeling

1. 10 μM oligonucleotides.
2. T4 polynucleotide kinase (T4 PNK).
3. 10× T4 PNK buffer.
4. γ[^{32}P]-ATP (PerkinElmer).
5. Illustra ProbeQuant G-50 micro columns (GE Healthcare).
6. EDTA: 0.5 M ethylenediaminetetraacetic acid disodium salt in water, pH to 8 with NaOH.
7. TE: 90 mM Tris–HCl (pH 8.0) and 2 mM EDTA (pH 8.0).
8. 2× Annealing buffer: 40 mM Tris–HCl (pH 8), 8% glycerol, 0.2 mM EDTA, 80 μg/mL BSA, 20 mM DTT, and 20 mM MgOAc.

2.2 Native Gel Electrophoresis

1. 1× TBE: 90 mM Tris base, 90 mM boric acid, and 2 mM EDTA (pH 8.0).
2. 10% Ammonium persulfate (APS) in water.
3. TEMED.
4. 5 mL DNA resolving gel: 8% 19:1 acrylamide–bis-acrylamide in 1× TBE, 80 μL APS, and 20 μL TEMED.
5. 3× DNA loading dye: 15% glycerol, 60 mM EDTA, 0.15% SDS, and 0.75% bromophenol blue.
6. Polyacrylamide gel electrophoresis rig.
7. Power supply for electrophoresis.
8. 0.34 mm Whatman paper.
9. Storage phosphor screen and cassette.

2.3 Biochemical Assays

1. 2× Helicase buffer: 50 mM Na-HEPES (pH 8.0), 10% glycerol, 100 mM NaOAc (pH 7.6), 300 μM NaCl, 15 mM MgOAc, and 0.02% Tween 20.
2. 50 mM adenosine triphosphate in water, pH to 7 with NaOH (ATP).
3. 4 mg/mL SDS–Proteinase K (SDS-PK) in 50 mM Tris–HCl (pH 8.0) and 3 mM $CaCl_2$.

3 Methods

3.1 DNA Substrate Preparation

1. Radiolabel 1 μL of 10 μM oligo 1 on the 5′ end with 1 μL T4 PNK, 2 μL 10× PNK buffer, 2 μL $\gamma[^{32}P]$-ATP, and 14 μL H_2O (*see* **Note 1**). This protocol uses a substrate with 5′ and 3′ ssDNA tails (i.e., fork) as an example, but a variety of substrates can be made with this method (Fig. 1a, b) (*see* **Note 2**). Incubate this reaction for 1 h at 37 °C. Stop the reaction with 1 μL 0.5 M EDTA and by incubating for 5 min at 95 °C.
2. Prepare an illustra ProbeQuant G-50 micro column by spinning at 700 × *g* for 1 min to remove the storage buffer. Remove unincorporated $\gamma[^{32}P]$-ATP by loading the entire labeling reaction onto a G-50 micro column. Place the G-50 micro column in a fresh 1.5 mL tube and spin for 2 min at 700 × *g* to collect the clean, labeled oligonucleotide, which is at ~450 nM assuming a 90% recovery from the G-50 micro column.
3. Set up the annealing reaction by incubating 5 μL of 450 nM labeled oligonucleotide, 5 μL of 500 nM complementary oligonucleotide (oligo 2), 25 μL 2× annealing buffer, and 15 μL H_2O (*see* **Note 3**). Incubate the reaction at 37 °C overnight

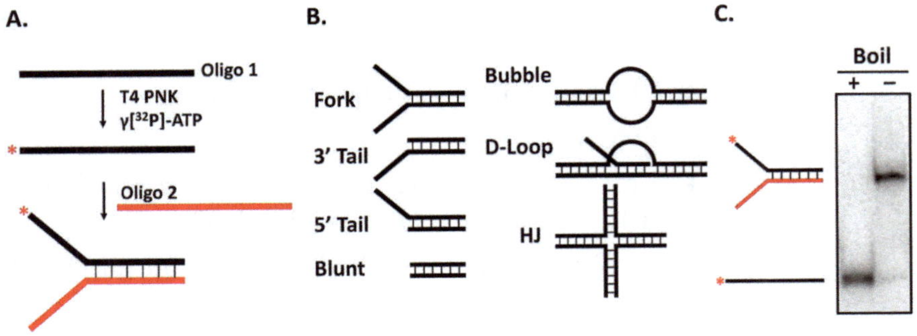

Fig. 1 Formation of DNA substrates. (**a**) Outline for labeling and annealing a DNA fork. Oligo 1 (black) is 5′ radiolabeled (indicated by a red asterisk) prior to annealing to the partially complementary oligo 2 (red). (**b**) Schematic examples of the types of substrates that can be made using this protocol. (**c**) Representative image of labeled fork DNA made using this protocol

[10] (*see* **Note 4**). The desired substrate will be at ~45 nM (*see* **Note 5**). Dilute the annealed DNA to 10 nM in 1× TE.

4. To ensure that the DNA was properly annealed, add 10 μL of the labeled, annealed DNA to two separate tubes and add 5 μL 3× DNA loading dye to each. Incubate one of these samples at 95 °C for 5 min, which will serve as the heat-denatured reference. Load 10 μL of each dye containing sample to an 8% 19:1 acrylamide–bis-acrylamide gel. Using 1× TBE running buffer, run the gel at 10 V/cm until the dye front is near the bottom of the gel (*see* **Note 6**).
5. Remove the gel from the rig and place it on a piece of 0.34 mm Whatman paper. Cover the top of the gel with plastic wrap and another piece of Whatman paper. Also, place two more pieces of Whatman paper below the gel. Dry the gel on a gel dryer for 20 min at 80 °C (*see* **Note 7**).
6. Remove the dried gel from the sandwich leaving the plastic wrap and Whatman paper that are in contact with the gel attached. Expose the gel to a storage phosphor plate for several hours or overnight (*see* **Note 8**). Visualize the DNA bands using a Typhoon Imager. The heat-denatured lane (ssDNA) should migrate faster than the annealed DNA substrate. The undenatured DNA will migrate slower, but there is usually some (<5%) contaminating denatured ssDNA in the annealed sample (Fig. 1c) (*see* **Note 9**).

3.2 Helicase Assays

1. Set up the following reactions on ice. Dilute protein from freezer stocks to 5× the desired final concentration in 1× helicase buffer. Mix 2 μL of protein with 1 μL 10 nM labeled DNA (from Subheading 3.1), 1 μL 50 mM ATP, 1 μL H_2O, and 5 μL 2× helicase buffer (*see* **Note 10**). For proteins that exhibit strong annealing activity, it helps to add 10 nM cold trap, that is, a molar excess of unlabeled ssDNA that is equivalent

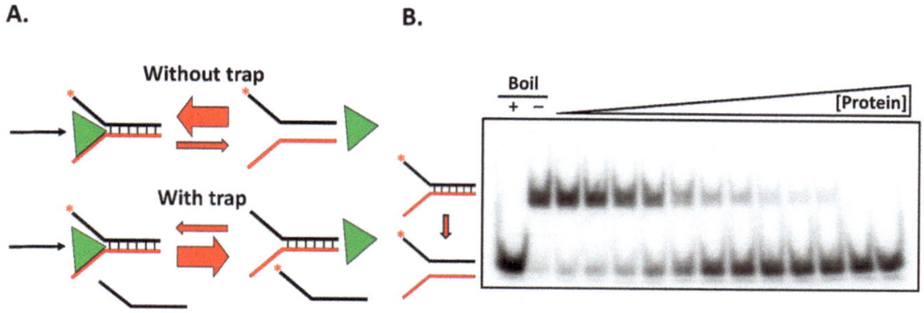

Fig. 2 DNA unwinding assay. (**a**) Complications when observing DNA unwinding by a protein with annealing activity (green triangle). In the absence of cold ssDNA trap, the unwound product can be reannealed resulting in lower than expected helicase activity. In extreme cases, the unwound product can be completely reannealed. Excess cold trap will be more likely to anneal with the unlabeled strand in the unwound product, resulting in observable labeled ssDNA. (**b**) Example of DNA unwinding. Increasing protein concentration should generate more ssDNA

to the labeled strand (oligo 1) in your substrate (Fig. 2a) (*see* **Note 11**). Incubate the reactions at 37 °C for 30 min (*see* **Note 12**). One reaction should omit protein to use as a negative control measuring spontaneously unwound DNA. Another reaction should be heat-denatured at 95 °C for 5 min to use as reference for ssDNA migration (Fig. 2b).

2. Place the unwinding reactions on ice after they are completed. To stop each reaction, add 5 μL of 3× DNA loading dye supplemented with 400 μg/mL SDS-PK to each reaction. Incubate the reactions at 37 °C for 1 min to release protein from the DNA (*see* **Note 13**). Load 10 μL of these reactions on an 8% 19:1 acrylamide–bis-acrylamide gel using 1× TBE running buffer and run at 10 V/cm until the dye front reaches the bottom of the gel.
3. Dry the gel as in Subheading 3.1, **step 5**.
4. Image the gel as in Subheading 3.1, **step 6**.
5. The percent of DNA unwound can be calculated with densitometry with the following equation: (ssDNA density/total × 100). The amount of DNA unwound from the no protein control should be subtracted from each of the reactions.

3.3 Annealing Assays

1. Set up annealing reactions on ice. Dilute protein concentration from stock to 5× the desired final concentration in 1× helicase buffer. Prepare complementary ssDNA by heat-denaturing the desired substrate made in 3.1 for 10 min at 95 °C (*see* **Note 14**). Mix together 2 μL protein, 1 μL 10 nM melted DNA, 2 μL H_2O, and 5 μL 2× helicase buffer (*see* **Note 15**). Incubate the reactions at 37 °C for 30 min. Omit protein from one reaction for a negative control for spontaneously annealed

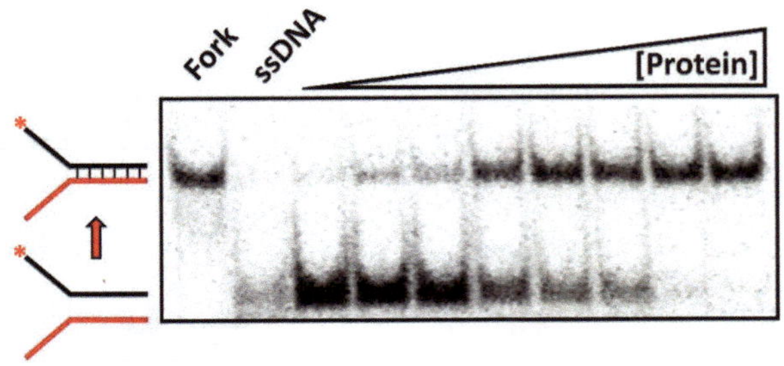

Fig. 3 Representative DNA annealing assay. This is an example of an enzyme with strong annealing activity. Increasing concentrations of proteins result in annealing of ssDNA to form the fork substrate

DNA. Another lane should contain no protein but an equimolar amount of annealed substrate as a reference for where annealed DNA migrates (Fig. 3).

2. Stop and load the reactions on a gel as in Subheading 3.2, **step 2**.
3. Dry the gel as in Subheading 3.1, **step 5**.
4. Image the gel as in Subheading 3.1, **step 6**.
5. The percentage of annealed DNA can be quantified using densitometry by calculating (annealed density/total × 100). The amount of spontaneously annealed DNA should be subtracted from each protein-containing reaction.

3.4 Strand Exchange Assays

1. The oligonucleotides should be designed such that the labeled, annealed DNA structure has a different migration rate than the strand exchange product (Fig. 4). These reactions should be set up on ice. Prepare protein by diluting to 5× the desired final concentration in 1× helicase buffer. Mix 2 μL protein with 5 μL 2× helicase buffer, 1 μL H_2O, 1 μL 10 nM starting substrate (e.g., fork), and 1 μL 10 nM ssDNA oligo that upon exchange makes a differentially migrating product (e.g., blunt). Incubate the reactions at 37 °C for 30 min. Prepare two additional reactions that omit protein, one that contains the starting substrate and the other with the exchange product.
2. Stop and load the reactions on a gel as in Subheading 3.2, **step 2** (*see* **Note 16**).
3. Dry the gel as in Subheading 3.1, **step 5**.
4. Image the gel as in Subheading 3.1, **step 6**.
5. Although there are multiple ways to quantify strand exchange, the simplest is to calculate the percent formation of the strand exchange product with the equation (exchange density/total × 100).

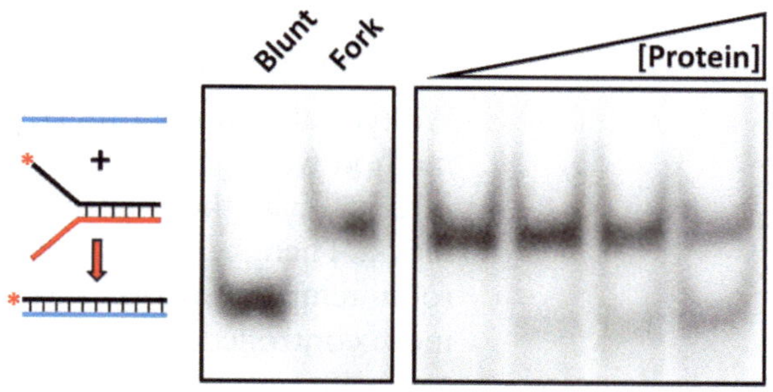

Fig. 4 Strand exchange to blunt dsDNA from a fork and a third ssDNA strand. Fork DNA (black and red) is incubated with complementary ssDNA (blue) along with increasing protein concentration. Formation of the blunt substrate (black and blue) increases in the presence of more protein

4 Notes

1. The 5′ labeled DNA method described here is a simple and highly efficient technique. For those seeking other methods of radiolabeling, New England BioLabs has a nice summary of alternative techniques that use different labeling enzymes and/or other radiolabels (https://www.neb.com/products/dna-modifying-enzymes-and-cloning-technologies/dna-labeling/dna-labeling). While our method utilizes radioactive probes, those without access to radioactive substrates can alternatively use fluorescently labeled DNA.
2. For the simplicity of describing this method, we refer to the formation of a commonly used fork substrate. However, more complicated substrates can be made as well, including those that involve annealing more than two strands (Fig. 1b). For these substrates, incubate all of the strands together at nearly equimolar amounts. This method can also be used to make dsRNA and DNA–RNA hybrids, although the efficiency is not as high as with DNA–DNA. RNA degrades more easily than DNA, so care should be taken when using these substrates. DEPC-treated water can be used to help prevent degradation. Finally, DNA sequence is an important factor to consider when designing substrates. For example, the fork we describe has ssDNA arms that can form secondary structures that can promote or disrupt the interaction of the desired protein with DNA [11].
3. Some labs that use variations of this protocol use a greater excess of the cold oligonucleotide (oligo 2) than we show here. This can be considered in order to drive the formation of annealing to the labeled oligonucleotide. However, this means there will be some amount of contaminating, unobservable cold oligonucleotide in subsequent reactions. We have

found that using nearly equimolar DNA is sufficient for driving the formation of clean substrate.

4. We typically anneal our DNA using the overnight isothermal method we suggest here. However, another common method is to incubate the annealing reaction for 5 min in a 95 °C heat block. Then, turn the heat block off, and let it slowly cool to room temperature over several hours. This can also be done in a more controlled setting in a thermocycler.
5. The annealing reaction can be done with a higher concentration of DNA. However, a limiting factor is the efficiency of radiolabeling. Adding more DNA than what can be labeled will result in inaccurate DNA concentrations and can confound results due to the presence of excess unlabeled DNA.
6. Gel conditions should be optimized based on the structure and size of the substrate. Smaller substrates (such as blunt dsDNA) will run at nearly the same position as its unwound ssDNA counterpart on typical gels. Thus, higher percentage gels are usually preferred for these substrates. The running buffer and gel composition can be changed as well for varying effects. For example, the addition of 10 mM MgOAc to the gel can improve the stability of duplex DNA in the gel.
7. Gel percentage and thickness will affect the conditions needed to dry without cracking. We recommend running and drying a gel without substrate to determine the best parameters for drying based on the equipment available before risking the loss of a gel containing important reactions. An alternative to drying under heat and vacuum is to freeze the gels at −20 °C during the overnight exposure in a storage phosphor cassette.
8. Storage phosphor exposure duration can vary. If using a high enough concentration of DNA that was made with fresh, hot probe, several hours is usually sufficient time for good signal. However, overnight exposures are often recommended, especially when using low concentrations of DNA in biochemical assays. Longer exposures may help with older probe, but typically, the improvements to the signal are insignificant even after multiple days.
9. If the substrates are consistently highly contaminated with ssDNA (>10%), gel extraction can be used to purify the annealed substrate away from the component ssDNA. To do this, load the whole sample of prepared DNA substrate onto a gel as described in Subheading 3.1, **step 4**. Do not dry this gel, but instead expose the gel to autoradiography film in a dark room for 1–3 min and develop on a X-OMAT film developer. Both ssDNA and dsDNA should be observable in the same lane. Align the gel to the film image, cut out the slower migrating product, crush and place the gel fragment in 0.5× TBE,

and let diffuse overnight. Remove the buffer to a fresh tube away from the gel chunks, and perform a standard DNA precipitation. Resuspend the DNA pellet in 50 μL TE. Continue with this substrate through Subheading 3.1, **step 4**. A substrate of known concentration will be required to determine the concentration of the gel extracted substrate, some of which will be lost during the extraction.

10. There is some flexibility in the amounts of reagents we recommend in all of these assays. For example, 5× helicase buffer is a viable option when more than 2 μL of protein is required to reach the desired final concentration. However, be cautious when adding protein directly from your storage buffer, especially if it contains high amounts of glycerol or is in drastically different buffer conditions than the reaction buffer. We recommend using a storage buffer similar to that used in the reaction buffer, and the helicase buffer itself can be optimized to the protein of interest. Lower (<5%) amounts of glycerol in the final reaction will improve gel images as well.
11. When using cold trap, it is helpful to set up the reactions with everything but cold trap and ATP. This will allow the protein to interact with, but not unwind, the labeled DNA of interest prior to competition by the trap. Adding the cold trap and ATP last will initiate the helicase reaction, but there are complications with method. We typically observe less overall unwinding with helicases that anneal DNA even under these conditions. Annealing is constantly in competition with unwinding, and the excess ssDNA competes for interaction with the protein. However, this method allows unwinding to be observed by helicases with exceptionally strong annealing activity [5]. For the more complex substrates with multiple points of annealing such as the bubble or Holliday junction, this method becomes even less efficient.
12. The temperature and duration of these biochemical reactions is just a recommended starting point. For uncharacterized enzymes, a range of temperatures should be tested to optimize the reaction based the protein. The method we describe is a helicase titration, which is a good place to start to determine a concentration for optimal helicase activity. However, it is often worth knowing the rate at which the helicase can unwind DNA substrates. Set up these reactions with a single concentration of protein and stop them at various time points. Most reactions are fast, so plan to take most data points from ~0 to 10 min.
13. The SDS-PK treatment we recommend here is very mild. Some enzymes may require longer treatments or higher temperatures (typically up to 50 °C). Depending on the protein, this step may be omitted entirely.

14. Alternatively, labeled ssDNA can be made as in Subheading 3.1, **step 1**. An equimolar amount of cold complementary oligo will need to be added to the reaction.
15. Most reports of annealing are ATP-independent, and in the case of helicases, adding ATP would inhibit annealing activity as the enzyme would unwind any annealed DNA.
16. The percentage of the gel is important here to be able to separate 3 DNA substrates: the starting substrate, the exchange product, and contaminating ssDNA.

Acknowledgments

Funding: This work was supported by the College of Arts and Sciences, Indiana University, and Indiana University Collaborative Research Grant fund of the Office of the Vice President for Research, a Collaboration in Translational Research Pilot Grant from the Indiana Clinical and Translational Sciences Institute, and the American Cancer Society [RSG-16-180-01-DMC].

References

1. Bochman ML (2014) Roles of DNA helicases in the maintenance of genome integrity. Mol Cell Oncol 1(3):e963429. https://doi.org/10.4161/23723548.2014.963429
2. Brosh RM Jr (2016) Special methods collection on DNA helicases. Methods 108:1–3. https://doi.org/10.1016/j.ymeth.2016.08.009
3. Hodeib S, Raj S, Manosas M, Zhang W, Bagchi D, Ducos B, Fiorini F, Kanaan J, Le Hir H, Allemand JF, Bensimon D, Croquette V (2017) A mechanistic study of helicases with magnetic traps. Protein Sci 26(7):1314–1336. https://doi.org/10.1002/pro.3187
4. Cejka P, Kowalczykowski SC (2010) The full-length *Saccharomyces cerevisiae* Sgs1 protein is a vigorous DNA helicase that preferentially unwinds Holliday junctions. J Biol Chem 285(11):8290–8301. https://doi.org/10.1074/jbc.M109.083196
5. Rogers CM, Wang JC, Noguchi H, Imasaki T, Takagi Y, Bochman ML (2017) Yeast Hrq1 shares structural and functional homology with the disease-linked human RecQ4 helicase. Nucleic Acids Res 45(9):5217–5230. https://doi.org/10.1093/nar/gkx151
6. Bhargava R, Onyango DO, Stark JM (2016) Regulation of single-strand annealing and its role in genome maintenance. Trends Genet 32(9):566–575. https://doi.org/10.1016/j.tig.2016.06.007
7. Macris MA, Krejci L, Bussen W, Shimamoto A, Sung P (2006) Biochemical characterization of the RECQ4 protein, mutated in Rothmund-Thomson syndrome. DNA Repair 5(2):172–180. https://doi.org/10.1016/j.dnarep.2005.09.005
8. Keller H, Kiosze K, Sachsenweger J, Haumann S, Ohlenschlager O, Nuutinen T, Syvaoja JE, Gorlach M, Grosse F, Pospiech H (2014) The intrinsically disordered amino-terminal region of human RecQL4: multiple DNA-binding domains confer annealing, strand exchange and G4 DNA binding. Nucleic Acids Res 42(20):12614–12627. https://doi.org/10.1093/nar/gku993
9. Daley JM, Kwon Y, Niu H, Sung P (2013) Investigations of homologous recombination pathways and their regulation. Yale J Biol Med 86(4):453–461
10. Kanter DM, Kaplan DL (2011) Sld2 binds to origin single-stranded DNA and stimulates DNA annealing. Nucleic Acids Res 39(7):2580–2592. https://doi.org/10.1093/nar/gkq1222
11. Rogers CM, Bochman ML (2017) *Saccharomyces cerevisiae* Hrq1 helicase activity is affected by the sequence but not the length of single-stranded DNA. Biochem Biophys Res Commun 486(4):1116–1121. https://doi.org/10.1016/j.bbrc.2017.04.003

Chapter 17

In Vitro Assay for Plasmid Length DNA Strand Exchange by Human DMC1

Steven D. Goodson, Russell B. Hawes, Sarah M. Waldvogel, and Michael G. Sehorn

Abstract

Meiosis is a specialized cell division that generates gametes. Meiotic recombination is essential not only to generate diversity in offspring, but also to hold homologous chromosomes together through chiasma allowing proper chromosome segregation. This process requires the meiosis-specific recombinase, DMC1. DMC1 facilitates the search for homology between the homologous chromosomes and is followed by DNA strand invasion and strand exchange to produce a linkage between the two homologous chromosomes. The development of biochemical in vitro assays and the purification of the requisite proteins factors has led to a better understanding of the molecular mechanisms of meiotic homologous recombination. In this chapter, a detailed in vitro assay to examine DNA strand exchange over 5000 bases of DNA catalyzed by human DMC1 is described. This method has proved to be valuable for examining the catalytic potential of hDMC1 and delineating the functional interaction with other HR factors.

Key words DMC1, Meiotic recombination, DNA strand exchange, Homologous recombination

1 Introduction

Meiosis is a type of cell division that produces haploid cells from diploid cells. To initiate meiosis, programmed DNA double-stranded breaks (DSBs) are introduced into the genome by the Spo11 enzyme [1]. The DSBs can be repaired to yield either a crossover or a non-crossover. Crossover formation results in physical linkages between homologous pairs of chromosomes. During meiosis, linkages between interhomolog chromatids are important to ensure proper segregation during meiotic prophase I [2]. DMC1, the meiosis specific RecA homolog, along with other accessory proteins is responsible for catalyzing the formation of crossover structures. To do this, the DSB is first processed to expose a 3′ ssDNA tail [3, 4] upon which DMC1 forms a right handed helical filament [5]. The DMC1-ssDNA filament is known as the presynaptic filament. The presynaptic filament searches for

Lata Balakrishnan and Jason A. Stewart (eds.), *DNA Repair: Methods and Protocols*, Methods in Molecular Biology, vol. 1999, https://doi.org/10.1007/978-1-4939-9500-4_17,

homology between the interhomolog chromatid and the ssDNA found within the presynaptic filament [2, 3]. Once homology has been found, the presynaptic filament strand invades the duplex DNA and promotes strand exchange to form a joint molecule that links the two interhomologs together. This chapter provides a detailed in vitro protocol to monitor human DMC1-catalyzed DNA strand exchange over five thousand bases of DNA.

2 Materials

2.1 Protein Purifications

1. Purified human RPA [6].
2. Purified human DMC1 [7].

2.2 Substrate Preparation

1. ϕX174 ssDNA (New England Biolabs).
2. ϕX174 Replicative Form I dsDNA (New England Biolabs).
3. ApaL1 (New England Biolabs).
4. Nuclease-free water.
5. 10× Cutsmart Buffer (New England Biolabs).
6. Phenol–chloroform–isoamyl alcohol (25:24:1).
7. M sodium acetate at pH 5.2.
8. 100% Isopropanol.
9. 70% ethanol.
10. 1× TE: 10 mM Tris–HCl pH 7.4, 1 mM EDTA.

2.3 Plasmid Length DNA Strand Exchange

1. 125 mM Tris–HCl at pH 7.4.
2. 25 mM ATP/30 mM $MgCl_2$.
3. 1 M creatine phosphate.
4. 1.5 mg/mL creatine kinase.
5. Nuclease-free water.
6. 0.14 μM ϕX174 ssDNA (New England Biolabs).
7. 3 M potassium chloride (KCl).
8. Buffer A: 20 mM KH_2PO_4, 10% glycerol, 0.5 mM EDTA, 100 mM KCl, pH 7.4.
9. ApaLI Linearized ϕX174 Replicative Form I dsDNA (0.14 μM) (New England Biolabs) (*see* **Note 1**).
10. 50 mM spermidine.
11. 173 μM purified hDMC1.
12. 50 μM purified hRPA.
13. 10% SDS.
14. 10 mg/mL proteinase K.

2.4 Product Resolution

1. Agarose.
2. 1× TAE buffer: 40 mM Tris–acetate, pH 7.5, 0.5 mM EDTA.
3. Agarose gel mold.
4. Agarose gel electrophoresis apparatus.
5. 0.25% Orange G in 50% glycerol.
6. 10 mg/mL ethidium bromide stain.
7. Gel documentation scanner.

3 Methods

In this assay, hDMC1 forms a presynaptic filament on the ϕX174 ssDNA circular stand (css). ϕX174 ssDNA is homologous to one strand contained in the ϕX174 Replicative Form I dsDNA (lds). hDMC1 presynaptic filament searches the linear duplex for homology. Once homology is located, a joint molecule intermediate (jm) is produced. Further strand exchange generates a linear single-stranded DNA (lss) and a nicked circular (nc) duplex DNA as final products (Fig. 1a). Reaction products can be resolved by using agarose gel electrophoresis (Fig. 1b, lanes 1–6).

3.1 ϕX174 Replicative Form I dsDNA Linearization

1. Digest 5 μg of ϕX174 Replicative Form I dsDNA in a buffer containing 1× Cutsmart, nuclease-free water, and 1 unit of ApaLI in a total of 100 μL for 1 h at 37 °C in a 1.5 mL microcentrifuge tube.
2. Add 150 μL of nuclease-free water.
3. Add equal volume of phenol–chloroform–isoamyl alcohol (25:24:1).
4. Invert the tube multiple times to mix.
5. Centrifuge the tube for 5 min at 18,000 ×*g* in a microcentrifuge (*see* **Note 2**).
6. Transfer the top phase of the separated solution to a new 1.5 mL microcentrifuge tube.
7. Discard the bottom phase.
8. Add equal volume of isopropanol to the separated top phase.
9. Add 1/10 volume of 3 M sodium acetate at pH 5.2.
10. Mix by inverting the tube.
11. Centrifuge the tube for 5 min at 18,000 × *g* in a microcentrifuge.
12. Remove and discard the liquid.
13. Allow the DNA pellet to air dry 5–10 min (*see* **Note 3**).
14. Resuspend the DNA pellet in 30 μL 1× TE, pH 7.4.

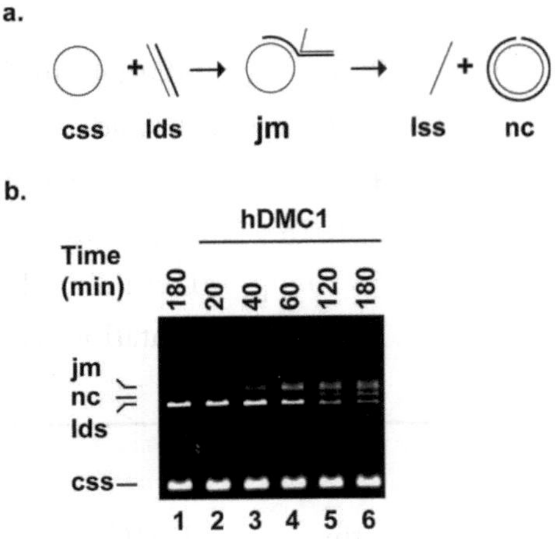

Fig. 1 Plasmid length DNA strand exchange. (**a**). The circular single-strand (css) bound by hDMC1 searches for homology in the linear double-stranded (lds) DNA. Once found, a joint molecule (jm) is formed and continued strand exchange yields nicked circular (nc) and linear single stranded (lss) DNA products. (**b**) A time course of DNA strand exchange in the absence (lane 1) or presence of hDMC1 (lanes 2–6)

3.2 Dilution of DNA Substrates

1. Dilute ApaLI-digested ϕX 174 dsDNA to a concentration of 0.14 μM in 1× TE, pH 7.4.
2. Dilute ϕX 174 ssDNA to a concentration of 0.14 μM in 1× TE, pH 7.4.

3.3 Plasmid Length DNA Strand Exchange

1. Set up the reaction buffer containing 25 mM Tris–HCl, pH 7.4, 5.5 nM of ϕX174 ssDNA, 1 mM DTT, 2.4 mM $MgCl_2$, 2 mM ATP, 16 mM creatine phosphate, 36 μg/mL creatine kinase in a final volume of 8.75 μL (*see* **Note 4**).
2. Add 1 μL of 173 μM hDMC1 and gently mix.
3. Incubate for 10 min at 37 °C (*see* **Note 5**).
4. Add 1 μL of 50 μM hRPA to the reaction.
5. Add 0.75 μL of 3 M KCl to the reaction and gently mix.
6. Incubate the reaction for 8 min at 37 °C (*see* **Note 6**).
7. Add 1 μL 0.14 μM bp ϕX174 dsDNA and 1 μL 50 μM spermidine to initiate the reaction (*see* **Note 7**).
8. Incubate the reactions at 37 °C for the desired reaction times (*see* **Note 8**).
9. Terminate the reactions through the addition of 0.7 μL of 10% SDS and 0.8 μL of 10 mg/mL proteinase K.
10. Incubate reactions at 37 °C for 30 min.

3.4 Product Resolution

1. Add 3.0 μL Orange G gel loading dye to each reaction and mix well. Load 12.5 μL of sample onto a 0.9% agarose gel.

2. Run samples for 3 h at 100 mA in 1× TAE buffer (*see* **Note 9**).
3. Stain gel in fresh ethidium bromide (4 μg/mL).
4. Destain gel for 10 min in 200 mL of fresh deionized water, three times.
5. Use a gel documentation imager to visualize and capture an image of the reaction products on the gel (*see* **Note 10**).

4 Notes

1. The concentration ϕX 174 dsDNA is reported as the concentration of DNA molecules.
2. Centrifugation causes the separation of the aqueous (top) and organic (bottom) phases. Nucleic acids are contained in the aqueous phase.
3. The pellet is considered "air-dry" when all of the residual liquid has evaporated and is no longer visible in the tube.
4. A reaction volume of 12.5 μL is sufficient to provide reactions products that can be easily visualized. Creatine phosphate and creatine kinase form an ATP regeneration system.
5. Incubating hDMC1 along with css in reaction buffer allows for the formation of the presynaptic filament.
6. The plasmid length DNA strand exchange reaction utilizes hDMC1 to catalyze DNA strand exchange and hRPA to prevent secondary structures in the ϕX174 ssDNA and to bind the liberated ssDNA. It is advised that hDMC1 be titrated to optimize for hDMC1 DNA strand exchange activity. Once the reactions have been optimized by titrating hDMC1 and hRPA, it is recommended that a titration with KCl be performed.
7. To increase the consistency of the reactions, both spermidine and ϕX174 Replicative Form I dsDNA can be combined in the correct proportion prior to the addition to the reaction.
8. A 3-h reaction time yields adequate amounts of nicked circular product to be visualized. In addition, the use of varied reactions times can elucidate the independent formation of joint molecule and nicked circular products.
9. Running the gel slowly helps resolve the distinct banding representing joint molecule and nicked circular. Running the gel too quickly can cause difficulty in distinguishing the joint molecule and nicked circular bands.
10. Any gel imager capable of detecting ethidium bromide fluorescence can be used to visualize and capture the image of the reaction products.

References

1. Keeney S, Giroux CN, Kleckner N (1997) Meiosis-specific DNA double-strand breaks are catalyzed by Spo11, a member of a widely conserved protein family. Cell 88(3):375–384
2. Krejci L, Altmannova V, Spirek M, Zhao X (2012) Homologous recombination and its regulation. Nucleic Acids Res 40(13):5795–5818
3. Sehorn MG, Sigurdsson S, Bussen W, Unger VM, Sung P (2004) Human meiotic recombinase Dmc1 promotes ATP-dependent homologous DNA strand exchange. Nature 429 (6990):433–437
4. Ohta K, Nicolas A, Furuse M, Nabetani A, Ogawa H, Shibata T (1998) Mutations in the MRE11, RAD50, XRS2, and MRE2 genes alter chromatin configuration at meiotic DNA double-stranded break sites in premeiotic and meiotic cells. Proc Natl Acad Sci U S A 95 (2):646–651
5. D'Amours D, Jackson SP (2002) The Mre11 complex: at the crossroads of DNA repair and checkpoint signaling. Nat Rev Mol Cell Biol 3 (5):317–327
6. Sigurdsson S, Trujillo K, Song B, Stratton S, Sung P (2001) Basis for avid homologous DNA strand exchange by human Rad51 and RPA. J Biol Chem 276(12):8798–8806
7. Sharma D, Say AF, Ledford LL, Hughes AJ, Sehorn HA, Dwyer DS, Sehorn MG (2013) Role of the conserved lysine within the Walker A motif of human DMC1. DNA Repair 12 (1):53–62

Chapter 18

In Vitro Assays for DNA Branch Migration

Andrew A. Kelso, Steven D. Goodson, and Michael G. Sehorn

Abstract

Homologous recombination is a high-fidelity DNA double-strand break repair pathway that uses a homologous template to repair the break. Recombinases are the central enzymes that facilitate the strand invasion step of homologous recombination, which forms a DNA joint molecule. These DNA joint molecules can be moved through branch migration activity. In this chapter, we describe two assays to determine the branch migration activity and directionality of an enzyme. Monitoring the branch migration activity of an enzyme can provide insight into the roles of these factors in homologous recombination.

Key words DNA branch migration, PX junction, Holliday junction, Homologous recombination, DNA double-strand breaks

1 Introduction

Homologous recombination (HR) is a DNA double-strand break (DSB) repair pathway that utilizes a homologous template to mend the DNA lesion [1]. Central to HR is the function of a specialized enzyme called a recombinase. Recombinases form nucleoprotein filaments on the end-resected ssDNA overhangs at the DSB site. These nucleoprotein filaments facilitate strand invasion into the homologous template (Fig. 1a). This invasion step results in a DNA joint molecule known as a displacement loop (D-loop) while the junction at the strand invasion point is termed a partial X (PX) junction (Fig. 1b). Primed by the invading strand, DNA synthesis occurs to repair the DNA lesion, after which the PX junction can be dissociated through branch migration in the $5' \rightarrow 3'$ direction (Fig. 1c). The dissociated strand anneals to the other processed end of the DSB, which is followed by DNA synthesis and ligation to fully repair the DNA DSB (Fig. 1c). This is the synthesis-dependent strand annealing (SDSA) pathway, and it is the dominant HR pathway in somatic cells because it results in non-crossover products (Fig. 1c). For SDSA, current evidence suggests that the human SWI2/SNF2 chromatin remodeler, RAD54 branch

Lata Balakrishnan and Jason A. Stewart (eds.), *DNA Repair: Methods and Protocols*, Methods in Molecular Biology, vol. 1999, https://doi.org/10.1007/978-1-4939-9500-4_18, © Springer Science+Business Media, LLC, part of Springer Nature 2019

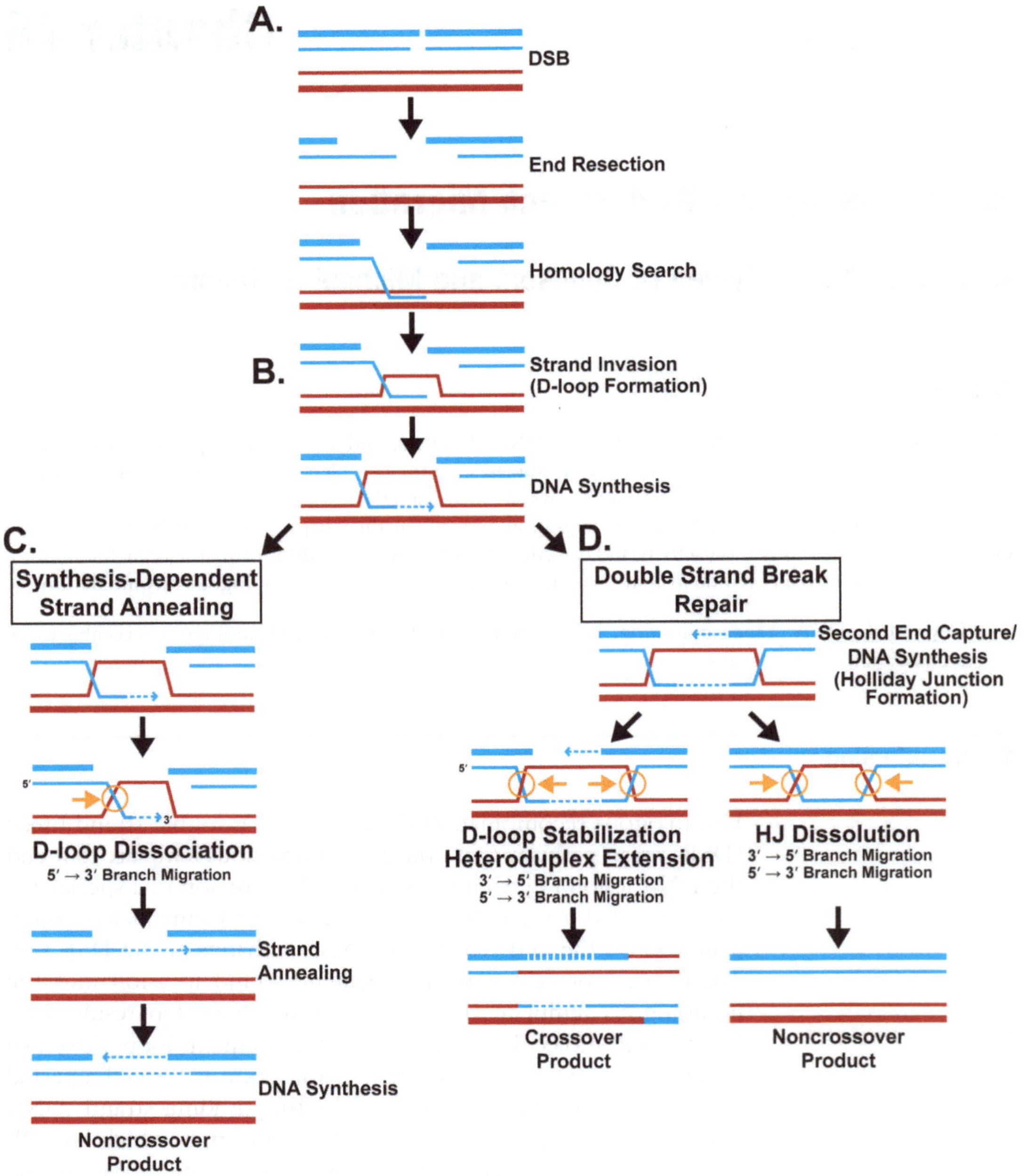

Fig. 1 Homologous recombination repair pathway. (**a**) When a double-strand break occurs, the ends of the chromosome are enzymatically processed. The resulting 3′ ssDNA overhang is recognized by a recombinase that searches for homology within a homologous template. (**b**) Once homology is recognized, strand invasion occurs resulting in a DNA joint molecule known as a D-loop. The invaded 3′ ssDNA primes DNA synthesis to repair the damaged DNA. (**c**) In the synthesis-dependent strand annealing pathway, resolution of the joint molecule occurs by dissociation of the invaded strand by branch migration. An annealing event occurs, followed by DNA synthesis and ligation to yield noncrossover products. (**d**) In the double-strand break repair pathway, the extension of a D-loop leads to second-end capture. A synthesis and ligation event results in the formation of a double Holliday junction (dHJ). This DNA intermediate can be extended or dissolved through branch migration. This pathway yields crossover or noncrossover products

migrates the DNA joint molecule (5′ → 3′), and RAD52 anneals the dissociated strand to the other end of the broken chromosome via single-strand annealing activity [2, 3].

Conversely, after strand invasion the D-loop can be extended resulting in second-end capture in which the D-loop anneals to the other end of the broken chromosome (Fig. 1d). A second synthesis and ligation event results in the formation of two DNA joint molecules called Holliday junctions (HJ) [4]. The resulting double HJs (dHJs) can be resolved to yield either crossover or noncrossover products. This is the double-strand break repair (DSBR) pathway which is the preferred HR pathway in meiosis due to the formation of crossover products. The length of the dHJs can be altered through branch migration activity (5′ → 3′ and/or 3′ → 5′) to either increase or decrease the amount of genetic information that is transferred in recombination events (Fig. 1d) [5]. Additionally, branch migration is associated with resolving the noncrossover pathway of DSBR [6] and with restarting stalled replication forks [7, 8]. Several branch migration enzymes (RAD54, SMARCAL1, RECQ1, and BLM) have been mechanistically characterized to have roles in these processes [2, 9–12].

This chapter provides two in vitro assays for analyzing the activity and directionality of branch migrating enzymes. This first assay is an oligonucleotide-based assay that analyzes branch migration activity using PX junction structures. The second assay utilizes plasmids to analyze D-loop dissociation via branch migration.

2 Materials

2.1 Protein Purification

1. Purified RAD51 recombinase [13].

2.2 Preparation of Oligonucleotides

2.2.1 Gel Purification of Oligonucleotides

1. The sequences of oligonucleotides used in these assays can be found in Table 1.
2. 1× TE: 10 mM Tris–HCl pH 7.5, 1 mM EDTA.
3. UV spectrometer.
4. 2× Denaturing loading buffer: 95% formamide, 20 mM Tris–HCl pH 7.5, 2 mM EDTA, 5% glycerol, and 0.1% bromophenol blue.
5. 1× Denaturing gel: 7 M urea, 10% polyacrylamide, 1× TAE.
6. 1× TAE: 40 mM Tris, 20 mM NaOAc, 2 mM EDTA, pH 7.5.
7. Protein (mini) gel electrophoresis apparatus.
8. Handheld UV lamp.
9. Dialysis membrane tubing 6 k–8 k Da MWCO.
10. Horizontal (mini) gel electrophoresis apparatus.
11. Centrifugal filter tube (3000 NMWL).

Table 1
List of oligonucleotides used in branch migration assays

Name	Purpose	Length	Sequence (5′ → 3′)
OL10	PX3-A	80 nt	TTGCCGTATTTCGAACTGCTAATGTTTGAATTCATGGCTAAC GACTTAGACCATTTTTTTTTTTTTTTTTTTTTTTTTTT
OL11	PX3-B	80 nt	GTATAAATTTTGTACAACCTAGGGTCGTGGTCTAAGTCGTT AGCCATGAATTCAAACATTAGCAGTTCGAAATACGGCAA
OL12	PX3-C	80 nt	TTGCCGTATTTCGAACTGCTAATGTTTGAATTCATGGCTCAC GACTTAGACCACGACCCTAGGTTGTACAAAATTTATAC
OL13	PX3-D	53 nt	TGGTCTAAGTCGTGAGCCATGAATTCAAACATTAGCAGTTC GAAATACGGCAA
OL14	PX5-A	80 nt	TTTTTTTTTTTTTTTTTTTTTTTTTTACTTAAGTACCGAATGC TGAATCTGGTGCTGGGATCCAACATGTTTTAAATATG
OL15	PX5-B	80 nt	CATATTTAAAACATGTTGGATCCCAGCACCAGATTCAGCATT CGGTACTTAAGTTTGTAATCGTCAAGCTTTATGCCGTT
OL16	PX5-C	80 nt	AACGGCATAAAGCTTGACGATTACAAACTTAAGTACCGACT GCTGAATCTGGTGCTGGGATCCAACATGTTTTAAATATG
OL17	PX5-D	54 nt	CATATTTAAAACATGTTGGATCCCAGCACCAGATTCAGCAG TCGGTACTTAAGT
OL24	5′-tailed DNA	100 nt	TTTTTTTTTTTTTTTTTTTTTTTTTTTTTTTTTTTGCCTCAC TGATTAAGCATTGGTAACTGTCAGACCAAGTTTACTCATA TATACTTTAGATTGATTT
OL25	3′-tailed DNA	100 nt	AAATCAATCTAAAGTATATATGAGTAAACTTGGTCTGACAGTT ACCAATGCTTAATCAGTGAGGCTTTTTTTTTTTTTTTTTT TTTTTTTTTTTTTTTTT
OL26	PolyA	35 nt	AAAAAAAAAAAAAAAAAAAAAAAAAAAAAAAAAAA
OL3	Nonmobile PX3-A and PX5-A	80 nt	TTGATAAGAGGTCATTTGAATTCATGGCTTAGAGCTTAATTGC TGAATCTGGTGCTGGGATCCAACATGTTTTAAATATG
OL4	Nonmobile PX3-B and PX5-B	80 nt	CATATTTAAAACATGTTGGATCCCAGCACCAGATTCAGCAT AGATGTCCATCTGTTGTAATCGTCAAGCTTTATGCCGTT
OL5	Nonmobile PX3-C and PX5-C	80 nt	AATAGACAGATCGCTGAGATAGGTGCCTCACTGATTAAGCA TTAAGCTCTAAGCCATGAATTCAAATGACCTCTTATCAA
OL6	Nonmobile PX3-D	40 nt	GCTTAATCAGTGAGGCACCTATCTCAGCGATCTGTCTATT
OL8	Nonmobile PX5-D	40 nt	AACGGCATAAAGCTTGACGATTACAACAGATGGACATCTA

2.2.2 5′-End Radiolabeling Oligonucleotides

1. 10× polynucleotide kinase buffer: 100 mM Tris–HCl pH 7.5, 100 mM $MgCl_2$, 50 mM DTT.
2. T4 polynucleotide kinase (5 U/μL).
3. [γ-^{32}P]-ATP 250 μCi.
4. Nuclease-free water.
5. 1× TE: 10 mM Tris–HCl pH 7.5, 1 mM EDTA.
6. Microspin column (~30 μm pore size).

2.2.3 Annealing of Oligonucleotide Complexes

1. 10× Annealing buffer: 100 mM Tris–HCl pH 8.0, 500 mM NaCl, 10 mM EDTA.
2. Parafilm M.
3. 2× nondenaturing loading buffer: 25 mM Tris–HCl pH 7.4, 50% glycerol, 0.1% Orange G.
4. 1× Native gel: 10% acrylamide, 1× TAE.
5. 1× TAE.
6. Protein (mini) gel electrophoresis apparatus.
7. Handheld UV lamp.
8. Dialysis membrane tubing 6 k–8 k Da MWCO.
9. 1× TE.
10. Horizontal (mini) gel electrophoresis apparatus.
11. Centrifugal filter tube (3000 NMWL).

2.3 BM of PX Junctions

1. 1× Reaction buffer: 25 mM Tris–HCl pH 7.5, 1.2 mM $MgCl_2$, 2 mM ATP, 1 μg/μL BSA, 1 mM DTT, 2 mM creatine phosphate, 45 μg/mL creatine kinase.
2. Nuclease-free water.
3. Buffer A: 20 mM KH_2PO_4 pH 7.5, 10% glycerol, 0.5 mM EDTA, 100 mM KCl.
4. 10% Sodium dodecyl sulfate (SDS): 10% (w/v) in nuclease-free water.
5. Proteinase K (20 μg/μL).
6. 2× Nondenaturing loading buffer.
7. 1× Native gel.
8. Protein (mini) gel electrophoresis apparatus.
9. 1× TAE.
10. Gel dryer.
11. Cellulose chromatography paper.
12. Phosphor screen.
13. Phosphorimager.

2.4 D-Loop Dissociation via Branch Migration

2.4.1 Preparation of D-Loop Structure

1. 1× Reaction buffer.
2. Nuclease-free water.
3. pBluescript II SK(+) purified from *Escherichia coli,*
4. 10% SDS.
5. Proteinase K (20 μg/μL).
6. Sephacryl S-500.
7. 1× TE.
8. 2× Nondenaturing loading buffer.
9. Agarose gel: 1% agarose (w/v) in 1× TAE.
10. 1× TAE.
11. Horizontal (mini) gel electrophoresis apparatus.
12. Gel dryer.
13. Ion exchange paper.
14. Phosphor screen.
15. Phosphorimager.
16. Centrifugal filter tube (3000 NMWL).

2.4.2 Branch Migration of the D-loop Structure

1. 1× Reaction buffer: 25 mM Tris–HCl pH 7.5, 1.2 mM $MgCl_2$, 2 mM ATP, 1 μg/μL BSA, 1 mM DTT, 2 mM creatine phosphate, 45 μg/mL creatine kinase.
2. Nuclease-free water.
3. Buffer A.
4. 10% SDS.
5. Proteinase K (20 μg/μL).
6. 2× nondenaturing loading buffer.
7. Agarose gel.
8. 1× TAE.
9. Horizontal (mini) gel electrophoresis apparatus.
10. Gel dryer.
11. DE81 Ion exchange paper.
12. Phosphor screen.
13. Phosphorimager.

3 Methods

3.1 Preparation of Oligonucleotides

3.1.1 Gel Purification of Oligonucleotides

1. Dissolve chemically synthesized oligonucleotides in the appropriate volume of 1× TE to reach a concentration of 1 μg/μL. DNA concentrations can be measured from absorbance at 260 nm using a UV spectrometer. Mix 75 μL of each oligonucleotide (1 μg/μL) with an equal volume of 2× denaturing loading buffer.
2. Incubate the samples at 90 °C for 5 min.
3. Load each sample onto a 1× denaturing gel and run at 100 V for 4 h at 55 °C. Use 1× TAE as the running buffer.
4. Remove the gel from the apparatus, and use a handheld UV lamp to shadow and locate the DNA band (*see* **Note 1**).
5. Excise the DNA band by cutting the gel with a razor blade, and place the excised gel in a dialysis membrane tube containing 2 mL of 1× TAE.
6. Electrophorese in a horizontal (mini) gel electrophoresis apparatus at 35 V overnight at 4 °C in 1× TAE (*see* **Note 2**). The next day increase to 70 V for 60 min. Then, switch the electrodes and run at 70 V for 2 min.
7. Remove the 2 mL of 1× TAE/DNA solution from the dialysis membrane and transfer to a centrifugal filter tube (3000 NMWL).
8. Centrifuge per the manufacturer's suggested guidelines at 4 °C until the final volume is reached (~50 μL).
9. Add 4 mL of 1× TE to the filter tube.
10. To facilitate buffer exchange (1× TAE for 1× TE) repeat **steps 8** and **9** two more times.
11. Centrifuge per the manufacturer's suggested guidelines at 4 °C until the final volume is reached (~50 μL).
12. Store at 4 °C for short-term use or −20 °C for long-term use (*see* **Note 3**).

3.1.2 5′-End Radiolabeling Oligonucleotides

1. To radiolabeled DNA substrates, set up the labeling reactions as follows:

   10 ng of purified DNA,

   2.5 μL 10× polynucleotide kinase buffer,

   1.0 μL T4 polynucleotide kinase,

   2.0 μL [γ-^{32}P]-ATP 250 μCi,

Nuclease-free water to bring the final volume to 25 μL (*see* **Note 4**).

2. Incubate at 37 °C for 45 min.
3. Add an additional 1.0 μL of T4 polynucleotide kinase.
4. Incubate at 37 °C for 45 min.
5. Add 50 μL of 1× TE.
6. Incubate at 80 °C for 15 min.
7. Use the microspin column to purify the radiolabeled oligonucleotide.
8. Store at −20 °C.

3.1.3 Annealing of Oligonucleotide Complexes

1. In a 1.5 mL microcentrifuge tube, mix equimolar amounts (*see* **Note 5**) of purified oligonucleotides in annealing buffer to a final volume of 50 μL (*see* **Note 6**); wrap the top of the closed tube with Parafilm M, and place the tubes in a floating tube rack.
2. Boil water in a beaker. Once a rolling boil is reached, place the floating tube rack containing the reaction tubes in the water and cover the top of the beaker with aluminum foil.
3. Boil for 10 min.
4. Place the beaker in a Styrofoam box, and move the box to 4 °C (refrigerator or cold room) to allow the reactions to cool slowly overnight (*see* **Note** 7).
5. After the reactions have cooled, add 50 μL of 2× nondenaturing loading buffer, and load the total reaction on a 1× native gel.
6. Run the gel at 100 V for 5 h at 4 °C using 1× TAE as the running buffer.
7. Follow **steps 4–12** from Subheading 3.1.1 *Gel purification of oligonucleotides.*

3.2 Branch Migration of PX

In this assay, the branch migration activity and directionality of an enzyme can be determined by using PX junctions that migrate in either the 3′ → 5′ (PX3) or 5′ → 3′ (PX5) direction. The first stage of this reaction is to anneal a radiolabeled fork and an unlabeled overhang (Fig. 2a, b). This annealing step is then followed by the introduction of a branch migrator enzyme/protein of interest. Branch migration of the PX structure results in the exchange of the radiolabeled oligonucleotide within the fork to the overhang (Fig. 2a, b), which is distinct when resolved on a nondenaturing polyacrylamide gel (Fig. 2c, d).

1. Set up the 1× reaction buffers as follows:

   2.0 μL of 125 mM Tris–HCl pH 7.5,

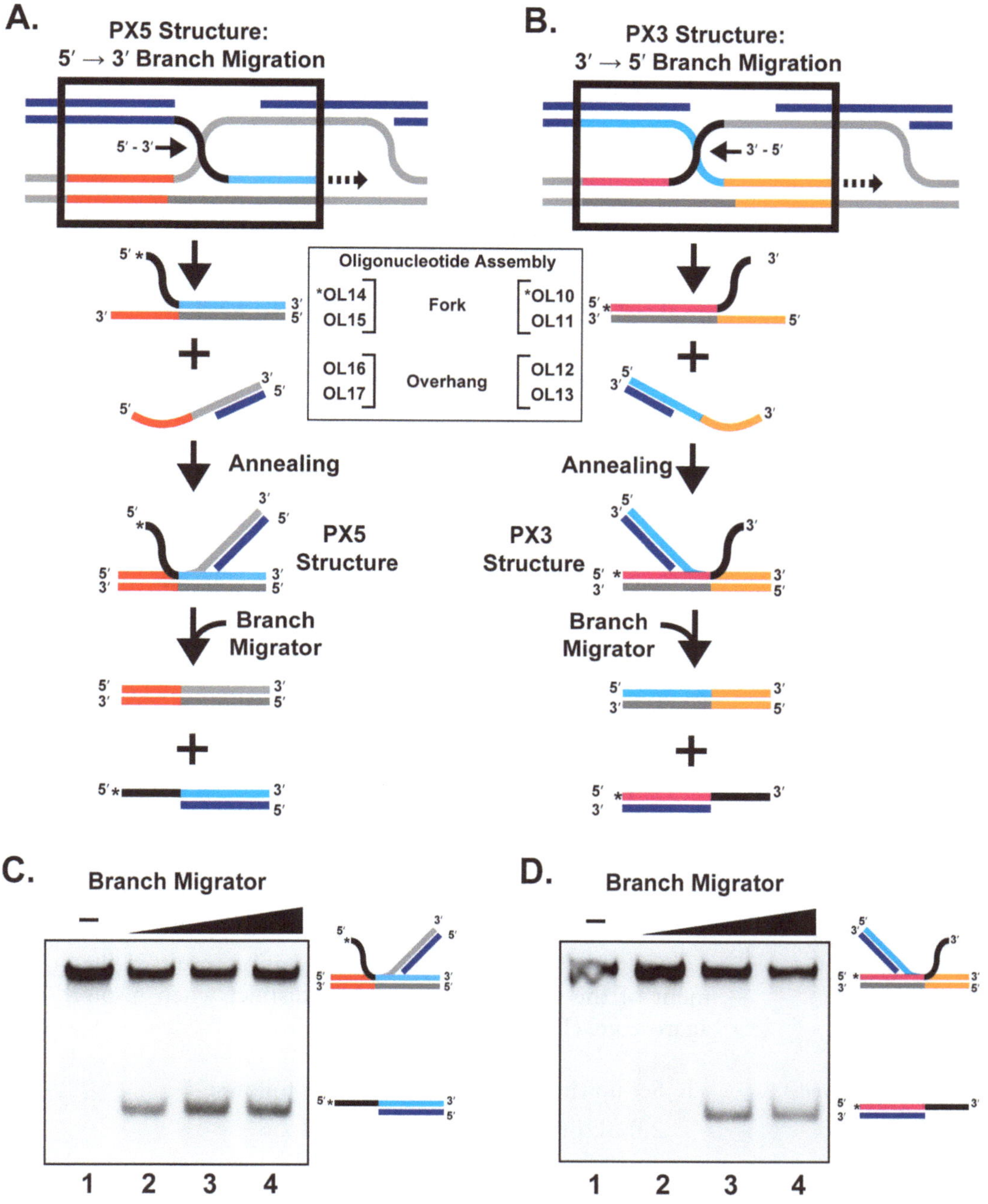

Fig. 2 Branch migration of PX junctions. Annealing and branch migration schematic for PX5 (**a**) and PX3 (**b**); Inset shows the assembly scheme and oligonucleotides used to develop the initial fork and overhang structures. * indicates radiolabeled oligonucleotide. Sample gels of PX5 (**c**) and PX3 (**d**) branch migration

0.8 μL 15 mM $MgCl_2$,
0.8 μL 25 mM ATP,
0.1 μL 100 μg/μL BSA,
0.1 μL 100 mM DTT,

0.2 μL 1 M creatine phosphate,

0.3 μL 1.5 mg/mL creatine kinase,

1.7 μL nuclease-free water.

2. Add 0.5 μL of 640 nM radiolabeled fork and 0.5 μL of 960 nM unlabeled overhang.
3. Anneal the fork and overhang of PX3 and PX5 junctions by incubating at 50 °C for 10 min, followed by 37 °C for 10 min, then 23 °C for 10 min. PX3: ^{32}P-OL10-OL11 fork + OL12-OL13 overhang and PX5: ^{32}P-OL14-OL15 fork + OL16-OL17 overhang (Fig. 2a, b) (*see* **Note 8**).
4. Add the branch migration enzyme/protein of interest (*see* **Note 9**). The addition of the protein should be balanced with buffer A to reach a final reaction volume of 10 μL.
5. Stop the reactions with the addition of 1 μL 10% SDS and 1 μL of Proteinase K (10 μg/μL).
6. Incubate for 20 min at 37 °C.
7. Add 3 μL of 2× nondenaturing loading buffer.
8. Load the reactions onto a 1× native gel.
9. Run the gel at 100 V for 90 min in 1× TAE.
10. Dry the gels by placing them on cellulose chromatography paper using a gel dryer for 30 min at 80 °C.
11. Expose the dried gel to a phosphor screen and analyze using a phosphorimager.

3.3 D-Loop Dissociation via Branch Migration

In this assay, the RAD51 recombinase is used to generate a plasmid-based D-loop structure with a 3′ or 5′ overhang (Fig. 3a). This structure is then purified and used to analyze D-loop dissociation activity and directionality. Branch migration results in the displacement of the tailed DNA, which is distinct when resolved on an agarose gel (Fig. 3b).

3.3.1 Preparation the D-Loop Structure

1. Set up the 1× reaction buffers as follows:

   2.0 μL of 125 mM Tris–HCl pH 7.5,

   0.8 μL 15 mM $MgCl_2$,

   0.8 μL 25 mM ATP,

   0.1 μL 100 μg/μL BSA,

   0.1 μL 100 mM DTT,

   0.2 μL 1 M creatine phosphate,

   0.3 μL 1.5 mg/mL creatine kinase,

   2.7 μL nuclease-free water.

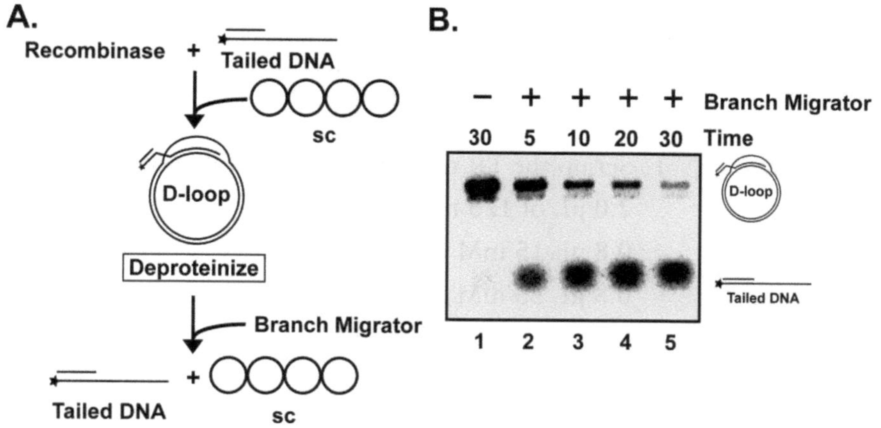

Fig. 3 D-loop dissociation assay via branch migration. (**a**) Schematic for generating tailed DNA/D-loop complex; *sc* supercoiled plasmid, * indicates radiolabeled oligonucleotide. (**b**) Sample gel of D-loop dissociation time course analysis

2. Add 1 μL of 20 μM RAD51 and either of 1 μL of 25 μM ^{32}P-OL24-OL26 (5′-tailed DNA) or 1 μL of 25 μM ^{32}P-OL25-OL26 (3′-tailed DNA), and incubate for 10 min at 37 °C.
3. Start the reaction by adding 1 μL of 700 μM base pairs pBluescript II and incubate for 60 min at 37 °C.
4. Stop the reactions with the addition of 1 μL 10% SDS and 1 μL of Proteinase K (10 μg/μL) and incubate for 20 min at 37 °C.
5. To isolate the tailed DNA/D-loop complexes, pass the D-loop reaction product over a 3.5 mL to Sephacryl S-500 size exclusion column in 1× TE. After 1.0 mL of flow through, collect each fraction in 200 μL aliquots.
6. Determine the fractions that contain the tailed DNA/D-loop complex by mixing 2.5 μL of the reaction product with 2.5 μL of 2× non-denaturing loading buffer and running the samples on a 1% agarose gel at 80 V for 70 min in 1× TAE.
7. Dry the gel using a gel dryer on ion exchange paper for 30 min at 80 °C.
8. Expose the dried gel to a phosphor screen and analyze using a phosphorimager.
9. Pool the fractions that contain the tailed DNA/D-loop complex, and transfer the pooled solution to a centrifugal filter tube (3000 NMWL).
10. Centrifuge the centrifugal filter tube containing the tailed DNA/D-loop complex per the manufacturer's suggested guidelines at 4 °C until the final volume is reached (~50 μL).
11. Add 4 mL of 1× TE to the filter tube to facilitate buffer exchange (1× TAE to 1× TE).

12. Repeat **steps 10** and **11** two more times.
13. Centrifuge per the manufacturer's suggested guidelines at 4 °C until the final volume is reached (~50 μL).

3.3.2 Branch Migration of D-loop Structure

1. Set up the 1× reaction buffers as follows:

   2.0 μL of 125 mM Tris–HCl pH 7.5,

   0.8 μL 15 mM $MgCl_2$,

   0.8 μL 25 mM ATP,

   0.1 μL 100 μg/μL BSA,

   0.1 μL 100 mM DTT,

   0.2 μL 1 M creatine phosphate,

   0.3 μL 1.5 mg/mL creatine kinase,

   1.7 μL nuclease-free water.
2. Add 1 μL of 140 nM 5′-tailed D-loop complex or the 3′-tailed D-loop complex.
3. Add the branch migration enzyme/protein of interest and incubate at 37 °C for 5–30 min (*see* **Note 9**). The addition of the protein should be balanced with buffer A to reach a final reaction volume of 10 μL.
4. Stop the reactions with the addition of 1 μL 10% SDS and 1 μL of Proteinase K (10 μg/μL) and incubate for 20 min at 37 °C.
5. Add 2.0 μL of 2× nondenaturing loading buffer to each sample.
6. Load the reactions on a 1% agarose gel and run at 80 V for 70 min in 1× TAE.
7. Dry the gel on ion exchange paper using a gel dryer for 30 min at 80 °C.
8. Expose the dried gel to a phosphor screen and analyze using a phosphorimager.

4 Notes

1. Place the gel on plastic wrap, and place the plastic wrap + gel on a white background to help visualize the DNA band.
2. Place the tubing in the center of the apparatus and submerge in running buffer (1× TAE). To prevent floating, use a buffer dam (or something similar) to hold the dialysis membrane tube in place.
3. The amount of purified oligonucleotide can vary depending on the efficiency of the purification, but typical efficiency ranges from 10 to 80% of the original amount of oligonucleotide.

4. The amount of T4 polynucleotide kinase can be adjusted to accommodate the instructions suggested by the manufacturer. If the amount changes, use nuclease-free water to bring the final volume to 25 μL.
5. A typical concentration for each oligonucleotide is 1–5 μM.
6. The initial oligonucleotide complexes that will be purified for the PX junctions consist of a fork and an overhang and for the D-loop structures is an overhang. See Table 1 and Fig. 2 for the oligonucleotides used and the assembly scheme.
7. Once the annealed oligonucleotide complexes have cooled to 4 °C, it is important to perform all subsequent steps (prior to use in the reactions) at 4 °C (or below when storing long term) to prevent denaturation of the complexes.
8. Nonmobile structures, or versions of the branch migration structures that are dissociated through helicase activity but not branch migration activity, can be assembled as follows: PX3 (^{32}P-OL3-OL4-OL5-OL6) and PX5 (^{32}P-OL3-OL4-OL5-OL8) (Table 1). These nonmobile structures are used as controls to help differentiate between helicase and branch migration activities.
9. Increasing concentrations of the enzyme are used to determine optimal amounts of the enzyme to be used in the reaction. Variation in reaction time and enzyme amounts can be used for kinetic analysis of the enzyme's activity.

References

1. San Filippo J, Sung P, Klein H (2008) Mechanism of eukaryotic homologous recombination. Annu Rev Biochem 77:229–257
2. Bugreev DV, Hanaoka F, Mazin AV (2007) Rad54 dissociates homologous recombination intermediates by branch migration. Nat Struct Mol Biol 14:746–753
3. Sugiyama T, Kantake N, Wu Y, Kowalczykowski SC (2006) Rad52-mediated DNA annealing after Rad51-mediated DNA strand exchange promotes second ssDNA capture. EMBO J 25:5539–5548
4. Holliday R (1964) A mechanism for gene conversion in fungi. Genet Res 5:282–304
5. Paques F, Haber JE (1999) Multiple pathways of recombination induced by double-strand breaks in *Saccharomyces cerevisiae*. Microbiol Mol Biol Rev 63:349–404
6. Wyatt HD, West SC (2014) Holliday junction resolvases. Cold Spring Harb Perspect Biol 6: a023192
7. Seigneur M, Bidnenko V, Ehrlich SD, Michel B (1998) RuvAB acts at arrested replication forks. Cell 95:419–430
8. McGlynn P, Lloyd RG, Marians KJ (2001) Formation of Holliday junctions by regression of nascent DNA in intermediates containing stalled replication forks: RecG stimulates regression even when the DNA is negatively supercoiled. Proc Natl Acad Sci U S A 98:8235–8240
9. Betous R, Mason AC, Rambo RP, Bansbach CE, Badu-Nkansah A, Sirbu BM, Eichman BF, Cortez D (2012) SMARCAL1 catalyzes fork regression and Holliday junction migration to maintain genome stability during DNA replication. Genes Dev 26:151–162
10. Bugreev DV, Mazina OM, Mazin AV (2006) Rad54 protein promotes branch migration of Holliday junctions. Nature 442:590–593
11. Pike AC, Gomathinayagam S, Swuec P, Berti M, Zhang Y, Schnecke C, Marino F, von Delft F, Renault L, Costa A et al (2015) Human RECQ1 helicase-driven DNA

unwinding, annealing, and branch migration: insights from DNA complex structures. Proc Natl Acad Sci U S A 112:4286–4291

12. Bizard AH, Hickson ID (2014) The dissolution of double Holliday junctions. Cold Spring Harb Perspect Biol 6:a016477

13. Chi P, Van Komen S, Sehorn MG, Sigurdsson S, Sung P (2006) Roles of ATP binding and ATP hydrolysis in human Rad51 recombinase function. DNA Repair 5:381–391

Chapter 19

Stabilization of the Human DMC1 Nucleoprotein Filament

Sarah M. Waldvogel, Steven D. Goodson, and Michael G. Sehorn

Abstract

The meiosis-specific recombinase, DMC1, is important for the generation of haploids during meiosis. DMC1 forms a helical nucleoprotein filament on ssDNA overhangs located at the processed double-stranded DNA break. The DMC1 filament performs a search for homology in homologous chromosome. Once homology is located, the DMC1 filament strand invades the homologous chromosome forming a displacement loop (D-loop). These connections are needed for accurate segregation to occur later in meiosis. Because DMC1 requires numerous accessory factors and specific ionic conditions to participate in this DNA repair process, in vitro assays were developed to understand how these accessory factors influence the biochemical properties of hDMC1. This chapter describes a method that can be used to investigate the stability of the human DMC1 nucleoprotein filament under various conditions and provides insight into an important early stage in DNA double-strand break repair by homologous recombination during meiosis.

Key words DMC1, Recombinase, Homologous recombination, Nucleoprotein filament, DNA double-stranded break repair

1 Introduction

DMC1 is a eukaryotic ortholog of the *E. coli* RecA recombinase and is a paralog of RAD51 [1]. While RAD51 participates in DNA double-stranded break (DSB) repair by homologous recombination during both mitosis and meiosis, DMC1 is specific to meiosis [1]. The protein SPO11 induces DSBs at the onset of meiosis [2] and the ends of the DSB are enzymatically processed to produce single-stranded 3′ overhangs [3]. These 3′ single-stranded overhangs are bound by replication protein A (RPA) to prevent formation of secondary structure and prevent degradation by nucleases [4]. While the role RPA plays in protecting the 3′ overhangs at DSBs is important, its presence inhibits the nucleation of RAD51 onto the 3′ ssDNA overhangs. Accessory protein factors are required to remove RPA and load the DMC1 recombinase onto the ssDNA overhangs [5]. DMC1 forms a right-handed helical nucleoprotein filament on ssDNA in the presence of ATP [6].

Lata Balakrishnan and Jason A. Stewart (eds.), *DNA Repair: Methods and Protocols*, Methods in Molecular Biology, vol. 1999, https://doi.org/10.1007/978-1-4939-9500-4_19,

The DMC1 presynaptic filament searches duplex DNA for homology. Once the DMC1 filament identifies a homologous sequence, it mediates DNA strand invasion of the homologous duplex sequence [7]. During meiosis, these DNA joints are important for proper segregation of homologous chromosomes during metaphase I and are preferentially resolved as crossover products [7].

The human DMC1 (hDMC1) nucleoprotein filament is a right-handed helix with 6 to 6.5 protomers per turn [8, 9], and each turn of the helical filament spans approximately 19 bp of ssDNA [9]. hDMC1 requires ATP binding, but not hydrolysis, for nucleoprotein filament assembly [6, 10]. Although electron microscopy has been used to elucidate many physical properties of the hDMC1 nucleoprotein filament, characterization of the biochemical properties of hDMC1 has relied on specific in vitro assays. Numerous interacting partners are attributed with regulating the activity of hDMC1 [8]. In the presence of Ca^{2+}, hDMC1 forms more stable filaments by attenuating the hydrolysis of ATP [11]. In addition to Ca^{2+}, hDMC1 filaments are stabilized by the HOP2-MND1 protein complex [12, 13]. The stabilization of the hDMC1 filament allows for more efficient homology search, strand invasion, and D-loop formation.

The assay described in this chapter provides a method for interrogating the stability of hDMC1 filaments. Briefly, hDMC1 is incubated with biotinylated dT oligonucleotide ssDNA bound to magnetic streptavidin, upon which it forms a filament. A competitor biotin-free dT oligonucleotide ssDNA trap is used to capture DMC1 that has dissociated from the ssDNA on the magnetic bead. The solution is then subjected to magnetic separation, and the supernatant is removed from the beads (Fig. 1). Resolving the supernatant and bead fractions by denaturing gel electrophoresis reveals the extent of the hDMC1 filament stabilization by comparing the amount of protein found in the supernatant and elution fractions. The assay described provides a versatile way to monitor the effect that changes in salt concentrations, addition of various hDMC1 interacting partners, or introducing mutations into hDMC1 itself have on hDMC1-ssDNA filament dynamics.

2 Materials

2.1 Protein Preparation

Purified human DMC1 (hDMC1) recombinase [10].

2.2 Preparation of ssDNA-Bound Magnetic Beads

1. Streptavidin-coated magnetic beads.
2. Buffer A: 10 mM Tris–HCl, pH 7.5, 1 mM EDTA, 100 mM NaCl.
3. Buffer B: 10 mM Tris–HCl, pH 7.5, 1 mM EDTA, 1 M NaCl.

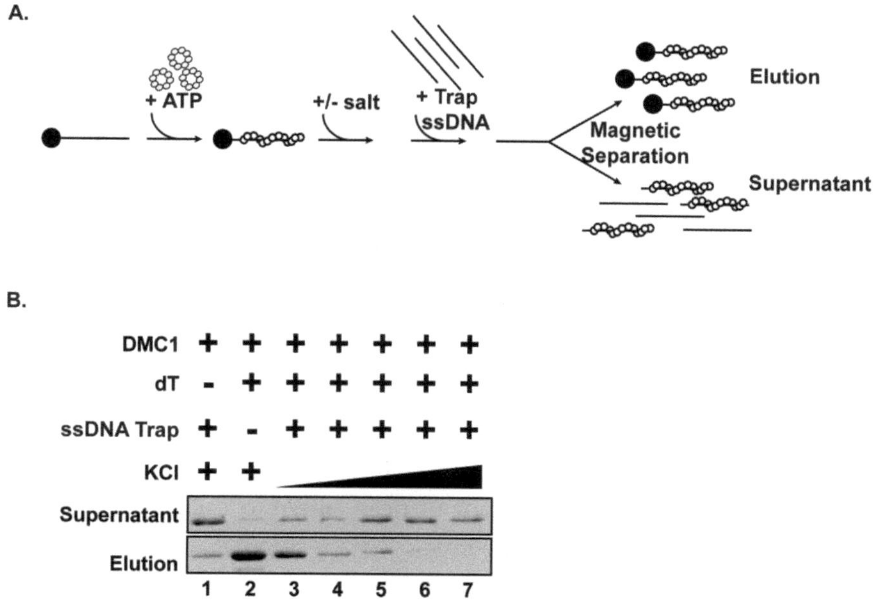

Fig. 1 (**a**) Schematic of the filament stability assay. Magnetic streptavidin beads bound to biotinylated dT oligonucleotide 90 mers are incubated with hDMC1, which forms filaments on the ssDNA. Unbound competitor ssDNA is added in excess to allow any unbound DMC1 to form a filament on the free DNA. The solution is subjected to magnetic separation, and the supernatant is removed from the beads. The supernatant and elution fractions can be resolved by SDS-PAGE, and the relative amounts of hDMC1 in the elution versus the supernatant represent the stability of the original filament. (**b**) A sample SDS-PAGE gel from a filament stability experiment. The supernatant lanes are displayed above the elution lanes. Lane 1 is a control with magnetic beads without bound oligonucleotide dT ssDNA. Most of the protein is found in the supernatant. Lane 2 is a control in which no free ssDNA trap is added. Most of the protein is found in the elution. For lanes 1 and 2, 0.1 M KCl was used. In lanes 3–7, increasing concentrations of KCl are titrated into the reaction (0.1–0.5 M), and the protein is increasingly found in the supernatant

4. 0.93 μM 5′ biotinylated dT 90-mer oligonucleotide (dT_{90}) (*see* **Note 1**).
5. Magnetic separator.

2.3 Stability of the hDMC1 Nucleoprotein Filament

1. ssDNA-bound magnetic beads,
2. 0.19 μM dT_{90} trap ssDNA.
3. 2× Buffer C: 40 mM Tris–HCl pH 7.4, 20% Glycerol 200 mM KCl, 5 mM ATP, 6 mM $MgCl_2$, 1 mM $CaCl_2$, and 0.5 μg/μ L BSA.
4. Buffer D: 20 mM KH_2PO_4, pH 7.5, 10% glycerol, 0.5 mM EDTA, 100 mM KCl.
5. 270 μM hDMC1 in Buffer D.
6. Buffer E: 20 mM KH_2PO_4, pH 7.5, 10% glycerol, 0.5 mM EDTA, 100 mM KCl, 0.1% Triton, 2.5 mM ATP, 3.0 mM $MgCl_2$, 0.5 mM $CaCl_2$.

7. 2× SDS loading buffer: 100 mM Tris–HCl, pH 6.8, 4% SDS (w/v), 0.2% (w/v) bromophenol blue, 20% (v/v) glycerol, 200 mM Β-mercaptoethanol (BME).
8. Magnetic separator.

2.4 SDS-PAGE

1. Bio-Rad Mini Protean 3 Cell gel electrophoresis apparatus.
2. SDS-PAGE gel: 10 × 7 cm, 0.75 mm thick, 12% polyacrylamide, 1% SDS.
3. Gel dryer.

3 Methods

3.1 Preparation of ssDNA-Bound Magnetic Beads

1. Resuspend streptavidin-coated magnetic beads and transfer a 400 μL aliquot of bead slurry to each of two 1.5 mL microcentrifuge tubes.
2. Use a magnetic separator to draw the beads to the bottom of the tube.
3. After complete separation where the beads are isolated on the bottom of the tube, remove and discard the supernatant.
4. Add 400 μL of Buffer A to the beads and gently mix the beads.
5. Use a magnetic separator to draw the beads to the bottom of the tubes.
6. After complete separation where the beads are isolated on the bottom of the tube, remove and discard the supernatant.
7. Repeat **steps 4–6** two additional times.
8. Add 800 μL of Buffer A to the beads and gently mix the beads.
9. Use a magnetic separator to draw the beads to the bottom of the tubes.
10. After complete separation where the beads are isolated on the bottom of the tube, remove and discard the supernatant.
11. Repeat **steps** 7–9 two additional times.
12. Add 6 μL of 0.93 μM biotinylated dT_{90} to one aliquot of beads (these beads will be the ssDNA-bound magnetic beads) and an equal volume of water (*see* **Note 2**) to the other aliquot of beads (these beads will be the mock-treated beads that do not have bound ssDNA).
13. Incubate both sets of beads at 25 °C for 10 min, gently mixing occasionally to prevent the beads from settling to the bottom of the tube.
14. Use a magnetic separator to draw the beads to the bottom of the tubes.
15. After complete separation where the beads are isolated on the bottom of the tube, remove and discard the supernatant.

16. Add 400 μL of Buffer B and gently mix the beads.
17. Use a magnetic separator to draw the beads to the bottom of the tubes.
18. After complete separation where the beads are isolated on the bottom of the tube, remove and discard the supernatant.
19. Add 400 μL of Buffer A to the beads.
20. Use a magnetic separator to draw the beads to the bottom of the tubes.
21. After complete separation where the beads are isolated on the bottom of the tube, remove and discard the supernatant.
22. Repeat **steps 18–20** two additional times.
23. Resuspend the beads in 400 μL of Buffer A.
24. Store at 4 °C in light-impenetrable tubes or wrap tubes with aluminum foil (*see* **Note 3**).

3.2 Stability of the hDMC1 Nucleoprotein Filament

1. Set up reactions in 1.5 mL microcentrifuge tubes as follows (*see* **Note 4**):

   15 μL Buffer C.

   5 μL ssDNA-bound magnetic beads for the experimental condition or 5 μL mock-treated magnetic beads to serve as the control (*see* **Note 5**).

   1 μL Buffer D.

   5.5 μL dH_2O.
2. Incubate at 37 °C for 5 min with occasional gentle mixing to prevent beads from settling to the bottom of the tube.
3. Add 1 μL of 270 μM hDMC1 and incubate at 37 °C for 10 min with occasional gentle mixing.
4. Add 2.5 μL of 0.19 μM dT_{90} oligonucleotide Trap ssDNA or 2.5 μL dH_2O (*see* **Note 5**) and incubate at 37 °C for 10 min with occasional gentle mixing.
5. Using a magnetic separator, draw the beads to the bottom of the tubes. After complete separation is achieved, remove and save the supernatant for analysis.
6. Wash the beads with 100 μL Buffer E (*see* **Note 6**). Using a magnetic separator, draw the beads to the bottom of the tubes. After complete separation is achieved, remove and save the supernatant for analysis.
7. Using the magnetic separator, remove the wash buffer from the beads. Discard the wash buffer.
8. Add 20 μL 2× SDS loading buffer to the beads and the supernatant. Heat the samples at 65 °C for approximately 1 min.
9. Resolve 8 μL samples of supernatant and beads on a 12% SDS-PAGE gel at 200 V for 50 min (Fig. 1b).

4 Notes

1. μM indicates the number of DNA molecules. The number of nanomoles is usually reported on tubes of lyophilized oligonucleotides ordered from a company or can be calculated from the OD_{260} and extinction coefficient of the oligonucleotide using Beer–Lambert's Law for an oligonucleotide in solution. The oligonucleotide can then be resuspended in or diluted to the appropriate volume to achieve the desired μM concentration.
2. An equal volume of the buffer in which the oligonucleotide is resuspended should be added to the second aliquot of beads, which represents a mock-treated bead condition lacking the oligonucleotide probe and will serve as a control for any nonspecific sticking of the protein to the magnetic beads.
3. The beads can also be stored at −20 °C, but avoid repeated thawing and refreezing. Even when stored at 4 °C, the beads have a usable lifetime up to 6 months.
4. Titrations of KCl, BSA, hDMC1, $CaCl_2$, and the ssDNA trap should be performed to find the optimum conditions for filament stabilization. To change the volume of hDMC1, or to add calcium or an accessory protein such as HOP2-MND1 to the assay, the newly added volume should be subtracted from the volume of Buffer D, as this is the buffer in which the proteins are suspended. For any volume changes to the other components, the volume of dH_2O should be adjusted appropriately.
5. At least two controls must be included: one in which mock-treated magnetic beads without a biotinylated ssDNA probe are used and one in which water is substituted for the ssDNA trap.
6. The concentrations of components in Buffer E during the wash should be the same as the final concentrations in the reaction mixture.

Acknowledgments

This work was supported in part by the Clemson University Creative Inquiry Program and by a Departmental Honors Research Grant from the Calhoun Honors College (SMW), and by the National Institutes of Health R01GM098510 (MGS).

References

1. Bishop DK, Park D, Xu L, Kleckner N (1992) DMC1: a meiosis-specific yeast homolog of *E. coli* recA required for recombination, synaptonemal complex formation, and cell cycle progression. Cell 69:439–456
2. Keeney S, Giroux CN, Kleckner N (1997) Meiosis-specific DNA double-strand breaks are catalyzed by Spo11, a member of a widely conserved protein family. Cell 88:375–384
3. Garcia V, Phelps SE, Gray S, Neale MJ (2011) Bidirectional resection of DNA double-strand breaks by Mre11 and Exo1. Nature 479:241–244
4. Sugiyama T, New JH, Kowalczykowski SC (1998) DNA annealing by RAD52 protein is stimulated by specific interaction with the complex of replication protein A and single-stranded DNA. Proc Natl Acad Sci U S A 95:6049–6054
5. Ferrari SR, Grubb J, Bishop DK (2009) The Mei5-Sae3 protein complex mediates Dmc1 activity in *Saccharomyces cerevisiae*. J Biol Chem 284:11766–11770
6. Sehorn MG, Sigurdsson S, Bussen W, Unger VM, Sung P (2004) Human meiotic recombinase Dmc1 promotes ATP-dependent homologous DNA strand exchange. Nature 429:433–437
7. Sansam CL, Pezza RJ (2015) Connecting by breaking and repairing: mechanisms of DNA strand exchange in meiotic recombination. FEBS J 282:2444–2457
8. Sheridan SD, Yu X, Roth R, Heuser JE, Sehorn MG, Sung P, Egelman EH, Bishop DK (2008) A comparative analysis of Dmc1 and Rad51 nucleoprotein filaments. Nucleic Acids Res 36:4057–4066
9. Yu X, Egelman EH (2010) Helical filaments of human Dmc1 protein on single-stranded DNA: a cautionary tale. J Mol Biol 401:544–551
10. Sharma D, Say AF, Ledford LL, Hughes AJ, Sehorn HA, Dwyer DS, Sehorn MG (2013) Role of the conserved lysine within the Walker A motif of human DMC1. DNA Repair (Amst) 12:53–62
11. Bugreev DV, Golub EI, Stasiak AZ, Stasiak A, Mazin AV (2005) Activation of human meiosis-specific recombinase Dmc1 by Ca^{2+}. J Biol Chem 280:26886–26895
12. Pezza RJ, Voloshin ON, Vanevski F, Camerini-Otero RD (2007) Hop2/Mnd1 acts on two critical steps in Dmc1-promoted homologous pairing. Genes Dev 21:1758–1766
13. Zhao W, Saro D, Hammel M, Kwon Y, Xu Y, Rambo RP, Williams GJ, Chi P, Lu L, Pezza RJ, Camerini-Otero RD, Tainer JA, Wang HW, Sung P (2014) Mechanistic insights into the role of Hop2-Mnd1 in meiotic homologous DNA pairing. Nucleic Acids Res 42:906–917

Part V

Methods to Assess Telomeric and Replication-Induced DNA Damage and Repair

Chapter 20

Measuring UV Photoproduct Repair in Isolated Telomeres and Bulk Genomic DNA

Elise Fouquerel, Ryan P. Barnes, Hong Wang, and Patricia L. Opresko

Abstract

Telomere repeats at chromosomal ends are essential for genome stability and sustained cellular proliferation but are susceptible to DNA damage. Repair of damage at telomeres is influenced by numerous factors including telomeric binding proteins, sequence and structure. Ultraviolet (UV) light irradiation induces DNA photoproducts at telomeres that can interfere with telomere maintenance. Here we describe a highly sensitive method for quantifying the formation and removal of UV photoproducts in telomeres isolated from UV irradiated cultured human cells. Damage is detected by immunospot blotting of telomeres with highly specific antibodies against UV photoproducts. This method is adaptable for measuring other types of DNA damage at telomeres as well.

Key words UV photoproducts, UV irradiation, Telomeres, Immunospot blot, Telomere isolation, Radiolabeled DNA probes

1 Introduction

Telomere caps at the ends of linear chromosome are essential for genome stability. Telomere loss triggers cellular senescence which contributes to aging-related diseases or induces chromosomal changes which drive carcinogenesis [1, 2]. Human telomeres consist of approximately 10–15 kilobases of tandem 5′-TTAGGG-3′ repeats in duplex DNA, and terminate in a 3′ single strand overhang of about 50–200 nucleotides. A complex of six proteins, termed shelterin, binds specifically to telomeric DNA sequences, and remodels the telomere into a cap-like structure [3]. DNA damage and repair at telomeres is uniquely impacted by several factors including telomeric DNA sequence, the ability of telomeric DNA to form alternate structures, and the ability of shelterin proteins to bind to and regulate various DNA repair enzymes (for review see [4]). The abundance of dipyrimidine residues in telomere repeats makes them susceptible to UV light induced DNA photoproducts. Cyclobutane pyrimidine dimers (CPD) and pyrimidine(6–4) pyrimidone

Lata Balakrishnan and Jason A. Stewart (eds.), *DNA Repair: Methods and Protocols*, Methods in Molecular Biology, vol. 1999, https://doi.org/10.1007/978-1-4939-9500-4_20,

photoproducts (6-4PP) are formed when UV irradiation induces covalent linkage between two adjacent pyrimidines [5]. Unrepaired DNA photoproducts can interfere with DNA replication and transcription, and can lead to mutations [6]. UV photoproducts are repaired by nucleotide excision repair (NER). Loss of NER causes the severe sunlight sensitivity and skin cancer prone genetic disorder xeroderma pigmentosum (XP) [7]. In the general population, UV exposure is associated with skin aging and increased risk for developing skin cancer [7]. UV exposure is also associated with accelerated telomere shortening in exposed skin from humans and mice [8, 9], indicating that UV damage can affect telomere maintenance. Human cell culture experiments reveal that UVC exposure induces CPDs and 6-4PPs at telomeres, which are removed by NER [10, 11]. Unrepaired CPDs prevent shelterin protein TRF1 from binding to telomeric DNA [11]. Therefore, DNA damage and repair at telomeres has important implications for telomere maintenance and function.

The ability to measure DNA damage and repair at telomeres is challenging due to the highly repetitive nature of the sequence, low abundance (less than 0.025% of the genome), and localization at chromosome ends. Here we describe a highly sensitive procedure for detecting UV photoproducts in telomeres isolated from UVC irradiated cells (Fig. 1) [11]. The telomere isolation procedure is based on a protocol originally developed to detect structure and sequence at chromosomal ends [12]. The annealing of a biotinylated oligonucleotide to the telomeric single strand overhang is used as a handle to capture telomeric restriction fragments with streptavidin-coated magnetic beads. Detection of UV photoproducts in the isolated telomeres employs the immunospot blot method and commercially available highly specific and sensitive antibodies against CPDs or 6-4PPs. This method is adaptable to detecting any form of DNA damage at telomeres for which sensitive and specific antibodies are available.

2 Materials

2.1 Mammalian Cell Culture and UV Exposure

1. Any suitable mammalian cell line can be used. This protocol describes culturing of BJ skin fibroblasts immortalized with hTERT (BJ-5ta, ATCC, Manassas, VA).
2. Dulbecco's Modified Eagle Medium, high glucose (DMEM) supplemented with 50 units/mL penicillin and 50 μg/mL streptomycin. 10% (v/v) fetal bovine serum is added to media.
3. 0.5% trypsin–EDTA.
4. Germicidal 254 nm UVC lamp and UVP UVX digital ultraviolet intensity meter/radiometer with 254 nm sensor.

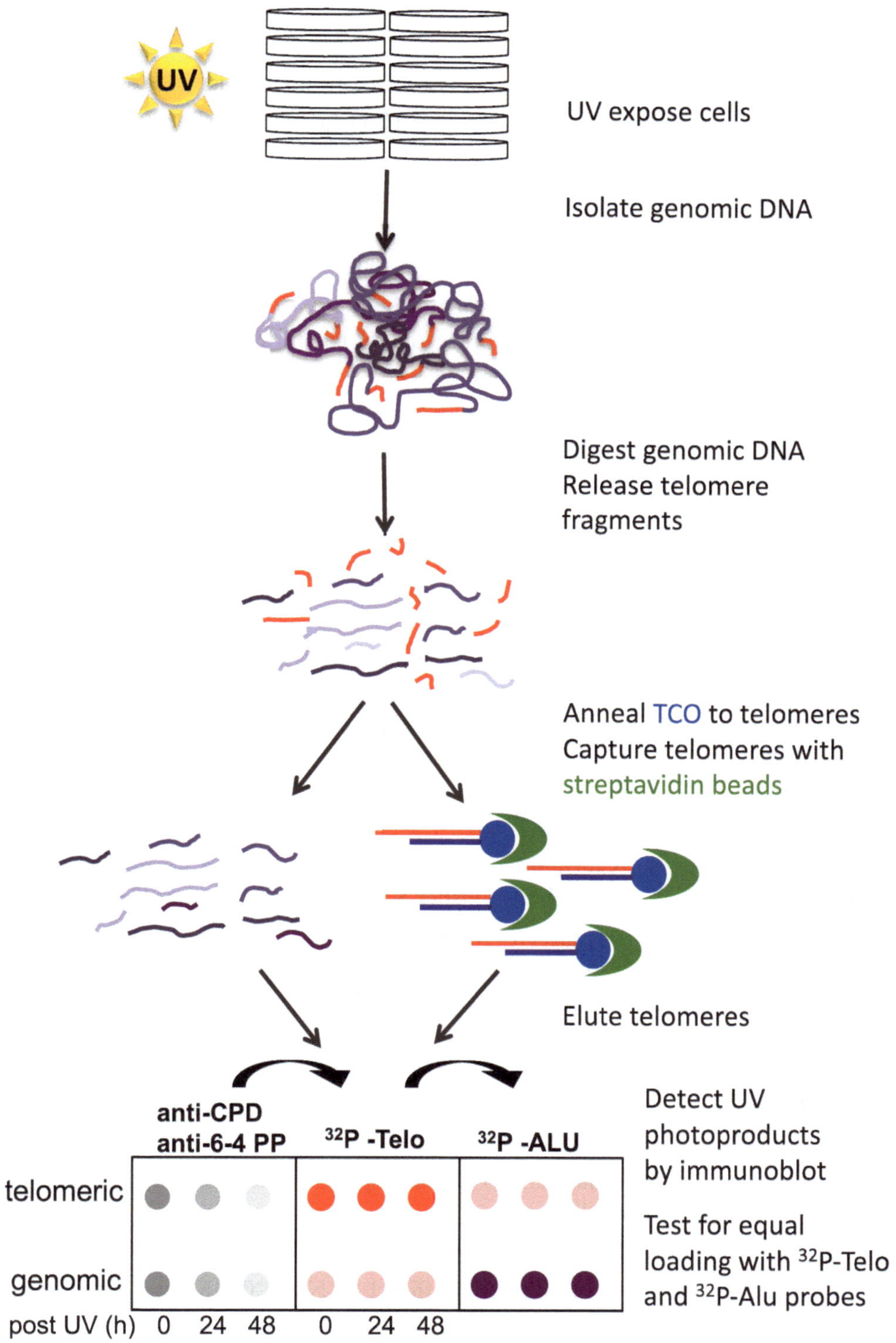

Fig. 1 Schematic of UV photoproduct detection in purified telomeres and genomic DNA assay. Cultured cells are irradiated with 10 J/m^2 UVC. Genomic DNA is isolated and digested with restriction enzymes to release intact telomere fragments, which are then annealed to a complementary biotinylated telomeric capture oligonucleotide (TCO). Telomere fragments are captured with streptavidin beads and then eluted by heating at 53 °C. Purified telomeres and digested genomic DNA are blotted on membranes, and then sequentially probed with antibodies against CPDs or 6-4PPs, radiolabeled telomeric probes, and then radiolabeled Alu DNA probes. Detection with radiolabeled probes serves as controls for equal loading

2.2 Genomic and Telomeric DNA Isolation

1. Qiagen 100/G DNA isolation kit (Qiagen, Valencia, CA).
2. Restriction enzymes 10 U/μL *Hinf*I, 10 U/μL *Alu*I, 5 U/μL *Mnl*I, 5 U/μL *Hph*I.
3. Restriction enzyme reaction buffer (10×): 330 mM Tris–HCl (pH 7.9), 100 mM magnesium acetate, 1 M LiCl, 5 mM DTT.
4. 20× SSC: 3.0 M NaCl and 0.3 M Sodium Citrate.
5. 10% Triton X-100.
6. Telomere capture oligonucleotide (TCO): Biotin-5′-ACTCC $(CCCTAA)_3$-3′.
7. M-280 streptavidin-coated magnetic beads.
8. 1×PBST: 1× phosphate buffered saline, 0.1% Triton X-100.
9. 5× Denhardt solution.
10. Elution buffer: 1 mM Tris–HCl (pH 7.5), 1 mM EDTA, 10 mM LiCl. Store at 4 °C.

2.3 Agarose Gel Electrophoresis

1. Agarose.
2. Gel Loading Dye, Purple (6×), no SDS.
3. 1× TAE: 40 mM Tris, 20 mM acetic acid, 1 mM EDTA.
4. Quick-Load Purple 2-Log DNA ladder (New England Biolabs, Ipswich, MA), or an equivalent DNA ladder.
5. SYBR green.

2.4 DNA Concentration Measurement

1. NanoDrop 3300 Fluorospectrometer (Nano Drop Technologies, ThermoFisher Scientific, Wilmington, DE).
2. PicoGreen (Invitrogen Molecular Probes, ThermoFisher Scientific, Carlsbad, CA).
3. *Hind*III digested lambda DNA.

2.5 DNA Immunospot Blot

1. Minifold I Spot Blot System (GE Healthcare, Piscataway, NJ).
2. Amersham Hybond™-N^+ membrane (GE Healthcare, Piscataway, NJ).
3. Whatman gel blot filter paper for capillary blotting and wicking applications.
4. Denaturation buffer: 1.5 M NaCl, 0.5 M NaOH.
5. Neutralization buffer: 1 M NaCl, 0.5 M Tris–HCl (pH 7.0).
6. TBST: 1× Tris Buffered Saline (pH 7.4), 0.3% Tween 20.
7. Blocking buffer: 5% nonfat dry milk, TBST.
8. Mouse monoclonal antibody against cyclobutane pyrimidine dimer (CPD) clone TDM2 (Cosmo Bio, Tokyo, Japan).
9. Mouse monoclonal antibody against pyrimidine(6–4) pyrimidone photoproduct (6–4 PP) clone 64 M-2 (Cosmo Bio, Tokyo, Japan).

10. Secondary antibody: peroxidase-conjugated AffiniPure Goat Anti-Mouse IgG (H + L).
11. Amersham ECL Prime Western Blotting Detection Reagent (GE Healthcare, Piscataway, NJ).
12. X-ray film.

2.6 Blotting with Radiolabeled Probes

1. T4 Polynucleotide Kinase 10 U/μL.
2. 10× T4 Kinase Buffer.
3. ATP, [γ-^{32}P]- 3000 Ci/mmol.
4. Telomere probe oligonucleotides: 5′-$(TTAGGG)_4$-3′ and 5′-$(CCCTAA)_4$-3′.
5. Alu probe oligonucleotide: 5′-GGCCGGGCGCGGTGGCT CACGCCTGTAATCCCAGCACTTTGGGAGGCCGAGGC GGGCGGA-3′.
6. Illustra MicroSpin G-25 Sephadex columns (GE Healthcare, Piscataway, NJ).
7. Hybridization buffer: 5×SCC, 5×Denhardt solution, 10 mM Na_2HPO_4, 1 mM $Na_2P_2O_7$.

3 Methods

3.1 Mammalian Cell Culture and UVC Exposure

Approximately 20 × 10^6 BJ-5ta cells (eleven 100 mm plates) are required to obtain 100 μg of genomic DNA to isolate telomeres for each repair time point. The number of cells depends on the cell line. The HeLa cell line requires approximately 12 × 10^6 cells from three to five 100 mm plates to obtain 100 μg genomic DNA. Each experiment should include an untreated control. Since CPDs are repaired slowly, appropriate time points for CPD repair are 0, 12, 24, and 48 h. Since 6-4PPs are repaired rapidly, appropriate time points for 6-4PP repair are 0, 3, 6, and 12 h.

1. Culture and passage BJ-5ta human cells according to the ATCC guidelines. For BJ-5ta cells seed approximately 1 × 10^6 cells in eleven 100 mm tissue culture dishes for each repair time point. Incubate at 37 °C and 5% CO_2 in humidified cell culture incubators.
2. UV irradiate the cells when the cultures are approximately 80% confluent. Turn on the 254 nm lamp and allow it warm for 20 min. Measure exposure time required to deliver 10 J/m^2 UVC by using a UVX digital ultraviolet intensity meter/radiometer with 254 nm sensor (*see* **Note 1**).
3. Prior to irradiation, wash the cells in 1× prewarmed PBS. Remove PBS and expose cells to the UVC lamp with the lid removed.
4. Add fresh media and place in the tissue culture incubator for desired recovery time.

3.2 Isolating Genomic and Telomeric DNA

1. Wash approximately 20 × 10^6 cells per time point in 1× PBS, then harvest by centrifugation at 335 × *g* at 4 °C. Discard the supernatant and proceed with genomic DNA isolation using the Qiagen 100/G kit according to the manufacturer's instructions.
2. Mix 100 μg of purified genomic DNA with 50 U each of *Hinf*I, *Alu*I, *Mnl*II, and *Hph*I in 1× restriction enzyme reaction buffer in a final volume of 200 μL. Incubate at 37 °C overnight. The reaction can be adjusted to 300 μL if necessary, or the genomic DNA sample can be concentrated.
3. To verify that genomic DNA was digested to completion, analyze an aliquot of the digestion reaction by agarose gel electrophoresis. Mix 1 μL of digested DNA or 1 μL of undigested DNA with 4 μL H_2O and 1 μL of 6× gel loading dye. Load DNA samples along with 5 μL of Quick-Load 2-Log DNA ladder onto a 0.8% agarose gel in 1× TAE. Electrophoreses for 1 h at 110 V. Visualize DNA fragments by ethidium bromide or SYBR green staining (Fig. 2a).
4. Add 11 μL of 20× SSC and 2.5 μL of 10% Triton X-100 to the 200 μL digestion reaction for a final concentration of 1×SCC and 0.1% Triton X-100. Add 3.5 pmols (1 μL) of TCO oligonucleotide. Place tube in a thermocycler and run the following

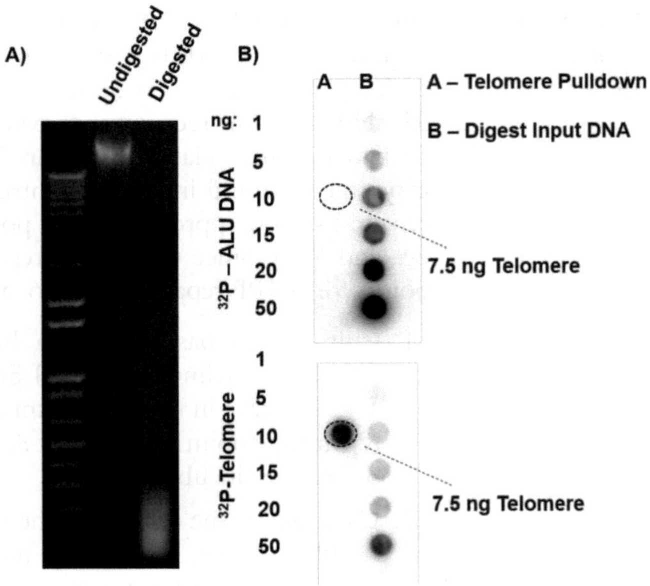

Fig. 2 Testing purity of isolated telomeres. (**a**) Agarose gel shows digestion of genomic DNA. (**b**) Telomere fragments (7.5 ng, eluent) and various amounts of digested genomic DNA (1–50 ng, input) are blotted on a two separate membranes. One membrane is probed with ^{32}-P radiolabeled oligonucleotides complementary to Alu repeat DNA, and the other is probed with ^{32}-P radiolabeled oligonucleotides complementary to telomeric DNA

program to anneal the TCO to the telomeric 3′ overhang by controlled stepwise cooling from 80 to 25 °C. The annealing reaction may be divided into several tubes.

Annealing program: Initial denaturation for 15 min at 80 °C. Cool from 80 to 65 °C in 15 cycles at 1 min each. Hold at 65 °C for 30 min in 1 cycle. Cool from 65 to 55 °C in 10 cycles at 1 min each. Hold at 55 °C for 20 min in 1 cycle. Cool from 55 to 45 °C in 10 cycles at 1 min each. Hold at 45 °C for 15 min in 1 cycle. Cool from 45 to 35 °C in 10 cycles at 1 min each. Hold at 35 °C for 15 min in 1 cycle. Cool from 35 to 25 °C in 10 cycles at 1 min each. Hold at 25 °C for 10 min in 1 cycle. Hold at 4 °C until sample is retrieved. This program requires 2 h and 40 min to run (*see* **Note 2**).

5. During the annealing reaction, prepare M-280 streptavidin-coated magnetic beads. After resuspending the beads, remove 20 μL of the beads slurry and add to 200 μL of PBST. Apply a magnet to collect beads against the side of the tube. Remove supernatant without disturbing the beads and repeat wash with 200 μL fresh PBST. Apply the magnet and remove the supernatant. Add 18 μL of 5× Denhardt solution to block nonspecific binding. Incubate for 2 h at room temperature and resuspend by flicking the tube every 20 min.
6. When the annealing reaction is complete, add all 200 μL to the prewashed and blocked beads. Rotate at 5 rpm overnight end over end at 4 °C.
7. Centrifuge the tube briefly at 233 × *g*. Apply the magnet for 2 min at room temperate to collect the beads on the side of the tube. Remove the supernatant and save for analysis.
8. Wash the beads three times with 200 μL of 1×SCC, 0.1% Triton X-100 solution. Then wash twice with 200 μL of 0.2×SCC solution. For each wash add the solution, resuspend the beads gently by tapping the tube, and apply the magnet for 1 min at room temperature. Then remove the wash without disturbing the beads. Save the washes for analysis later.
9. To elute the telomere fragments, resuspend the beads in 30 μL of elution buffer. Incubate for 20 min at 53 °C. Resuspend by gently tapping the tube after 10 min. Gentle heating at this step breaks the hydrogen bonds between the TCO oligonucleotide and the telomeres to release the telomere fragments.
10. Apply the magnet and collect the eluent. This is elution 1. Store samples at −20 °C.
11. Resuspend the beads in 15 μL of elution buffer. Incubate for 20 min at 53 °C. Resuspend by gently tapping the tube after 10 min. This is elution 2 and can be pooled with elution 1. Store samples at −20 °C. Yields are typically 10–15 ng of purified telomere fragments from 100 μg of genomic DNA.

3.3 Measuring Concentrations of Purified Telomeres

The NanoDrop 3300 fluorospectrometer allows for measurement of very low DNA concentrations. This instrument can measure PicoGreen bound duplex DNA in a range of 1 to 1000 ng/mL (*see* **Note 3**).

1. Prepare standards by serial dilution of *Hind*III digested lambda DNA to concentrations of 4000 ng/mL, 2000 ng/mL, 1000 ng/mL, 400 ng/mL, and 200 ng/mL in sterile ddH_2O. The standards can be stored at −20 °C for use in future experiments.
2. Dilute PicoGreen dye stock 1:200 (v/v) in DMSO solution.
3. Mix 2 μL of each standard with 2 μL of diluted PicoGreen dye. Mix 2 μL of purified telomeres, or 2 μL of genomic DNA diluted about 1:500 in ddH_2O with 2 μL of diluted PicoGreen dye for a 2× concentration. Prepare a reference sample lacking DNA. Incubate for 5 min at room temperature.
4. Blank with the reference sample. Prepare a standard curve by reading 2 μL of each standard twice according to the manufacturer's protocol titled "PicoGreen Assay for dsDNA."
5. Measure 2 μL of each sample of purified telomeres twice and determine the concentration based on the standard curve. Concentrations of purified telomeres are typically 200 to 400 ng/mL (*see* **Note 4**).
6. Testing the purity of the telomere fractions can be done by blotting various amounts of input genomic DNA (0.25–10 ng) and 10 ng of purified telomeres as described in Subheading 3.4, **steps 1–9**. Prepare two separate membranes with the identical samples. Incubate membrane 1 with Alu repeat probe DNA as described in Subheading 3.5, which serves as a marker for genomic DNA. This allows for the analysis of the amount of contaminating Alu repeat DNA in the purified telomere fractions. Incubate membrane 2 with the telomeric probe as described in Subheading 3.5. This allow for analysis of the enrichment of telomeric DNA (Fig. 2b).

3.4 Immunospot Blot for UV Photoproducts

1. To prepare the samples mix 10–20 ng of purified genomic DNA or isolated telomere fragments with 200 μL 2× SCC solution. 7.5 ng DNA is required for CPD detection, whereas 15 ng DNA is required for 6–4 PP detection.
2. To prepare the manifold spot blot apparatus, cut a piece of Whatman filter paper and a piece of Hybond-N+ membrane to fit the manifold. Presoak the filter paper and membrane in 2× SCC.
3. Assemble spot blot apparatus according to the manufacturer's directions. Place the filter support plate on top of the vacuum piece, aligning the pins. Lay down the filter paper and place the

membrane on top, ensuring there are no bubbles. Next add the spot-sample well plate, aligning the pins. Finish off the sandwich with the metal clamping plate and clamp with the steel latches. Apply the vacuum for 1 min and then turn off.

4. Load samples while noting the position and apply the vacuum until liquid is no longer visible in the wells. Turn off the vacuum. Samples can also be loaded while the vacuum is on.
5. Disassemble the apparatus and remove the membrane. Place membrane face down onto a piece of filter paper presoaked in denaturation buffer and incubate at room temperature for 8 min.
6. Remove membrane and lay face down onto a piece of filter paper presoaked in neutralizing buffer, and incubate at room temperature for 8 min.
7. Remove membrane and place it face up on a dry piece of filter paper. Incubate at room temperature for 5 min.
8. Place membrane on another piece of filter paper. Wrap in Saran wrap and bake under vacuum for 2 h at 80 °C. This process serves to fix the DNA onto the membrane.
9. Remove the membrane and cool to room temperature. Incubate membrane in blocking solution for 1 h at room temperature with gentle rocking.
10. Dilute the primary antibody 1:5000 either anti-CPD or anti-6-4PP in TBST. Add to the membrane and incubate overnight at 4 °C.
11. Wash membrane three times for 5 min each with TBST using gentle rocking.
12. Dilute the secondary antibody, peroxidase-conjugated anti-mouse, 1:5000 in blocking buffer. Incubate at room temperature for 1 h.
13. Wash membrane three times for 5 min each with TBST using gentle rocking.
14. Lay membrane on a piece of Saran wrap. Apply the peroxidase substrate using the ECL Primer Western Blotting Detection Reagent according to the manufacturer's protocol.
15. Wrap membrane in Saran wrap and place in a cassette. Expose membrane to a piece of X-ray film in the dark room. Develop the film to visualize the blot (Fig. 3).
16. Antibody signal intensities can be quantified using ImageJ software.

3.5 Spot Blot Detection of Telomeric and Alu Repeat DNA

To control for equal loading of telomeric DNA and genomic DNA, the membrane can be subsequently blotted with radiolabelled oligonucleotide probes that are complementary to telomeric DNA or

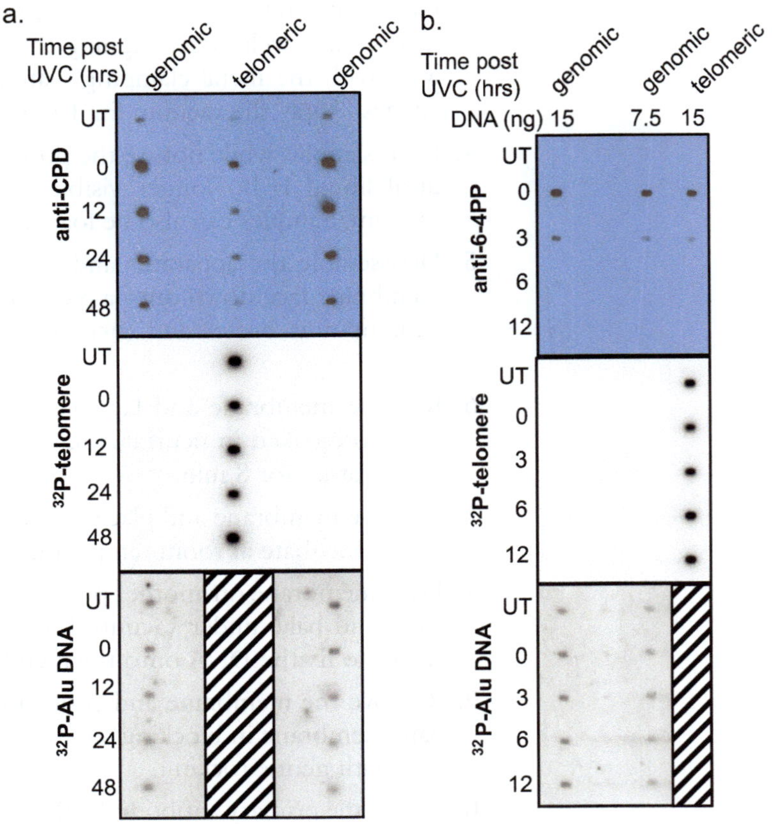

Fig. 3 UV photoproduct detection in telomeric and genomic DNA. BJ-hTERT cells are untreated or exposed to 10 J/m^2 UVC and harvested at various repair time points (0–48 h). (**a**) Isolated telomeres (7 ng) and genomic DNA (7 ng) are blotted and sequentially probed with antibodies against CPDs, radiolabeled probes against telomeric DNA, and then radiolabeled probes against Alu repeat DNA. (**b**) Isolated telomeres (15 ng) and genomic DNA (7.5 or 15 ng) are blotted and sequentially probed with antibodies against 6-4PPs , radiolabeled probes against telomeric DNA, and then radiolabeled probes against Alu repeat DNA. (reproduced from ref. 11 with permission from *Nature Communications*)

Alu repeat sequences. Stripping the membrane between probes leads to loss of DNA. Therefore, probes need to be applied consecutively (*see* **Note 5**).

1. In a total volume of 25 μL, mix 10 pmols of each telomeric oligonucleotide probe or 10 pmol of the Alu repeat oligonucleotide probe with 10 units of T4 polynucleotide kinase (PNK) and 30 μCi ^{32}P-γATP in the 1× T4 PNK buffer provided by the manufacturer. Incubate for 1 h at 37 °C. Heat inactivate at 65 °C for 20 min.
2. Remove unincorporated ^{32}P-γATP using a MicroSpin column according to the manufacturer's directions.

3. Incubate the membrane for 30 min at 42 °C in 10 mL hybridization buffer.
4. Add the 25 μL radiolabelled telomere probe reaction to the hybridization buffer and membrane. Incubate overnight at 42 °C.
5. Remove hybridization buffer with probe and store. It can be reused 2–3 times.
6. Wash the membrane two times each with 2×SCC, 0.1% SDS for 10 min at room temperature.
7. Wash the membrane two times each with 2×SCC, then two times each with 0.2×SCC, as in #6.
8. Wrap membrane in Saran wrap and expose to a phosphorimager screen for 1–2 h. Incubate overnight if the signal is too weak. Visualize and quantify signal intensities using a Phosphorimager (Fig. 3).
9. Repeat **steps 1–8** with the Alu repeat probe. The signal with the telomeric probe will remain, but the Alu probe is required to visualize loading of bulk genomic DNA.
10. Quantification of CPD and 6-4PP signals can be normalized to the signal obtained for telomeric and Alu repeat DNA to control for potential variation in sample loading.

4 Notes

1. UVC irradiation is damaging. Take precautions to ensure protection of eyes and skin. It is best to expose cells in an enclosed system.
2. We also obtained good results annealing the TCO to telomeric fragments using a water bath. Place the microfuge tubes containing the annealing reaction in a 500 mL beaker that is filled with ddH_2O and heated to 90 °C, with a stir bar spinning at a moderate speed (400 rpm). Allow the beaker to cool to room temperature on the stir plate.
3. PicoGreen dye is light sensitive. Perform experiments in amber tubes or foil covered tubes to minimize light exposure.
4. The Qubit™ Fluorometer also allows for the measurement of very low concentration DNA samples. This instrument has the advantage of requiring only two supplied standards and has several kits available for different purposes. Allow Qubit reagents to equilibrate to room temperature, and prepare a Qubit tube for each sample and standard. Use clear, unmarked tubes in fluorometer, as described by the manufacturer. Mix 10 μL of each supplied standard with 190 μL Qubit dsDNA HS working solution. Mix 1–10 μL of each sample with Qubit reagent, bringing the total volume to 200 μL. In our

experience, using 2–10 μL of sample yields consistent readings. After vortexing briefly, incubate tubes for 2 min at room temperature and read on the fluorometer according to the manufacturer.

5. Special protocols and safety precautions should be followed when using radioactive materials, and when disposing of solid and liquid radioactive waste.

Acknowledgments

The authors thank Dr. Woodring Wright (University of Texas Southwestern Medical Center) for helpful discussion regarding the telomere isolation protocol and acknowledge Dr. Dhvani Parikh for her contributions toward developing and optimizing the method for UV photoproduct detection at telomeres. This work was supported by an NIH grant (R01ES022944, ES28242, ES025606 to P.L.O, R21ES027641, R01GM107559, R01GM123246 to H.W.), **P30 ES025128 (through a pilot project grant to H.W. by CHHE at NCSU)**, and NIH grant (K99ES027028 to E.F.).

References

1. Armanios M, Blackburn EH (2012) The telomere syndromes. Nat Rev Genet 13:693–704
2. Artandi SE, DePinho RA (2010) Telomeres and telomerase in cancer. Carcinogenesis 31:9–18
3. Palm W, de Lange T (2008) How shelterin protects mammalian telomeres. Annu Rev Genet 42:301–334
4. Fouquerel E, Parikh D, Opresko P (2016) DNA damage processing at telomeres: the ends justify the means. DNA Repair (Amst) 44:159–168
5. Yoon JH, Lee CS, O'Connor TR, Yasui A, Pfeifer GP (2000) The DNA damage spectrum produced by simulated sunlight. J Mol Biol 299:681–693
6. Hoeijmakers JH (2009) DNA damage, aging, and cancer. N Engl J Med 361:1475–1485
7. DiGiovanna JJ, Kraemer KH (2012) Shining a light on xeroderma pigmentosum. J Invest Dermatol 132:785–796
8. Ikeda H, Aida J, Hatamochi A, Hamasaki Y, Izumiyama-Shimomura N, Nakamura K, Ishikawa N, Poon SS, Fujiwara M, Tomita K, Hiraishi N, Kuroiwa M, Matsuura M, Sanada Y, Kawano Y, Arai T, Takubo K (2014) Quantitative fluorescence in situ hybridization measurement of telomere length in skin with/without sun exposure or actinic keratosis. Hum Pathol 45:473–480
9. Stout GJ, Blasco MA (2013) Telomere length and telomerase activity impact the UV sensitivity syndrome xeroderma pigmentosum C. Cancer Res 73:1844–1854
10. Kruk PA, Rampino NJ, Bohr VA (1995) DNA damage and repair in telomeres: relation to aging. Proc Natl Acad Sci U S A 92:258–262
11. Parikh D, Fouquerel E, Murphy CT, Wang H, Opresko PL (2015) Telomeres are partly shielded from ultraviolet-induced damage and proficient for nucleotide excision repair of photoproducts. Nat Commun 6:8214
12. Wright WE, Tesmer VM, Huffman KE, Levene SD, Shay JW (1997) Normal human chromosomes have long G-rich telomeric overhangs at one end. Genes Dev 11:2801–2809

Chapter 21

Single-Molecule DNA Fiber Analyses to Characterize Replication Fork Dynamics in Living Cells

Srijita Dhar, Arindam Datta, Taraswi Banerjee, and Robert M. Brosh Jr.

Abstract

Understanding the molecular dynamics of DNA replication in vivo has been a formidable challenge requiring the development of advanced technologies. Over the past 50 years or so, studies involving DNA autoradiography in bacterial cells have led to sophisticated DNA tract analyses in human cells to characterize replication dynamics at the single-molecule level. Our own lab has used DNA fiber analysis to characterize replication in helicase-deficient human cells. This work led us to propose a model in which the human DNA helicase RECQ1 acts as a governor of the single-stranded DNA binding protein RPA and regulates its bioavailability for DNA synthesis. We have also used the DNA fiber approach to investigate the interactive role of DDX11 helicase with a replication fork protection protein (Timeless) in human cells when they are under pharmacologically induced stress. In this methods chapter, we present a step-by-step protocol for the single-molecule DNA fiber assay. We describe experimental designs to study replication stress and staining patterns from pulse–chase labeling experiments to address the dynamics of replication forks in stressed cells.

Key words DNA fiber, Replication, Genomic instability, DNA synthesis

1 Introduction

The first demonstration of DNA fiber autoradiography was reported by the Cairns lab in 1963 to characterize replication in bacterial cells using radiolabeled thymidine [1]. Five years later, a modified procedure involving pulse labeling and radiolabeled thymidine incorporation was used to visualize replicating units in mammalian cells [2]. In 1993, DNA capture on Millipore filters was replaced by DNA fiber spreading to adhere "stretched" DNA molecules from lysed cells onto glass plate; DNA sequences were detected by fluorescent hybridization to construct a restriction endonuclease-based DNA map of the hamster dihydrofolate reductase gene [3]. Technical advances to further align and straighten DNA molecules using a molecular combing strategy was described in 1994 [4]. The modern conventional fluorescent labeling of

Lata Balakrishnan and Jason A. Stewart (eds.), *DNA Repair: Methods and Protocols*, Methods in Molecular Biology, vol. 1999,
https://doi.org/10.1007/978-1-4939-9500-4_21,

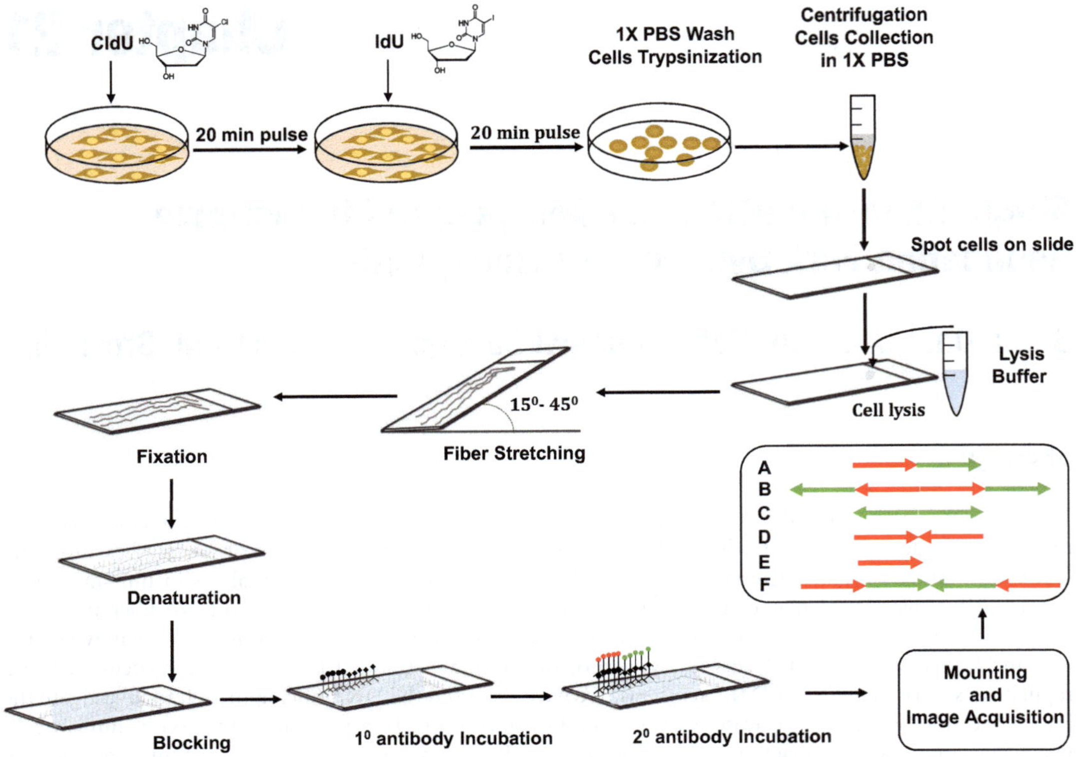

Fig. 1 Schematic showing the procedural steps in the DNA fiber assay. Structures of halogen nucleotides 5-chloro-2′-deoxyuridine (CldU) and 5-iodo-2′-deoxyuridine (IdU) are shown. The primary antibodies against CldU and IdU are incubated with the denatured DNA fibers fixed to the slide. Secondary antibodies conjugated with red and green fluorescent dyes are then added. Possible replication structures are shown: (*A*) Ongoing replication fork; (*B*) Bidirectional replication origin fired during first labeling (CldU); (*C*) Replication origin fired during second labeling (IdU); (*D*) Replication fork termination during first labeling; (*E*) Replication fork stalling during first labeling; (*F*) Termination of replication fork during second labeling. Red and green arrow represents Cldu and IdU labeled tracks respectively

DNA fibers with modified nucleotides and fluorescently labeled secondary antibodies was achieved in 1998, allowing researchers to propose replicon clusters responsible for propagation of S-phase in human cells [5]. An overview of the procedural steps in the pulse-labeling approach for DNA fiber studies, shown in Fig. 1, will be elaborated upon in subheadings 2 and 3 below. Further advances using the DNA fiber approach were made to study replication stress [6], and this topic remains of great interest [7]. Chromatin fiber technology yielded insight into the importance of changes in nucleosome occupancy that prevail during embryogenesis and carcinogenesis [8]. Furthermore, recent studies using the DNA fiber technique have provided evidence for increased replication initiation as an important property of cancer cells harboring p53 gain-of-function mutants [9, 10]. Modern developments in DNA and chromatin fiber analysis continue to emerge, emphasizing its powerful strategy and utility for studying complex questions in nucleic acid metabolism in vivo.

In the following two sections, we describe illustrative examples of how DNA fiber analyses in our own laboratory were used to illuminate novel mechanisms whereby human cells maintain normal replication fork dynamics, a determining factor for genomic stability. Following this, we will provide step-by-step procedures with technical notes that can be used or modified to apply DNA fiber methodology to other interesting cell biological problems involving genomic DNA synthesis.

1.1 RECQ1's Paramount Role in Replication Fork Dynamics

Of the five human RecQ helicases known to exist and play important roles in chromosomal stability of human cells, RECQ1 may very well be the least understood [11]. We employed DNA fiber analyses to investigate replication fork dynamics in RECQ1-deficient human cells and cells expressing site-directed mutants of RECQ1 which severely compromised their ability to unwind a preferred forked duplex DNA substrate of just 19 base pairs [12]. The experimental design involved first incubating the cells with 20 μM CldU for 20 min followed by 100 μM IdU for 20 or 45 min. This would allow for visualization of DNA synthesized within the first 20 min by red labeling and the next 20 or 45 min by green labeling. After lysing the cells, the DNA was allowed to spread on tilted glass slides, then fixed and neutralized, and immunostained using rat anti-BrdU (CldU) and mouse anti-BrdU (IdU) antibodies and the appropriate Alexa Fluor-conjugated secondary antibodies. Imaging was done with a Zeiss Axiovert microscope and Axio Vision software.

The microfluidic replication studies were revealing. Shortened fibers characterized by red staining followed by shortened lengths of green staining were observed in a greater percentage of the forks for RECQ1-depleted HeLa cells (2–6 μM) compared to those same cells exogenously expressing wild-type RECQ1 (4–10 μM) [12]. Similarly, RECQ1-depleted cells expressing either the RECQ1-W227A or RECQ1-F231A mutants also displayed a greater percentage of forks with shortened fibers (2–6 μM). From the DNA fiber analyses, it could be concluded that neither RECQ1 mutant was able to restore normal replication.

The slower replication fork rates inferred by the shorter DNA fibers in cells deficient of RECQ1 or expressing either RECQ1 site-directed mutant led us to analyze the DNA fibers from the microfluidic tract analyses in another way. We hypothesized that a greater number of dormant origins would be fired in the more slowly replicating RECQ helicase-deficient cells. So we quantified the forks with only IdU (green) signals, which represented nascent origins fired during the second labeling period. The percentage of dormant origins fired for RECQ1-depleted cells or those cells expressing either RECQ1-W227A or RECQ1-F231A were twofold greater than control (wild-type RECQ1) cells or

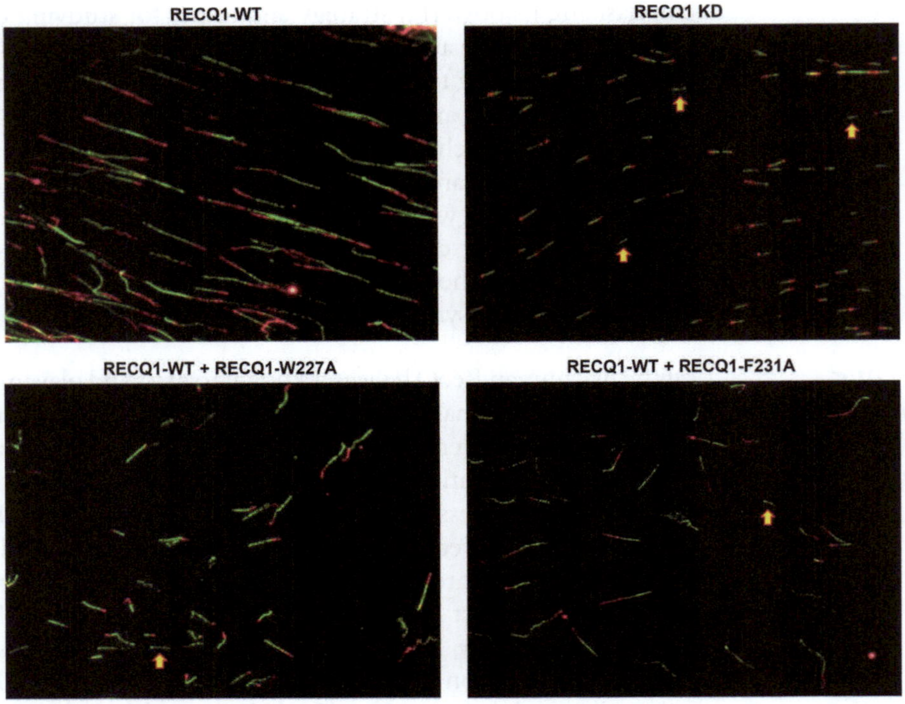

Fig. 2 Representative results from DNA fiber assays with HeLa cells. The effect of RECQ1 depletion (RECQ1 KD) or exogenous expression of dominant negative catalytic mutant RECQ1 proteins (RECQ1-W227A & RECQ1-F231A) on the rate of DNA synthesis in HeLa cells is visualized by the length of DNA fibers. Yellow arrows indicate dormant origins that are fired to compensate for the reduced DNA replication rate. Key: RECQ1-WT—HeLa cells expressing endogenous wild-type RECQ1; RECQ1 KD—HeLa cells in which endogenous wild-type RECQ1 has been depleted using a RNAi; RECQ1-WT + RECQ1-W227A—HeLa cells expressing endogenous wild-type RECQ1 and exogenous RECQ1-W227A; RECQ1-WT + RECQ1-F231A—HeLa cells expressing endogenous wild-type RECQ1 and exogenous RECQ1-F231A. The yellow arrows indicate DNA fibers synthesized from dormant origins

RECQ1-depleted cells exogenously expressing RECQ1-WT. Based on the DNA fiber analyses, we concluded that RECQ1 helicase activity plays an essential role in conferring normal fork dynamics as reflected by a prolific replication rate and nonelevated dormant origin firing [12]. From further studies, we determined that the abnormal fork dynamics had dire consequences for the cells in terms of genomic instability and DNA damage that could be suppressed by elevated RPA expression levels. Based on the results, we proposed a model in which RECQ1 acts as a governor of RPA to ensure that normal replication dynamics are managed in vivo to maintain genomic stability. Interestingly, expression of either RECQ1 aromatic loop mutant in HeLa cells endogenously expressing wild-type RECQ1 caused shorter replication tract lengths and dormant origin firing, consistent with their dominant negative effect on cell proliferation and DNA damage induction [12]. Representative results from DNA fiber assays are shown in Fig. 2.

1.2 Collaboration of a Fork Protection Complex Protein and a DNA Helicase During Replication Stress

Proteins of the fork protection complex are known to play an important role in sister chromatid cohesion, a process that cells use to tether newly replicated sister chromatids until their timely separation in metaphase [13]. Sister chromatid defects are observed in cells of persons with the chromosomal instability disease Warsaw breakage syndrome linked to biallelic mutations in the *DDX11* gene encoding a DNA helicase [14, 15]. In recently published work, we found in collaboration with the Pisani lab that the fork protection complex protein Timeless (Tim) interacts physically and functionally with DDX11 [16]. To determine if the in vitro interaction was meaningful in a biological setting, we employed DNA fiber track analysis to assess if Tim and DDX11 collaborate during periods of replication stress. To address this, we wanted to assess the outcome when a replication inhibitor was introduced during a period of advancing replication forks in vivo. Essentially, HeLa cells depleted of Tim, DDX11, or both by RNA interference were first pulse-labeled with CldU for 20 min, then the replication inhibitor hydroxyurea (HU) was added (or not), and finally cells were labeled with IdU for an additional 20 min. We observed that in HU-treated cells either Tim deficiency or DDX11 caused an approximately twofold reduction in IdU tract lengths, and the combined deficiency of both Tim and DDX11 resulted in a similar twofold reduction with no additive effect, suggesting that Tim and DDX11 operate in the same pathway for efficient progression of replication forks when cells are faced with an agent that induces replication stress by nucleotide depletion [16]. These findings helped to substantiate a model in which Tim and DDX11 collaborate with one another during periods of replication stress to ensure smooth replisome progression.

A basic flowchart for the DNA fiber assay is shown in Fig. 3. While the examples above provide relatively straightforward information about the status of DNA replication in stressed or unstressed cells, there are additional ways to set up the experiments with respect to labeling conditions that can provide insight to other aspects of fork dynamics. In the following sections, we will describe the procedures for DNA fiber analyses that we have use in our lab at the National Institute on Aging, NIH and discuss the parameters of fork dynamics that can be achieved by modifications of the procedure and careful analysis of the staining patterns.

2 Materials

1. 6-well tissue culture dish.
2. Dulbecco's Modified Eagle's medium (DMEM).
3. Streptomycin and penicillin.
4. DNA damaging agent.

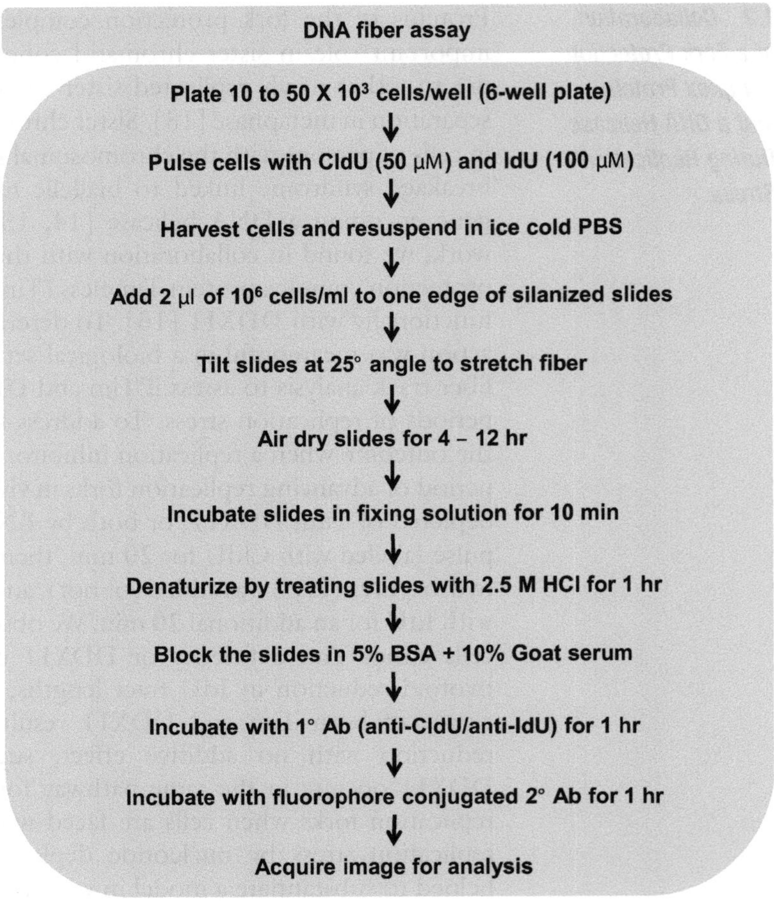

Fig. 3 Flowchart representing the DNA fiber assay. Diagrammatic representation for DNA fiber assay used to study replication dynamics in vitro. Cells are treated with two kinds of thymidine analog (CldU and IdU), which are incorporated differently in the DNA forks depending on the drug treatment and culture conditions. The two thymidine analogs are labeled with different fluorophores and can be visualized as separate colors, allowing us to study new, ongoing, stalled and resected forks

5. 5-Chloro-2′-deoxyuridine.
6. 5-Iodo-2′-deoxyuridine.
7. PBS, pH 7.4.
8. Trypsin.
9. Lysis buffer:
   (a) 200 mM Tris pH 7.4 (Stock solution)
   (b) 50 mM EDTA (Stock solution)
   (c) 0.5% SDS (Stock solution)

10. Fixing solution:
    (a) Methanol.
    (b) Glacial acetic acid.
11. 2.5 M HCl (Stock solution).
12. 400 mM Tris–HCl pH 7.4 (Stock solution).
13. Blocking buffer:
    (a) 5% BSA
    (b) 10% goat serum
14. Anti-CldU antibody.
15. Tween 20.
16. Anti-IdU antibody.
17. Alexa Fluor 488-AffiniPure goat anti-mouse IgG.
18. Alexa Fluor 647-AffiniPure goat anti-rat IgG.
19. Silanized slide.
20. Humidity Chamber.
21. ProLong Gold antifade mountant.
22. Premiere brand coverslips.
23. Zeiss Axiovert 200 M microscope.
24. Axio Vision software packages.

3 Methods

Single-molecule DNA fiber assay:

1. For human cell lines including Hela and U2OS, plate 10,000–50,000 cells in 2 mL of Dulbecco's Modified Eagle's medium with 10% fetal bovine serum, 100 μg/mL streptomycin, and 100 U/mL penicillin in each well of a 6-well plate (*see* **Note 1**).
2. Incubate the cells overnight in a 5% CO_2 incubator at 37 °C.
3. Aspirate media and add 2 mL of prewarmed fresh media containing 50 μM CldU to each well. Incubate for 20 min at 37 °C (*see* **Notes 2** and **3**).
4. While the cells are incubating in CldU containing media prepare DMEM media containing 100 μM IdU (*see* **Note 4**) and prewarm it to 37 °C in a water bath. Maintaining the constant temperature is critical for accurate results.
5. Aspirate the media containing CldU after 20 min and add the media with IdU immediately and incubate for an additional 20 min. It is critical to maintain the time accurately (*see* **Note 5**).

6. To study the effect of replication stress on replication dynamics, cells can be treated with specific DNA damaging agents for a predetermined period of time (*see* **Notes 6** and 7). According to the type and timing of the drug treatment we can study the effect of replication stress on fork dynamics. They are as follows:
   (a) Replication stress induced before CldU and IdU pulse: This method is used to study the effect of drug treatment on the number of origin firing, dormant origin firing, interorigin distance, replication tract length, and fork speed.
   (b) Replication stress induced in between CldU and IdU pulse: This approach can be utilized to study changes in the ongoing replication dynamics. Changes in replication tract length and fork speed can identify the effect of drug in the maintenance and processivity of forks. Reactivation of stalled forks and new origin firing as an effect of DNA damage can also be studied.
   (c) Replication stress induced after IdU pulse: with this approach we can identify resection of forks upon replication stress by measuring the IdU-labeled tract length before and after drug treatment.
7. Aspirate the media and wash the cells with ice-cold 1× PBS twice. Trypsinize the cells and count them. The cells are the pelleted by centrifuging them in a 1.5 mL Eppendorf tube at 450 × *g* for 1–2 min. Resuspend the cells in ice-cold 1× PBS to a final concentration of 10^6/mL.
8. Mark the slides in duplicate (*see* **Note 8**). Ideally, start with two independent sets of the same experiment and then make slides for each set of experiment in duplicate (*see* **Note 9**).
9. Add a drop of 2 μL from 10^6cells/mL solution on one edge of the slide (next to the slide label) and let it set for 1 min.
10. Add 10 μL of lysis buffer (*see* **Notes 10** and **11**) in drops onto the cells. Make sure that the tip does not touch the initial drop of cells on the slide. Wait till the sides of the lysis buffer is slightly dried out (usually 1–2 min). Then rest the slide in a slight tilted (around 10° to 25° angle) position to allow the liquid to flow down the length of the slide. The angle at which the slide is tilted is critical for success of the DNA spreading (*see* **Note 12**).
11. Fix the slides by immersing them in reservoirs with fixing solution (*see* **Note 13**). Make sure that the slides are immersed completely in the fixing solution and incubate for 10 min. After incubation is completed take the slides out and let them air-dry.

12. Wash the slides in reservoir with 1× PBS at room temperature. Dry out the extra PBS by touching the edge of the slides on an absorbent tissue paper (*see* **Note 14**).
13. Denature the DNA by treating the slides with 2.5 M HCl for 1 hr at room temp in a Coplin jar.
14. Neutralize the slides in 400 mM Tris–HCl pH 7.4 for 10 min. Keep it on a rocker in a reservoir the entire time of neutralization.
15. Wash the slides with 1× PBS at room temperature (3–5 min on a rocker). Wash once with PBST (0.1% Tween 20) for 3–5 min and dry the slides just enough to get rid of excess liquid.
16. Block the slides in 5% BSA + 10% goat serum (*see* **Note 15**). 75 μL of blocking solution is carefully placed on each slide in three drops along the length of the slide and evenly distributed over the entire slide using a strip of parafilm. The parafilm will cover the slide and will have the thin layer of the blocking solution between itself and the slide (*see* **Note 16**).
17. Incubate the slides with blocking solution inside a prewarmed humidity chamber at RT for 1 hr (*see* **Note 17**).
18. Add 75 μL of rat Anti-CldU (1:200 dilution) primary antibody, following the procedure described in **step 16**. Incubate for 1 hr at room temperature (*see* **Note 18**).
19. Wash the slides in a reservoir with PBST (0.1% Tween 20) for 5 min at room temperature with constant rocking. Repeat 3 times.
20. Add Alexa Fluor 647 conjugated goat anti-rat secondary antibody at a dilution of 1:100 and incubate at room temperature for 45 min. This procedure has to be performed in the dark.
21. Wash as mentioned in **step 19**.
22. Add mouse anti-IdU second primary antibody at a dilution of 1:40 and incubate at room temperature for 1 hr (*see* **Note 19**).
23. Wash as mentioned in **step 19**.
24. Add goat anti-mouse secondary antibody conjugated with Alexa Fluor 488 at a dilution of 1:100 and incubate for 45 min at room temperature.
25. Wash as mentioned in **step 19**.
26. Drain the excess liquid by touching the edge of the slide on a tissue paper. Add 75 μL of ProLong Gold mountant and carefully place the coverslip. Let the slides cure overnight (*see* **Note 20**).
27. Images of the slides can be taken by using 63× oil immersion objective of a fluorescence microscope (e.g., Zeiss Axiovert 200 M) with the appropriate filters. The images can be

analyzed using the Axio Vision software packages by Zeiss. For each sample, we suggest that a minimum of 100 forks be analyzed. Below listed are some replication dynamics that can be studied using DNA fiber assay:

(a) Origin firing:
   - During the first (CldU) pulse: green (IdU)–red (CldU) –green (IdU) labeled fibers indicating a bidirectional replication fork. The origin is presumed to be located in the center of the green (IdU)–red (CldU)--green (IdU) labeled fiber.
   - During the second (IdU) pulse/dormant origin firing: green (IdU) labeled fibers.

(b) Replication tract length and fork speed: Measuring the length of a unidirectional fork labeled red (CldU) -green (IdU) fibers. The tract length measured in μm can be converted to kilobase (kb) by using the multiplication factor of 2.59 [5]. Fork speed can be determined by dividing the replication tract length (kb) by time (min).

(c) Interorigin distance: Distance between the origins of two consecutive green (IdU)–red (CldU)–green (IdU) labeled fibers.

(d) Stalled fork: Indicated by red (CldU) labeled fibers (drug treatment after CldU and before IdU pulse).

(e) Restarting of stalled forks: Depicted by red (CldU) –green (IdU) labeled fibers (drug treatment after CldU pulse and before IdU pulse).

(f) Resection of fork: Shortening of the tract length in a unidirectional fork labeled as red (CldU)–green (IdU) DNA fibers, treated with drug after IdU (second) pulse.

4 Notes

1. Depending on cell type the culture conditions should be adjusted and optimized to grow to no more than 70% confluency and be in their logarithmic growth phase at the time of harvesting.
2. Make a 10 mM stock concentration of CldU and then add 10 μL of the stock CldU to 2 mL DMEM to get the final concentration of 50 μM CldU.
3. Grow the cells overnight in the incubator on top a Styrofoam piece. This helps with the maintenance of temperature while adding/changing media and minimizes cold shock to the cells. HeLa cells are more resistant but more sensitive cells might change their replication pattern when exposed to cold shock.

4. Always use higher concentration of label for the second labeling to dilute out any residual label from the first treatment.
5. Try not to delay between the additions of the second label in order for the two labeling to be continuous.
6. In order to determine the effect of replication stress on replication fork dynamics the cells we recommend standardizing the dose and duration of DNA damaging agent treatment by performing pilot experiment.
7. The "control" wells are recommended to be treated only with the solvent that the DNA damaging drug is dissolved in. The slides made from the cells in "control" wells can be utilized to analyze any change in the replication dynamics.
8. Mark the slides with pencil to avoid washing away with alcohol.
9. If there are five samples: plate them in duplicate for initial processing that will give ten wells. For each well make slides in duplicate, so finally there should be 20 slides, alternatively, there can be multiple DNA/cell lines made on the same slide.
10. Lysis buffer: Add 2 mL of 1 M Tris pH 7.4, 1 mL of 0.5 M EDTA, and 0.5 mL of 10% SDS to 6.5 mL dH_2O.
11. Make sure the lysis buffer gradually slides down and does not flow too quickly in order to get nicely stretched DNA fibers along the length of the slide. The angle at which the slide is tilted should be optimized for better results.
12. The thickness of a hardbound notebook usually used to keep notes should be ideal to rest the slide to achieve the optimal angle.
13. Fixing solution: Add 3-parts methanol to 1-part acetic acid.
14. The slides do not need to be completely dry. At this stage, any additional slides can be stored in 70% ethanol at 4 °C for later use.
15. The serum used depends on the animal in which the secondary antibody is raised; for example, if the secondary antibodies against CldU and IdU primary antibodies are raised in goat then blocking solution should contain goat serum.
16. It is critical to avoid bubbles in this step.
17. Add some water in the humidity chamber and place it in a 37 °C incubator prior to blocking step.
18. The choice of the primary and related secondary antibodies depends on the experimental setup. Pilot experiments need to be conducted in order to optimize the type and dilution of the antibodies used.
19. All steps after and including **step 19** should be done in dark.
20. The overlaying of coverslip should be done carefully, avoiding the formation of any air bubbles.

Acknowledgments

This work is supported by the National Institutes of Health, National Institute on Aging.

References

1. Cairns J (1963) The bacterial chromosome and its manner of replication as seen by autoradiography. J Mol Biol 6:208–213
2. Huberman JA, Riggs AD (1968) On the mechanism of DNA replication in mammalian chromosomes. J Mol Biol 32(2):327–341
3. Parra I, Windle B (1993) High resolution visual mapping of stretched DNA by fluorescent hybridization. Nat Genet 5(1):17–21. https://doi.org/10.1038/ng0993-17
4. Bensimon A, Simon A, Chiffaudel A, Croquette V, Heslot F, Bensimon D (1994) Alignment and sensitive detection of DNA by a moving interface. Science (New York, NY) 265(5181):2096–2098
5. Jackson DA, Pombo A (1998) Replicon clusters are stable units of chromosome structure: evidence that nuclear organization contributes to the efficient activation and propagation of S phase in human cells. J Cell Biol 140(6):1285–1295
6. Merrick CJ, Jackson D, Diffley JF (2004) Visualization of altered replication dynamics after DNA damage in human cells. J Biol Chem 279(19):20067–20075. https://doi.org/10.1074/jbc.M400022200
7. Quinet A, Carvajal-Maldonado D, Lemacon D, Vindigni A (2017) DNA fiber analysis: mind the gap! Methods Enzymol 591:55–82. https://doi.org/10.1016/bs.mie.2017.03.019
8. Cohen SM, Chastain PD 2nd, Rosson GB, Groh BS, Weissman BE, Kaufman DG, Bultman SJ (2010) BRG1 co-localizes with DNA replication factors and is required for efficient replication fork progression. Nucleic Acids Res 38(20):6906–6919. https://doi.org/10.1093/nar/gkq559
9. Datta A, Ghatak D, Das S, Banerjee T, Paul A, Butti R, Gorain M, Ghuwalewala S, Roychowdhury A, Alam SK, Das P, Chatterjee R, Dasgupta M, Panda CK, Kundu GC, Roychoudhury S (2017) p53 gain-of-function mutations increase Cdc7-dependent replication initiation. EMBO Rep 18(11):2030–2050. https://doi.org/10.15252/embr.201643347
10. Singh S, Vaughan CA, Frum RA, Grossman SR, Deb S, Palit Deb S (2017) Mutant p53 establishes targetable tumor dependency by promoting unscheduled replication. J Clin Invest 127(5):1839–1855. https://doi.org/10.1172/jci87724
11. Brosh RM Jr (2013) DNA helicases involved in DNA repair and their roles in cancer. Nat Rev Cancer 13(8):542–558. https://doi.org/10.1038/nrc3560
12. Banerjee T, Sommers JA, Huang J, Seidman MM, Brosh RM Jr (2015) Catalytic strand separation by RECQ1 is required for RPA-mediated response to replication stress. Curr Biol 25(21):2830–2838. https://doi.org/10.1016/j.cub.2015.09.026
13. Uhlmann F (2016) SMC complexes: from DNA to chromosomes. Nat Rev Mol Cell Biol 17(7):399–412. https://doi.org/10.1038/nrm.2016.30
14. Capo-Chichi JM, Bharti SK, Sommers JA, Yammine T, Chouery E, Patry L, Rouleau GA, Samuels ME, Hamdan FF, Michaud JL, Brosh RM Jr, Megarbane A, Kibar Z (2013) Identification and biochemical characterization of a novel mutation in DDX11 causing Warsaw breakage syndrome. Hum Mutat 34(1):103–107. https://doi.org/10.1002/humu.22226
15. van der Lelij P, Chrzanowska KH, Godthelp BC, Rooimans MA, Oostra AB, Stumm M, Zdzienicka MZ, Joenje H, de Winter JP (2010) Warsaw breakage syndrome, a cohesinopathy associated with mutations in the XPD helicase family member DDX11/ChlR1. Am J Hum Genet 86(2):262–266. https://doi.org/10.1016/j.ajhg.2010.01.008
16. Cali F, Bharti SK, Di Perna R, Brosh RM Jr, Pisani FM (2016) Tim/Timeless, a member of the replication fork protection complex, operates with the Warsaw breakage syndrome DNA helicase DDX11 in the same fork recovery pathway. Nucleic Acids Res 44(2):705–717. https://doi.org/10.1093/nar/gkv1112

Chapter 22

Direct Visualization of DNA Replication at Telomeres Using DNA Fiber Combing Combined with Telomere FISH

Himabindu Gali, Emily Mason-Osann, and Rachel Litman Flynn

Abstract

The ability to analyze individual DNA fibers undergoing active DNA synthesis has emerged as a powerful technique in the field of DNA replication. Much of the initial analysis has focused on replication throughout the genome. However, more recent advancements in this technique have allowed for the visualization of replication patterns at distinct loci or regions within the genome. This type of locus-specific resolution will greatly enhance our understanding of the dynamics of DNA replication in regions that provide a challenge to the replication machinery. Here, we describe a protocol that will allow for the visualization of DNA replication through one of the most structurally complex regions in the human genome, the telomeric DNA.

Key words DNA fiber assay, Telomere FISH, Telomere replication, SMARD, DNA combing

1 Introduction

The repetitive nature of the telomeric DNA sequence, coupled with the natural topological constraints of the telomere structure itself poses a challenge for the DNA replication machinery. These natural impediments require a specialized set of proteins to ensure faithful progression of the replication fork through the ends of linear chromosomes [1, 2]. To better understand the dynamics of telomere replication, techniques have emerged to study the dynamics of telomere replication at the single molecule level. The single-molecule analysis of replicating DNA (SMARD) was the first single-molecule approach described to visualize replication specifically at telomeric DNA [3]. SMARD involves labeling of the replicating DNA using halogenated nucleoside analogs, isolation of telomeric DNA by pulse-field gel electrophoresis, and combined immunofluorescent and fluorescent in situ hybridization detection of the labeled telomeric DNA. Since the initial publication of the technique, SMARD has been applied in several studies assaying telomere replication and telomere replication stress [4–8]. Here, we

Lata Balakrishnan and Jason A. Stewart (eds.), *DNA Repair: Methods and Protocols*, Methods in Molecular Biology, vol. 1999, https://doi.org/10.1007/978-1-4939-9500-4_22,

describe a variation of SMARD that does not require the enrichment of telomere fibers through pulse field gel electrophoresis but instead relies on DNA fiber assay followed by high throughput imaging to increase the number of telomeric events captured for analysis. Logarithmically growing cells are pulsed with 10 μM CldU and 100 μM IdU sequentially for 20 min each, followed by a 3 h chase where cells are grown in the absence of the analogs. This pulse–chase cycle was repeated 4 times to maximize the number of telomeres labeled during replication and ensure sufficient events for the subsequent analysis. Labeled cells are harvested, counted, and embedded into agarose plugs using agarose with a low melting point. DNA fibers are then extracted from the plugs by detergent lysis and protein digestion before being mechanically stretched onto coverslips. Telomeric fibers undergoing replication are identified by telomere specific DNA probe hybridization and immunodetection of the incorporated nucleoside analogs (Fig. 1).

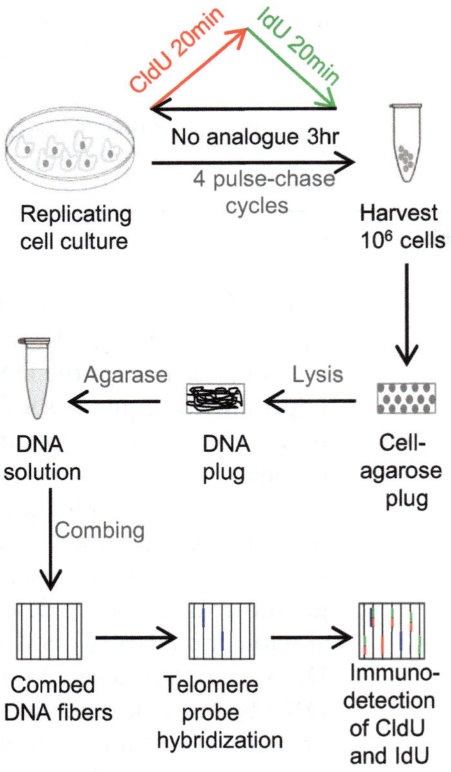

Fig. 1 Schematic representation of DNA fiber assay and analysis by telomere FISH and immunofluorescent detection of nucleoside analogs. The bottom row of the figure represents coverslips with combed DNA fibers in black lines, hybridized telomere probes in blue, CldU and IdU detection in red and green respectively

2 Materials

2.1 Incorporation of Nucleoside Analogs

1. Culture Media (Cell type specific).
2. 1× PBS: Dissolve 8 g NaCl, 0.2 g KCl, 1.44 g Na_2HPO_4, 0.24 g KH_2PO_4 in 800 mL distilled water. Adjust pH to 7.4 using HCl and make up the volume to one liter with distilled water.
3. 5-Chloro-2′-deoxyuridine (CldU) (Sigma, C6891): For 10 mM stock solution dissolve 262.7 mg of CldU in 100 mL of 1× PBS and store aliquots at −20 °C
4. 5-Iodo-2′-deoxyuridine (IdU) (Sigma, I7125): For 10 mM stock solution dissolve 354.1 mg of IdU in 100 mL of 1× PBS and store aliquots at −20 °C.

2.2 Preparation of DNA Plugs and Lysis

1. 1.2% low-melting point agarose (Lonza,50111): dissolved in PBS and kept at 50 °C.
2. Gel plug mold (Bio-Rad, 1703713).
3. 0.5 M EDTA (pH 8.0)
4. N-Lauroylsarcosine sodium salt solution (Sigma, L7414).
5. Proteinase K solution (Life Technologies 25530049).
6. TE—Tris–EDTA solution (10 mM Tris, 1 mM EDTA, pH 8.0).

2.3 Preparation of DNA Solution

1. β-Agarase I (NEB, M0392S).
2. 0.5 M MES Buffer (pH 5.5): Dissolve 97.62 g 2-morpholinoethanesulfonic acid in 750 mL water. Adjust pH to 5.5 using 10 N NaOH. Adjust final volume to 1 L, filter sterilize and store at 4 °C (*see* **Note 1**).
3. Water bath set at 42 °C, 50 °C, and 68 °C.

2.4 DNA Combing

1. Fiber Comb Molecular Combing System (Genomic Vision, MCS-001).
2. Silanized Coverslips (Genomic Vision, COV-002-RUO).
3. Disposable DNA reservoirs (Genomic Vision, RES-001).

2.5 Probe Hybridization and Immunofluorescence Staining

1. 0.1 N NaOH.
2. 0.1% β-Mercaptoethanol.
3. Glutaraldehyde solution (Sigma, G5882).
4. 70%, 90% and 100% Ethanol.
5. Blocking reagent (Sigma, 11096176001).
6. TelC-Biotin probe (PNA Bio Inc., F2001).

7. 2× SSC: Dissolve 17.53 g NaCl and 8.82 g sodium citrate in 800 mL distilled water. Adjust pH to 7 using HCl and make up the volume to one liter with distilled water.
8. Formamide (Fisher Bioreagents, BP228-100).
9. 1× PBS containing 0.05% Tween 20 (1× PBS-T)
10. BlockAid blocking solution (Life Technologies, B10710).
11. ProLong Gold without DAPI (Life Technologies, P36931).
12. Glass slides.

2.6 Antibodies

1. Rat anti-BrdU (recognizes CldU) (Abcam, ab6326).
2. Mouse anti-BrdU (recognizes IdU) (BD Biosciences, 347580).
3. Goat anti-rat Alexa Fluor 568 (Life Technologies, A11077).
4. Goat anti-mouse Alexa Fluor 488 (Life Technologies, A11001).
5. Biotinylated anti-streptavidin (Vector Laboratories BA-0500).
6. Streptavidin Alexa Fluor 647 conjugate (Life Technologies, S32357).

3 Methods

3.1 Pulse Labeling Nascent DNA with Nucleoside Analogs

1. Exponentially growing cells are sequentially pulse-labeled with 10 μM CldU for 20 min followed by 100 μM IdU for 20 min in CO_2 incubator (*see* **Note 2**).
2. Cells are washed 3 times with PBS and incubated in media without CldU/IdU for 3 h.
3. Pulse–chase cycle is carried out 4 times before harvesting 1,000,000 cells by trypsinization from each sample.

3.2 DNA Plug Preparation by In-Gel Lysis

1. Collect cells by centrifugation at 1200 rpm (135 × *g*) for 5 min and resuspend the cell pellet in 45 μL PBS for each plug.
2. Add equal volume of 1.2% low melting point agarose solution and mix gently. Immediately, pour 90 μL of the mixture into the gel plug mold.
3. Let the suspension solidify at 4 °C for 30 min (*see* **Note 3**) in gel plug mold before transferring to tubes containing lysis solution.
4. Cells in agarose plugs are lysed overnight at 50 °C in 1% Sarcosyl solution, containing 0.5 M EDTA and 2 mg/mL Proteinase-K.
5. Next day, discard lysis solution and wash the plugs in TE 3 times for 1 h each at room temperature with gentle rocking.

3.3 DNA Solution Preparation

1. Remove TE, add 1 mL 0.5 M MES (pH 5.5) to each plug and dissolve at 68 °C in a heat block for 20 min to release DNA into solution.
2. Cool down melted plug at 42 °C in a heat block for 10 min and gently add 1.5 μL β-agarase, but do not mix the solution (*see* **Note 4**).
3. Incubate at 42 °C overnight making sure the tube is well fixed without agitation.

3.4 DNA Combing

1. The following day, gently transfer the β-agarase treated plug to DNA combing reservoir already filled with 1 mL of 0.5 M MES buffer (pH 5.5) to comb DNA onto a silanized coverslip with the Fiber Comb molecular combing system according to manufacturer's instructions (*see* **Note 5**).

3.5 Dehydration and Denaturation of Combed Coverslips

1. Dehydrate coverslips containing DNA fibers in an oven at 60 °C for 2 h.
2. Next, denature DNA fibers by immersing coverslips in alkali denaturing buffer containing 0.1 N NaOH in 70% ethanol and 0.1% β-mercaptoethanol for 12 min.
3. Fix DNA fibers to coverslips by adding 0.5% glutaraldehyde to alkali denaturing buffer for 5 min at room temperature.
4. Wash with 1× PBS 3 times for 5 min each at room temperature with gentle rocking.

3.6 Telomere Probe Hybridization

1. Dehydrate the coverslips by immersing in 70%, 85% and 100% ethanol for 2 min each.
2. Let the coverslips dry at 37 °C for 20 min.
3. While the coverslips dry, prepare 10 nM TelC-Biotin PNA probe in 25 μL hybridization solution (70% formamide, 0.25% blocking reagent, 10 mM Tris (pH 7.5), 5 mM $MgCl_2$).
4. Flip the coverslip onto 25 μL hybridization mixture containing the telomere probe spotted on a glass slide, denature at 85 °C for 10 min and place in a humidified chamber at 37 °C overnight.
5. The following day wash the coverslip with 2× SSC containing 50% formamide 3 times at 37 °C for 5 min each, 3 times in 2× SSC alone at 37 °C for 5 min each, and one final wash in 1× PBS at room temperature. All washes should be performed with gentle rocking.

3.7 Immunofluorescence Detection

1. Block the coverslip using Block Aid for 30 min at 37 °C by flipping the coverslip onto a glass slide spotted with BlockAid in a humidified chamber.

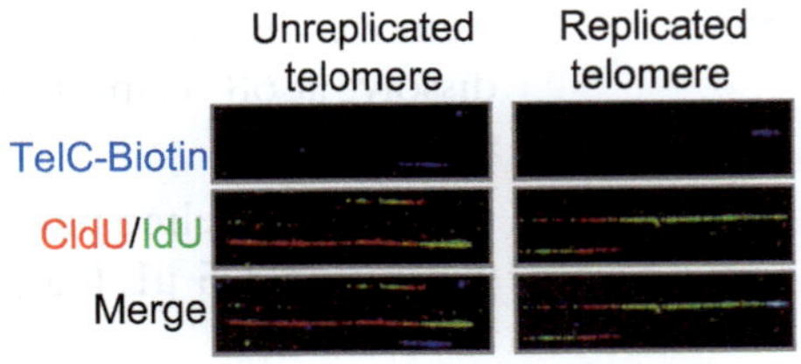

Fig. 2 Telomere targeted analysis of replication: Representative unreplicated (left) and replicated telomere (right) during pulse–chase cycles

2. Incubate the coverslip with 2 μL anti-rat BrdU + 2 μL streptavidin Alexa Fluor 647 conjugate + 21 μL Block Aid for 1 h at 37 °C by flipping the coverslip onto a glass slide in a humidified chamber. Wash the coverslip 3 times with 1× PBS-T for 5 min each at room temperature with gentle rocking followed by incubation with 4 μL anti-mouse BrdU +2 μL biotinylated anti-streptavidin +19 μL Block Aid for 1 h at 37 °C also by flipping the coverslip onto a glass slide in a humidified chamber.
3. Wash the coverslip 3 times with 1× PBS-T for 5 min each at room temperature with gentle rocking. Incubate the coverslip with 2 μL goat anti-mouse Alexa Fluor 488 + 2 μL goat anti-rat Alexa Fluor 568 + 2 μL streptavidin Alexa Fluor 647 conjugate +19 μL of BlockAid by flipping the coverslip onto a glass slide in a humidified chamber.
4. Wash the coverslip 3 times with 1× PBS-T for 5 min each at room temperature with gentle rocking. Dehydrate the coverslip by immersing into 70%, 85%, and 100% ethanol for 2 min each.
5. Let the coverslips dry at room temperature for 10 min followed by mounting onto glass slides with ProLong gold antifade reagent.

3.8 Image Acquisition and Analysis

Images were acquired using NIS element software (Nikon) and a Nikon eclipse Ti microscope with a 63× objective. The length of replication tracts and telomeres were measured using ImageJ. 1 μm of DNA represents 2.59 ± 0.24 kb for conversion of microns to kilobases (Fig. 2) (*see* **Note 6**).

4 Notes

1. To improve the quality and efficiency of DNA coming it is important to maintain the pH of MES buffer at 5.5.
2. IdU does not easily dissolve into solution; briefly warm the solution to 60 °C to dissolve completely.

3. Melted low-melting point agarose is kept at 50 °C in advance to maintain the solution state. Cell-agarose mixture was thoroughly mixed at 50 °C to avoid initial cell clumping and the eventual formation of bundled DNA fibers during combing. We also allow the cell-agarose plugs to solidify firmly at 4 °C for 15 to 30 min before lysis.
4. During β-agarase digestion do not mix and make sure the tubes are not agitated to avoid DNA becoming entangled.
5. We use molecular combing system from Genomic Vision to comb DNA onto silanized coverslips. Alternatively, DNA can be stretched on glass slides manually by directly lysing cells on the glass slide [9]. If insufficient telomeric tracts are detected, telomeric DNA can be enriched using pulse field gel electrophoretic separation and isolation of telomeric DNA before manually stretching on silanized glass slides [4–8, 10]
6. Alternative to manual analysis Genomic Vision provides Easy Scan services for high throughput imaging of whole coverslips to analyze replication patterns.

References

1. Martínez P, Blasco MA (2015) Replicating through telomeres: a means to an end. Trends Biochem Sci 40(9):504–515. https://doi.org/10.1016/j.tibs.2015.06.003
2. Maestroni L, Matmati S, Coulon S (2017) Solving the telomere replication problem. Genes 8(2):55. https://doi.org/10.3390/genes8020055
3. Sfeir A, Kosiyatrakul ST, Hockemeyer D, MacRae SL, Karlseder J, Schildkraut CL, de Lange T (2009) Mammalian telomeres resemble fragile sites and require TRF1 for efficient replication. Cell 138(1):90–103. https://doi.org/10.1016/j.cell.2009.06.021
4. Arnoult N, Schluth-Bolard C, Letessier A, Drascovic I, Bouarich-Bourimi R, Campisi J et al (2010) Replication timing of human telomeres is chromosome arm-specific, influenced by subtelomeric structures and connected to nuclear localization. PLoS Genet 6(4): e1000920. https://doi.org/10.1371/journal.pgen.1000920
5. Drosopoulos WC, Kosiyatrakul ST, Yan Z, Calderano SG, Schildkraut CL (2012) Human telomeres replicate using chromosomespecific, rather than universal, replication programs. J Cell Biol 197(2):253–266. https://doi.org/10.1083/jcb.201112083
6. Drosopoulos WC, Kosiyatrakul ST, Schildkraut CL (2015) BLM helicase facilitates telomere replication during leading strand synthesis of telomeres. J Cell Biol 210(2):191–208. https://doi.org/10.1083/jcb.201410061
7. Dilley RL, Verma P, Cho NW, Winters HD, Wondisford AR, Greenberg RA (2016) Break-induced telomere synthesis underlies alternative telomere maintenance. Nature 539 (7627):54–58. https://doi.org/10.1038/nature20099
8. Pan X, Drosopoulos WC, Sethi L, Madireddy A, Schildkraut CL, Zhang D (2017) FANCM, BRCA1, and BLM cooperatively resolve the replication stress at the ALT telomeres. Proc Natl Acad Sci 114(29): E5940–E5949. https://doi.org/10.1073/pnas.1708065114
9. Jackson DA, Pombo A (1998) Replicon clusters are stable units of chromosome structure: evidence that nuclear organization contributes to the efficient activation and propagation of S phase in human cells. J Cell Biol 6:1285–1295. https://doi.org/10.1083/jcb.140.6.1285
10. Norio P (2001) Visualization of DNA replication on individual Epstein-Barr virus episomes. Science 294(5550):2361–2364. https://doi.org/10.1126/science.1064603

Chapter 23

Telomere and G-Quadruplex Colocalization Analysis by Immunofluorescence Fluorescence In Situ Hybridization (IF-FISH)

Miaomiao Zhang, Rui Liu, and Feng Wang

Abstract

Four-stranded G-quadruplexes consists of tracks of guanines that are stabilized by Hoogsteen base pairing. The formation of G-quadruplexes in genomic DNA contribute to numerous biological processes in vivo, including replication, transcription, and telomere maintenance. Here, we present a detailed method to detect the colocalization of G-quadruplexes with telomeres in vivo, using the BG4 antibody developed from Dr. Shankar Balasubramanian's lab.

Key words Telomere, G-quadruplex, IF-FISH

1 Introduction

G-quadruplexes are noncanoical secondary structures formed in DNA or RNA sequences, which contain runs of guanines that form a cyclic Hoogsteen hydrogen-bonding arrangement of four guanines [1, 2]. It has been reported that G-quadruplexes can stall replication forks, regulate transcription and translation and are implicated in genomic and telomeric DNA integrity [3–6]. Moreover, given that the formation of G-quadruplexes within telomere overhangs could block telomerase recognition, the induction of telomeric G-quadruplexes stablizers have also been considered as potential anticancer therapy reagents [7–9]. Genome-wide analysis revealed over 700,000 potential G-quadruplex motifs, with the sequence $G_{\geq 3}N_xG_{\geq 3}N_xG_{\geq 3}N_xG_{\geq 3}$, in the human genome [10]. Due their G-rich property, the most abundant G-quadruplexes are located at specific chromosomal regions, including origins, promoters, rDNA, immunoglobulin switch regions, and telomeres [7, 11, 12].

Telomeres are DNA–protein complex found at the chromosome ends that protect chromosomes from improper degradation

Lata Balakrishnan and Jason A. Stewart (eds.), *DNA Repair: Methods and Protocols*, Methods in Molecular Biology, vol. 1999, https://doi.org/10.1007/978-1-4939-9500-4_23,

and chromosomal fusions [13]. The structure and stability of human telomeres are important for cancer, aging, and overall genetic stability [14, 15]. Human telomeric DNA consists of TTAGGG repeats that end in a single-stranded 3′ G-overhang [16]. Studies have provided evidence that G-quadruplexes form on the telomeres overhang in vivo and have elucidated the potential mechanism by which these structures are resolved during telomere replication [4, 17]. However, the visualization of G-quadruplexes remained a challenge until the development of a G-quadruplex structural specific antibody by Dr. Shankar Balasubramanian's lab [18]. Here, we described an optimized protocol to detect telomeres and G-quadruplexes on the same slide and show their colocalization in vivo using the BG4 antibody.

2 Materials

2.1 Cell Culture

1. 1× PBS: 137 mM NaCl, 2.7 mM KCl, 4.3 mM Na_2HPO_4, 1.4 mM KH_2PO_4, pH 7.2–7.4.
2. Trypsin–EDTA solution.
3. Clean and sterile coverslips (*see* **Note 1**).
4. Six-well plate.

2.2 Fluorescence in Situ Hybridization for Telomere (FISH)

1. Hybridization solution: Make fresh, 70% formamide, 10 mM Tris–HCl pH 7.2 with 1× blocking reagent (diluted from 10× blocking reagent stock solution).
2. 10× blocking reagent stock solution (Roche, made according to the manufacturer's instructions).
3. Peptide nucleic acid (PNA) TelC-FITC probe: FITC-OO-5′ CCCTAACCCTAACCCTAA 3′ (*see* **Note 2**).
4. Washing buffer 1: 10 mM Tris–HCl pH 7.2, 70% formamide, 0.1% BSA (BSA must be dissolved before adding formamide).
5. Washing buffer 2 (TBST): 20 mM Tris–HCl pH 8.0, 150 mM NaCl, 0.1% Tween 20, make fresh.
6. 37 °C water bath.
7. Clean glass slide.

2.3 Immunofluorescence (IF) for G-quadruplex

1. Extraction Buffer: 20 mM HEPES pH 7.0, 20 mM NaCl, 5 mM $MgCl_2$, 30 mM saccharose, 0.5% NP-40, 1× PBS.
2. PBST: 0.1% Tween 20 in 1× PBS.
3. Fixative solution: 2% paraformaldehyde in 1× PBS.
4. Pepsin solution: 1 mg/mL, dissolve in 10 mM glycine.
5. Permeabilization solution: 0.5% Triton X-100 in 1× PBS.
6. RNase A: 50 μg/mL working solution (*see* **Note 3**).

7. 70%, 90% and 100% ethanol.
8. Blocking buffer (PBG): 1× PBS, 0.5% BSA (Albumin Bovine V), 1% fish gelatin.
9. Antibodies: BG4 antibody [18], anti-Flag (MA4) mouse monoclonal antibody (Ray Antibody), and goat anti-mouse IgG (H + L) cross-absorbed secondary antibody Alexa Fluor 555.
10. Mounting Medium with DAPI (0.2–0.5 μg/mL).

2.4 Image Capture and Analysis

1. UV epifluorescence microscope equipped with filters of appropriate wavelengths, typically for blue: excitation, 360 nm—emission, 460 nm, red: excitation, 545 nm—emission, 610 nm, and green: excitation, 490 nm—emission, 520 nm.
2. ImageJ analysis software.

3 Methods

3.1 Cell Culture and Climbing

Plate cells the day before the experiment at a density of 10^5 cells per well on sterile coverslips that are placed in a six-well plate with 2 mL of complete growth medium (*see* **Note 4**).

3.2 FISH Procedure

1. Aspirate the medium from the six-well plate and wash each well twice with 1× PBS.
2. Extraction: Gently add 2 mL of Extraction Buffer to each well and incubate for 15 min at room temperature (RT) without shaking (*see* **Note 5**).
3. Wash cells 3 times for 5 min each with 2 mL PBST, then wash 3 times for 5 min with 2 mL 1× PBS.
4. Fixation: fix the cells with 2 mL Fixative solution for 20 min at RT (*see* **Note 6**).
5. Wash 3 times for 5 min each with 2 mL 1× PBS with shaking.
6. Incubate the slides in pepsin solution (prewarmed to 37 °C) for 5 min at 37 °C (*see* **Note 7**).
7. Wash twice for 5 min each with 2 mL 1× PBS with shaking.
8. Fix cells again with Fixative solution for 2 min at RT (*see* **Note 8**).
9. Wash 3 times for 5 min each with 2 mL 1× PBS with shaking.
10. Permeabilization: add 2 mL Permeabilization solution and gently shake at RT for 20 min.
11. Wash 3 times for 5 min each with 2 mL 1× PBS with shaking.

12. Add 100 μL 50 μg/mL RNase at 37 °C overnight.
13. Wash cells 3 times for 5 min each with 2 mL PBST.
14. Dehydrate the cells in 70%, 90%, 100% EtOH, 2 min each.
15. Aspirate the ethanol completely and let the coverslips dry for a few minutes. During this step, prepare the PNA probe by diluting it to 0.3 ng/μL in hybridization solution.
16. Add at least 25 μL of the PNA hybridization solution on a clean glass slide and then overlay the coverslip containing the cells onto the slide, avoiding bubbles.
17. Denature the DNA by heating the slides at 80 °C for 3 min.
18. Incubate for 2.5 h at RT in the dark.
19. Place coverslip back in the six-well dish and wash twice for 15 min with the wash buffer 1 with shaking (*see* **Note 9**).
20. Wash three times for 5 min each with the wash buffer 2.

3.3 IF Procedure

1. Blocking: Treat slides with the blocking buffer (PBG) for 1 h at RT. All subsequent step should be performed in the dark, unless otherwise indicated.
2. Dilute the BG4 antibody at a 1:1000 ratio in 50–100 μL blocking buffer. Put the antibody solution on flat strip of Parafilm and gentle place the coverslip, cell side down on the solution. Incubate for 1 h at RT.
3. Place coverslip back in the six-well dish and wash 4 times for 5 min each in PBST with shaking.
4. Incubate 1 h with anti-Flag (MA4) Mouse Monoclonal Antibody. Prepare the antibody as above in **step 2**.
5. Place coverslip back in the six-well dish and wash 4 times for 5 min each in PBST with shaking.
6. Incubate 30 min in 1:1000 Alexa Fluor 555 conjugated goat anti-mouse IgG (H + L) secondary antibody (*see* **Note 10**).
7. Place coverslip back in the six-well dish and wash 4 times for 5 min each in PBST with shaking.
8. Dehydrate the cells in 70%, 90%, 100% EtOH, 2 min each.
9. Let the coverslips air-dry in the dark.
10. For each slide, use 10 μL of Mounting Medium with DAPI to mount the coverslip onto a slide.
11. Slides can be stored at −20 °C.

3.4 Image Capture and Analysis

Search for cell nuclei under the microscope using the DAPI channel with an oil immersion lens. Visualize telomeric and G-quadruplex signals using the appropriate filter (red for Cy3/Alexa Fluor 555, green for FITC/Alexa Fluor 488). In addition, once you have chosen the proper exposure time, this should be used for all pictures. 25–30 images should allow a statistical analysis of number of telomere and G-quadruplex.

4 Notes

1. The cells are cultured on the coverslip in the six-well plates, so the coverslip should be clean and sterile. In general, the coverslips are treated with moderate acid solution overnight and then stored in 100% ethanol. Or, the ultrasonic cleaner can be used to clean the coverslips.
2. PNA TelC-Cy3 probe can also be used, but the TelC-Cy3 PNA probe working solution concentration should be reduced compared to the TelC-FITC. Disolve the PNA probe at 0.2 μg/mL in 0.1% trifluoroacetic acid (TFA) to elongate the storage time. PNA probe can be ordered from PANAGENE or other companies offering telomere FISH probes.
3. This protocol is used to detect the intramolecular telomere G-quadruplexes. The RNase digests the RNA G-quadruplex and the intermolecular G-quadruplex that formed between DNA and RNA.
4. These studies can be combined with small-molecule compounds, such as G-quadruplex stabilizers, or other functional analysis, in which case the number of days the cells are cultured may be varied.
5. The Extraction buffer should be made fresh and added slowedly because the cells may come off from the coverslip before fixation.
6. Other fixative solution such as methanol can also be used for fixation. If methanol is used, the cells shoule be fixed for 15 min on ice.
7. Pepsin should be prepared fresh and prewarmed at 37 °C. This step eliminates the cytoplasmic material and greatly reduces background.
8. Fixation by 2% paraformaldehyde again helps to protect the morphology and structure of nucleus after pepsin digestion.
9. Prewarming washing buffer 1 at 37 °C and increasing the shaking speed can help decrease the background signal, especially with the PNA TelC-Cy3 probe.
10. To distinguish the telomere and G-quadruplex signals, a red signal (Alexa Fluor 555) is used to stain G-quadruplexes, and a green fluorophore dye (FITC) is used to label PNA probe, or vice versa.

Acknowledgments

We thank Dr. Shankar Balasubramanian (University of Cambridge, UK) for providing BG4 antibody (plasmid). This work was supported by Ministry of Science and Technology (2017YFC1001904), National Natural Science Foundation of China (91649102, 31771520, 21177091, 21647008, 31471293, 81501386, 81671054, 81771135), the Key Project of Tianjin Research Program of Application Foundation and Advanced Technology (15JCZDJC35100).

References

1. Lipps HJ, Rhodes D (2009) G-quadruplex structures: in vivo evidence and function. Trends Cell Biol 19(8):414–422. https://doi.org/10.1016/j.tcb.2009.05.002
2. Li W, Wu P, Ohmichi T, Sugimoto N (2002) Characterization and thermodynamic properties of quadruplex/duplex competition. FEBS Lett 526(1–3):77–81
3. Bochman ML, Paeschke K, Zakian VA (2012) DNA secondary structures: stability and function of G-quadruplex structures. Nat Rev Genet 13(11):770–780. https://doi.org/10.1038/nrg3296
4. Rhodes D, Lipps HJ (2015) G-quadruplexes and their regulatory roles in biology. Nucleic Acids Res 43(18):8627–8637. https://doi.org/10.1093/nar/gkv862
5. Bugaut A, Balasubramanian S (2012) 5′-UTR RNA G-quadruplexes: translation regulation and targeting. Nucleic Acids Res 40(11):4727–4741. https://doi.org/10.1093/nar/gks068
6. Beaudoin JD, Perreault JP (2013) Exploring mRNA 3′-UTR G-quadruplexes: evidence of roles in both alternative polyadenylation and mRNA shortening. Nucleic Acids Res 41(11):5898–5911. https://doi.org/10.1093/nar/gkt265
7. Balasubramanian S, Neidle S (2009) G-quadruplex nucleic acids as therapeutic targets. Curr Opin Chem Biol 13(3):345–353. https://doi.org/10.1016/j.cbpa.2009.04.637
8. Buket O, Clement L, DanZhou Y (2014) DNA G-quadruplex and its potential as anticancer drug target. Sci China Chem 57(12):1605–1614. https://doi.org/10.1007/s11426-014-5235-3
9. Zahler AM, Williamson JR, Cech TR, Prescott DM (1991) Inhibition of telomerase by G-quartet DNA structures. Nature 350(6320):718–720. https://doi.org/10.1038/350718a0
10. Huppert JL, Balasubramanian S (2005) Prevalence of quadruplexes in the human genome. Nucleic Acids Res 33(9):2908–2916. https://doi.org/10.1093/nar/gki609
11. Balasubramanian S, Hurley LH, Neidle S (2011) Targeting G-quadruplexes in gene promoters: a novel anticancer strategy? Nat Rev Drug Discov 10(4):261–275. https://doi.org/10.1038/nrd3428
12. Besnard E, Babled A, Lapasset L, Milhavet O, Parrinello H, Dantec C, Marin JM, Lemaitre JM (2012) Unraveling cell type-specific and reprogrammable human replication origin signatures associated with G-quadruplex consensus motifs. Nat Struct Mol Biol 19(8):837–844. https://doi.org/10.1038/nsmb.2339
13. de Lange T (2002) Protection of mammalian telomeres. Oncogene 21(4):532–540. https://doi.org/10.1038/sj.onc.1205080
14. Podlevsky JD, Chen JJ (2012) It all comes together at the ends: telomerase structure, function, and biogenesis. Mutat Res 730(1–2):3–11. https://doi.org/10.1016/j.mrfmmm.2011.11.002
15. Blackburn EH, Greider CW, Szostak JW (2006) Telomeres and telomerase: the path from maize, Tetrahymena and yeast to human cancer and aging. Nat Med 12(10):1133–1138. https://doi.org/10.1038/nm1006-1133
16. Neidle S (2010) Human telomeric G-quadruplex: the current status of telomeric

G-quadruplexes as therapeutic targets in human cancer. FEBS J 277(5):1118–1125. https://doi.org/10.1111/j.1742-4658.2009.07463.x

17. Lemmens B, van Schendel R, Tijsterman M (2015) Mutagenic consequences of a single G-quadruplex demonstrate mitotic inheritance of DNA replication fork barriers. Nat Commun 6:8909. https://doi.org/10.1038/ncomms9909

18. Biffi G, Tannahill D, McCafferty J, Balasubramanian S (2013) Quantitative visualization of DNA G-quadruplex structures in human cells. Nat Chem 5(3):182–186. https://doi.org/10.1038/nchem.1548

Chapter 24

FISHing for Damage on Metaphase Chromosomes

P. Logan Schuck and Jason A. Stewart

Abstract

Fluorescence in situ hybridization (FISH) is used to examine chromosomal abnormalities and DNA damage. Developed in the early 1980s, this technique remains an important tool for understanding chromosome biology and diagnosing genetic disease and cancer. Use of FISH on metaphase chromosomes allows the visualization of chromosomal abnormalities at specific loci. Here, we describe methods for creating metaphase chromosome spreads and the use of telomere FISH probes to detect chromosome ends.

Key words FISH, Microscopy, Chromosome, Telomere, Metaphase spread

1 Introduction

In 1842, Carl Wilhelm von Nägeli first observed chromosomes in dividing plant cells [1]. While some herald this as the beginning of karyotyping, it was not until 1955 that the correct number of human chromosomes was determined by metaphase spread analysis, helping to usher in the field of chromosome biology [2]. This field was rapidly advanced in 1968 when the first in situ hybridization was successfully performed on *Xenopus* chromosomes [3]. This first approach used radiolabeled ribosomal RNA probes to detect stationary chromosomal DNA. The subsequent creation of biotin-labeled polynucleotides allowed for easy capture and fluorescein labeling of specific RNA/DNA sequences, making the technique more accessible [4].

Early fluorescent-labeled probes were large and relied on highly repetitive DNA sequences to obtain a detectable signal. However, this was a major improvement over autoradiographs, which required long exposure times and handling of radioactive material [3]. In the early 1980s, the quality and synthesis of probes continued to improve along with the detection of fluorescent dyes, leading to what we now know as fluorescence in situ hybridization (FISH) [5–8].

Lata Balakrishnan and Jason A. Stewart (eds.), *DNA Repair: Methods and Protocols*, Methods in Molecular Biology, vol. 1999, https://doi.org/10.1007/978-1-4939-9500-4_24,

Early FISH analysis was used to probe chromosome structure and stability, and is now employed to detect aneuploidy, gene deletions, gene duplications, translocations, and chromosome breakage. For example, this technique has been extensively employed in the mapping and analysis of common chromosomal fragile sites [9, 10]. FISH is also regularly used to study the replication and maintenance of repetitive DNA regions of the genome, such as centromeres, rDNA and telomeres [11–14]. More recently, a modified FISH technique was developed to more closely examine double-strand breaks at specific loci, using split-apart or breakaway probes. This technique relies on two colored FISH probes that flank the double-strand break [15, 16].

In the early 1990s, cytogeneticists realized the utility of this technique as a clinical diagnostic and began using FISH for karyotyping and the detection of chromosome abnormalities in diseases such as Fragile X and Down's syndrome [17–19]. The use of FISH as a clinical diagnostic continued to expand in intervening years becoming a powerful tool to detect and treat a number of genetic diseases and cancers. For example, FISH is used to identify whether breast cancers are HER2 positive or negative, which aids in the choice of drug treatment [20, 21]. It is also used in the treatment of leukemia [22]. In recent years, FISH has been combined with flow cytometry (Flow-FISH) for rapid detection of chromosome aberrations in clinical settings. Yet Flow-FISH is not capable of detecting chromosome fusions, chromosome fragility and structural aberrations [23]. With such limitations, traditional FISH remains a powerful tool for understanding basic chromosome biology and the molecular mechanisms of genetic disease and cancer.

Here, we describe the use of FISH to detect telomeric DNA on metaphase chromosomes. This technique can detect end-to-end chromosomal fusions, telomere fragility, and telomere loss [11, 13]. This protocol will describe the preparation of metaphase spreads and telomeric DNA with telomere FISH probes. However, this protocol can also be modified for use with FISH probes to other chromosomal loci [24, 25].

2 Materials

2.1 Cell Culturing

1. 6 cm sterile culturing dishes.
2. Cell growth media: specific to the cells being cultured (*see* **Note 1**).
3. Phosphate buffered saline (PBS): 137 mM sodium chloride, 2.7 mM potassium chloride, 10 mM disodium phosphate, 1.8 mM monopotassium phosphate pH 7.4.
4. Trypsin–EDTA Solution: 0.25% trypsin with 0.01% EDTA.

2.2 Cell Collection

1. 0.5 mg/mL colcemid in 100% ethanol. Solution should be stored at −20 °C. (*see* **Note 2**).
2. 15 mL conical tubes.
3. 0.075 M KCl.
4. 37 °C water bath.
5. Carnoy's fixative solution: three parts 100% methanol to 1 part glacial acetic acid (vol:vol), prepare fresh (*see* **Note 3**).

2.3 Metaphase Spreads

1. 1.5 mL microcentrifuge tubes.
2. Plain unfrosted slides stored in ultrapure water at 4 °C (*see* **Note 4**).
3. Heat block capable of reaching 70–80 °C.
4. Ultrapure water.
5. Paper towel or cloth.
6. Waste beaker.

2.4 FISH

1. Coplin jars or equivalent microscope slide staining rack/reservoir system.
2. PBS.
3. 4% formaldehyde in PBS.
4. 1 mg/mL pepsin in 10 mM glycine pH 2.0, prewarmed to 37 °C.
5. 70%, 90%, and 100% ethanol.
6. Telomere FISH probe (*see* **Note 5**).
7. 22 mm × 30 mm coverslip glass, No 1.5.
8. Hybridization oven or heat block capable of reaching 80 °C.
9. 10% blocking reagent: 10% Blocking reagent (Roche) (weight–volume) dissolved in 100 mM maleic acid, 150 mM NaCl. Adjust pH to 7.5 with NaOH. Autoclave and store at −20°C.
10. Hybridization buffer: 10 mM Tris–HCl, pH 7.2, 70% formamide, 1% blocking reagent. Make fresh and filter-sterilize.
11. Hybridization chamber: light blocking, humidified chamber (*see* **Note 6** and Fig. 3).
12. Wash solution #1: 10 mM Tris–HCl, pH 7.2, 70% formamide, 0.1% BSA. Dissolve the BSA prior to adding formamide.
13. Wash solution #2: 0.1 M Tris–HCl pH 7.2, 0.15 M NaCl, 0.08% Tween 20.
14. Mounting media containing 0.5 μg/mL DAPI.

3 Methods

3.1 Cell Collection

1. Grow cells overnight to 40–50% confluency in 6 cm culture dish. For HeLa cells, a good starting point is to plate ~5×10^5 cells the day prior to collection.
2. Add 4 μL of 0.5 mg/mL colcemid to each 6 cm dish and incubate for 2 h in a 37°C CO_2 incubator (*see* **Note 7**).
3. Collect the media from the 6 cm plate in a 15 mL tube.
4. Wash cells once with PBS. Add the PBS wash to the 15 mL tube with the media (*see* **Note 8**).
5. Add 0.5 mL trypsin–EDTA to the cells and incubate at 37 °C until cells detach from the plate (~3 min).
6. Add 3 mL of complete media to neutralize the trypsin and add the cells suspension to the tube with the media and PBS wash.
7. Spin down the cells in the 15 mL conical tube at 300 rcf (relative centrifugal force) for 4 min.
8. Wash the cells in PBS by removing the supernatant and resuspending the cell pellet in 5 mL of PBS (*see* **Note 9**).
9. Spin down the cells in the 15 mL conical tube at 300 rcf for 5 min and remove the supernatant.
10. Resuspend the cell pellet in the residual liquid and add 5 mL of prewarmed (37 °C) 0.075 M KCl dropwise to the cells, while gently vortexing the tube (*see* **Note 10**).
11. Immediately place the 15 mL conical with cells in a 37 °C water bath for 15–20 min. Ensure the cells are suspended by inverting the tubes occasionally (*see* **Note 11**).
12. Spin down the cells at 300 rcf for 5 min.
13. Remove the 0.075 M KCl. Resuspend the cell pellet in the residual liquid and then add 5 mL of Carnoy's fixative solution dropwise as in Subheading 3.1, **step 9** (*see* **Note 10**).
14. Place tubes at 4 °C at least overnight (*see* **Note 12**).

3.2 Slide Preparation

1. Ensure that the heat block is between 70 and 80 °C. Flip the heat block upside down so the flat surface is face up. Place a folded paper towel to fit over the heat block and dampen the paper towel with water so that it is wet throughout (*see* **Note 13** and Fig. 1).
2. Spin down the 15 mL conical tube containing the fixed cells at 800 rcf for 5 min.
3. Remove the supernatant and resuspend the cell pellet in 0.5–1 mL of fresh Carnoy's fixative solution (*see* **Note 14**).
4. Transfer the resuspended fixed cells to a 1.5 mL microcentrifuge tube and keep on ice or at 4 °C.

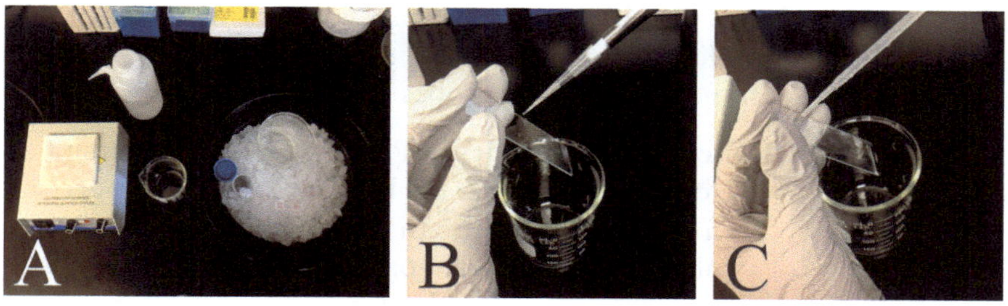

Fig. 1 (**a**) Setup for making metaphase spreads. The ice bucket contains samples in microcentrifuge tubes, fresh fixative with a transfer pipette and glass slides in a beaker of water. The heat block on the left has a paper towel folded over the top and is soaked with water regularly. The beaker in the middle of the image is a waste beaker to catch fresh fixative used to wash the slides. (**b**) Fixed cells are dropped onto a slide at an ~45° angle. (**c**) The slides are washed with fresh fixative before being placed onto the moist paper towel on the heat block

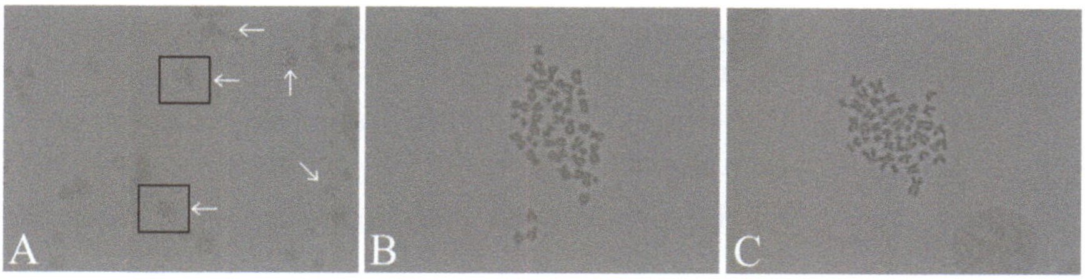

Fig. 2 (**a**) Chromosome spreads from HeLa cells under 10× magnification. Arrows indicate metaphase spreads. (**b**, **c**) Close-up of metaphase spreads seen in (**a**), denoted by boxes

5. Take a prechilled slide and hold it at an ~45° angle over the waste beaker (*see* **Note 15** and Fig. 1). From a few inches above the slide, gently drop 50 μL of the fixed cells across the middle of the slide so that the drops travel to the bottom of the slide (*see* **Note 16**).
6. Immediately rinse the slide with 1–2 mL of fresh Carnoy's fixative, running the solution down the slide.
7. Place the slide, with the cell side up, on the heat block with the wet paper towel for 30–60 s. Most or all of the fixative should be evaporated at this time (*see* **Note 17**).
8. Wipe the residual water off the back of the slide and let cool to room temperature.
9. Check the first slide of each sample under a light microscope to ensure that cell concentration and spreading are adequate (*see* Fig. 2) and then make additional slides, as needed.
10. Label and place the slides into a slide storage box or folder at least overnight before beginning the FISH protocol outlined below. The slides can also be stained with Giemsa or DAPI at this point to look for gross chromosomal abnormalities (*see* **Note 18**).

3.3 FISH Probe Application

Here, we will describe the procedure of telomere FISH (*see* **Note 5**). However, this method can be modified for use with other chromosome-specific probes (*see* **Note 19**).

1. Rehydrate the slides in PBS for 5 min by placing the slide in a Coplin jar, such that the fixed cells are submerged in the liquid (*see* **Note 20**). Ensure that enough liquid is used to completely submerge the region on the slide containing the cells. All subsequent steps are performed at room temperature, unless otherwise indicated.
2. Transfer the slides into a new Coplin jar with 4% formaldehyde in PBS for 2 min to fix the slides.
3. Wash the slides in Coplin jars with PBS for 5 min. Repeat twice with fresh PBS each time.
4. Treat slides with prewarmed pepsin at 37 °C for 5–8 min and no more than 10 min. It is recommended that the Coplin jar containing the pepsin solution be prewarmed to 37 °C (*see* **Note 21**).
5. Wash the slides in a PBS containing Coplin jar for 2 min. Repeat twice with fresh PBS each time.
6. Fix the spreads again in 4% formaldehyde as in Subheading 3.3, **step 2**.
7. Wash the slides in a PBS containing Coplin jar for 5 min. Repeat twice with fresh PBS each time.
8. Dehydrate the spreads by placing them in increasing ethanol concentrations for 1–2 min each. Start with 70% ethanol, move to 90% ethanol, and finally 100% ethanol.
9. Remove the slides from the Coplin jars and place them cell side up on paper towel. Allow the slides to air-dry. Ensure the slides are protected from dust and other debris as they dry.
10. Prepare the telomere FISH probe by diluting it to 10 nM in hybridization buffer (*see* **Note 22**). From this point on, minimize the amount of light exposure to the slides.
11. Prepare 35 μL of the probe + hybridization solution per slide. Place a small drop on the slide near the bottom where the cells are located. The remaining solution is spread in a line across a 22 mm × 30 mm glass coverslip. Take the slide and gently flip it over in your hand. Carefully fuse the solution on the slide with the that on the coverslip. The coverslip should then get pulled up onto the slide due to surface tension (*see* **Note 23**).
12. Incubate the slide, coverslip side up, at 80°C for 5 min (*see* **Note 24**).
13. Place the slide in a prewarmed hybridization chamber (37 °C) and allow the probe to hybridize with the DNA for at least two hours at room temperature. Slides can also be left to hybridize overnight (*see* **Note 25** and Fig. 3).

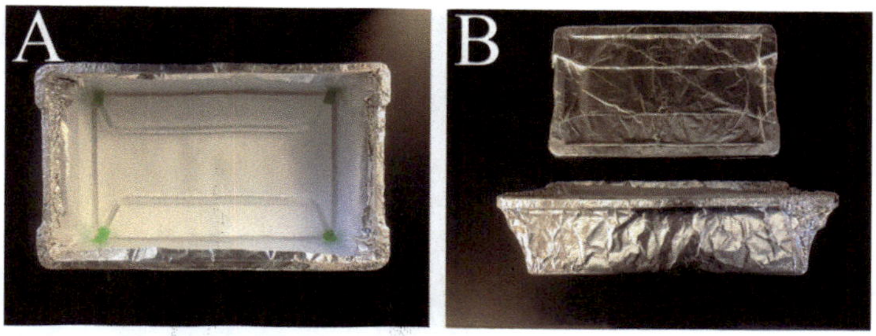

Fig. 3 Example of a homemade hybridization chamber. (**a**) A plastic storage container covered in aluminum foil with filter paper placed in the bottom of the container. Glass rods were bent to fit into the bottom of the container and hold microscope slides. (**b**) The lid and container are covered in aluminum foil to protect the samples from light

14. Place a small amount of the hybridization wash solution #1 in a petri plate or similar container. Gently dip the slides with the coverslip side down into the solution and allow the coverslips to slide off, when necessary gently help the coverslip off with a gloved finger or the side of the dish, being careful not to scrape the slide.
15. Wash slides in an aluminum foil covered, or other light protected, Coplin jar filled with wash solution #1 for 15 min. Repeat with a second wash step. All wash steps from this point on should be performed in light-protected Coplin jars.
16. Wash the slides with wash solution #2 for 5 min. Repeat twice (*see* **Notes 26** and **27**).
17. Dehydrate the slides as in Subheading 3.3, **step 8**.
18. Air-dry the slides in the dark.
19. Prepare 25 μL of mounting media with DAPI per slide. Place one drop on a 22 mm × 30 mm coverslip and apply the remaining mounting media on the slide, being careful not to scrape the pipette tip across the slide. Gently and slowly lay the coverslip down starting at one end of the mounting media on the slide and moving toward the other, avoiding the introduction of bubbles into the media.

3.4 Imaging FISH-Labeled Spreads

1. If necessary, allow the mounting media to harden overnight.
2. Image slides between 40 and 100× magnification, using the correct light filters. Ensure the proper immersion media is used, as microscopes objectives 40× and above typically require water or oil.
3. Take images of the FISH signals on metaphase chromosomes and overlay them with the DAPI. Save and export images (*see* Fig. 4 and **Note 28**).

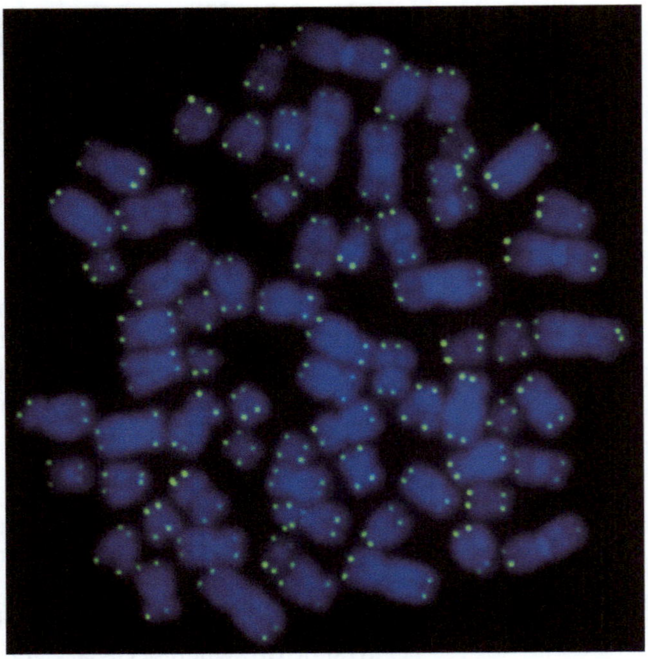

Fig. 4 Example image of telomere FISH on a metaphase spread. Blue, DAPI; Green, FITC-labeled telomere FISH probe

4 Notes

1. The protocol described is optimized for HeLa and HCT116 cells. Steps and conditions may need to be optimized for other cell lines.
2. Colcemid is a spindle poison that prevents the formation of microtubules, which causes the cells to arrest in prometaphase [26]. Other spindle poisons can be used in place of colcemid, such as colchicine and nocodazole. Nocodazole has the benefit of being reversible, whereas colcemid and colchicine are irreversible. The chemotherapeutic drug paclitaxel, sold under the name Taxol, also acts as a spindle poison.
3. The fixative solution should be kept cold (4 °C or on ice) and made fresh on the day of its use. Over time, methanol and acetic acid will form the ester, methyl acetate, which prevents chromosome spreading [27].
4. The slides should be stored in water at least overnight prior to their use. Each slide should be separated and individually submerged, which increases spreading efficiency during metaphase spread preparation.
5. Telomere FISH probes can target either the C-rich or G-rich strand and typically contain three telomere repeats (C-strand

probe: [TTAGGG]$_3$; G-strand [CCCTAA]$_3$) labeled with a fluorophore. A centromere probe can be combined with the telomere FISH protocol to help detect chromosomal fusions. Premade and custom probes are available for purchase from several companies and can be labeled with a variety of different fluorophores. The telomere probe used in this protocol has a peptide nucleic acid (PNA) backbone, which are more stable than DNA probes. Locked nucleic acid (LPA) and bridged nucleic acid (BNA) probes are also available, and they provide greater stability and binding than DNA probes.

6. A dark, humidified chamber is used to prevent the FISH probe solution from drying out during hybridization. Homemade chambers can be made out of a plastic box that has been blacked out using aluminum foil (*see* Fig. 3). In the bottom of the box, an absorbent material should be placed, along with a way to elevate slides. We find that parallel glass rods on top of filter paper works well. The filter paper is soaked with water to humidify the chamber.
7. Colcemid is used to enrich the mitotic cell population. Concentration and incubation time should be optimized for each cell line and to ensure no unintended side effects. Prolonged colcemid treatment has been shown to condense sister chromatids to the point they separate and attempt to reconstitute a nucleus [26].
8. Cells can also be collected by mitotic shake-off, which enriches the number of mitotic cells but yields lower total cell numbers. To collect cells by mitotic shake-off, tap the side of the dish to dislodge mitotic cells from the plate. Use of a screw-top, non-vented capped cell culture flask allows you to increase the strength at which you can tap the flask, increasing the number of mitotic cells obtained. Next, wash media across the cells several times and collect the media. Add PBS and repeat the steps above to dislodge and collect cells. Proceed to Subheading 3.1, **step 6**.
9. One way to minimize the loss and clumping of cells is to resuspend the cell pellet in the residual supernatant. This can be done by flicking the 15 mL conical tube several times or dragging it along a ridged surface several times such as a 1.5 mL centrifuge tube rack.
10. On a typical vortex, the speed should be set just below half full speed. The tube is held in one hand loosely to allow a sample to be gently vortexed, while the other hand is used to add the solution dropwise to the sample. Only the first 1–2 mL of the solution needs to be added to the sample dropwise.
11. This step is used to swell the cells. The 0.075 M KCl is a hypotonic solution and will cause the cells to take in water.

Incubation times may need to be adjusted for different cell types. There is a risk of having the cells burst if they are allowed to incubate in the 0.075 M KCl solution for too long or the solution is too hypotonic [28].

12. The fixed cells can be saved for at least several years.
13. Ensure that the paper towel on the heat block is wet each time before placing slides on it. When making multiple slides, water will need to be continually added to the paper towel at regular intervals. For safety, make sure that the heat block used is able to get wet.
14. The volume of Carnoy's fixative added will depend on the number of cells. For proper spreading, the cell density on the slide should be ~20–50% in most areas. To check this, make one metaphase slide and then check it under a light microscope. If the cell density is too high, then dilute the samples. If it is too low, then spin the cells down and resuspend them in a lower volume.
15. The slide should have a film of water on it when removed.
16. A common misconception in making metaphase spreads is that cells burst due to the force of being dropped from a long distance. However, the cells burst due to heating, humidity, and expansion of the volatile liquid inside, so the distance that you drop the solution does not need to be more than a few inches, which helps the solution spread across the slide.
17. This step influences the quality of the metaphase spreads. The spreads and interphase cells will be barely visible under the microscope. As you prolong the incubation time on the heat block, the cells will become more visible under the microscope, and, in our experience, the quality of the metaphase spreads decrease.
18. If you want to perform metaphase spreads analysis without FISH, the slides may be stained the next day using Giemsa stain or DAPI [28, 29]. This is useful for determining gross chromosomal abnormalities, such as chromosomal breaks or gaps, aneuploidy, or chromatid cohesion loss.
19. Chromosome- and locus-specific probes are available from a number of manufacturers. Many of these probes include their own hybridization buffer and instructions from the manufacturer, which should be used as a starting point for FISH analysis with the probe.
20. Between steps a lint-free tissue is used to blot off the excess liquid at the bottom of the slide.
21. Pepsin is used to remove protein from the chromosomes to provide better access for the probe. Other proteases such as trypsin can also be used.

22. The telomere probe concentration may need to be empirically determined. A range of 4–55 nM has been reported to work for telomere FISH. Telomere length, probe quality, fluorophore brightness, and hybridization conditions can affect signal detection.
23. This step must be done carefully to prevent bubbles from forming between the coverslip and slide. The volume can also be scaled for smaller or larger coverslips. The volume should be sufficient for the coverslip to slightly float in the liquid on the slide.
24. A hybridization oven is ideal, but a heat block or water bath may also be used for this step.
25. Ensure that an excess of water is on the filter paper to prevent the hybridization buffer drying out on the slide.
26. Instead of using mounting media containing DAPI in Subheading 3.3, **step 19**, DAPI (0.5 μg/mL final concentration) can be added to the second wash step and the wash extended to 10 min.
27. If the FISH probe signal is weak, the signal can be amplified with antibodies conjugated with biotin followed by an incubation with an avidin-conjugated fluorophore on the telomere FISH probe. Additional amplification steps can also be added.
28. Telomere FISH can be used to assess chromosome fusions, telomere fragility, telomere signal loss, and telomere length [30, 31]. There are several free software packages that are available for telomere length analysis, including TFL-Telo and telometer (an ImageJ plug-in) [32].

Acknowledgments

We would like to thank members of the Stewart lab for critical reading of the manuscript and helpful discussions. This work was supported by the National Institutes of Health [R00GM104409 to J.A.S.] and startup funds from the University of South Carolina.

References

1. Scott DH (1891) Carl Wilhelm von Nägeli. Nature 44:580. https://doi.org/10.1038/044580a0
2. Tjio JH (1978) The chromosome number of man. Am J Obstet Gynecol 130(6):723–724
3. Pardue ML, Gall JG (1969) Molecular hybridization of radioactive DNA to the DNA of cytological preparations. Proc Natl Acad Sci U S A 64(2):600–604
4. Langer PR, Waldrop AA, Ward DC (1981) Enzymatic synthesis of biotin-labeled polynucleotides: novel nucleic acid affinity probes. Proc Natl Acad Sci U S A 78 (11):6633–6637

5. Arndt-Jovin DJ, Robert-Nicoud M, Kaufman SJ, Jovin TM (1985) Fluorescence digital imaging microscopy in cell biology. Science 230(4723):247–256
6. Kricka LJ, Fortina P (2009) Analytical ancestry: "firsts" in fluorescent labeling of nucleosides, nucleotides, and nucleic acids. Clin Chem 55(4):670–683. https://doi.org/10.1373/clinchem.2008.116152
7. Bauman JG, Wiegant J, Borst P, van Duijn P (1980) A new method for fluorescence microscopical localization of specific DNA sequences by in situ hybridization of fluorochrome-labelled RNA. Exp Cell Res 128(2):485–490
8. Gall JG (2016) The origin of in situ hybridization – a personal history. Methods 98:4–9. https://doi.org/10.1016/j.ymeth.2015.11.026
9. Ma K, Qiu L, Mrasek K, Zhang J, Liehr T, Quintana LG, Li Z (2012) Common fragile sites: genomic hotspots of DNA damage and carcinogenesis. Int J Mol Sci 13(9):11974–11999. https://doi.org/10.3390/ijms130911974
10. Mrasek K, Wilhelm K, Quintana LG, Theuss L, Liehr T, Leskovac A, Filipovic J, Joksic G, Joksic I, Weise A (2015) BAC-probes applied for characterization of fragile sites (FS). Methods Mol Biol 1227:289–298. https://doi.org/10.1007/978-1-4939-1652-8_15
11. Brault ME, Autexier C (2011) Telomeric recombination induced by dysfunctional telomeres. Mol Biol Cell 22(2):179–188. https://doi.org/10.1091/mbc.E10-02-0173
12. Amakawa G, Ikemoto K, Ito H, Furuya T, Sasaki K (2013) Quantitative analysis of centromeric FISH spots during the cell cycle by image cytometry. J Histochem Cytochem 61(10):699–705. https://doi.org/10.1369/0022155413498754
13. Pinzaru AM, Hom RA, Beal A, Phillips AF, Ni E, Cardozo T, Nair N, Choi J, Wuttke DS, Sfeir A, Denchi EL (2016) Telomere replication stress induced by POT1 inactivation accelerates tumorigenesis. Cell Rep 15(10):2170–2184. https://doi.org/10.1016/j.celrep.2016.05.008
14. Koo DH, Zhao H, Jiang J (2016) Chromatin-associated transcripts of tandemly repetitive DNA sequences revealed by RNA-FISH. Chromosom Res 24(4):467–480. https://doi.org/10.1007/s10577-016-9537-5
15. Cuthbert G, Thompson K, Breese G, McCullough S, Bown N (2000) Sensitivity of FISH in detection of MLL translocations. Genes Chromosomes Cancer 29(2):180–185
16. Gascoyne RD, Lamant L, Martin-Subero JI, Lestou VS, Harris NL, Muller-Hermelink HK, Seymour JF, Campbell LJ, Horsman DE, Auvigne I, Espinos E, Siebert R, Delsol G (2003) ALK-positive diffuse large B-cell lymphoma is associated with Clathrin-ALK rearrangements: report of 6 cases. Blood 102(7):2568–2573. https://doi.org/10.1182/blood-2003-03-0786
17. Piper J, Fantes J, Gosden J, Ji LA (1990) Automatic detection of fragile X chromosomes using an X centromere probe. Cytometry 11(1):73–79. https://doi.org/10.1002/cyto.990110109
18. Ried T, Landes G, Dackowski W, Klinger K, Ward DC (1992) Multicolor fluorescence in situ hybridization for the simultaneous detection of probe sets for chromosomes 13, 18, 21, X and Y in uncultured amniotic fluid cells. Hum Mol Genet 1(5):307–313
19. Jalal SM, Law ME (1997) Detection of newborn aneuploidy by interphase fluorescence in situ hybridization. Mayo Clin Proc 72(8):705–710
20. Petroni S, Caldarola L, Scamarcio R, Giotta F, Latorre A, Mangia A, Simone G (2016) FISH testing of HER2 immunohistochemistry 1+ invasive breast cancer with unfavorable characteristics. Oncol Lett 12(5):3115–3122. https://doi.org/10.3892/ol.2016.5125
21. Agersborg S, Mixon C, Nguyen T, Aithal S, Sudarsanam S, Blocker F, Weiss L, Gasparini R, Jiang S, Chen W, Hess G, Albitar M (2018) Immunohistochemistry and alternative FISH testing in breast cancer with HER2 equivocal amplification. Breast Cancer Res Treat 170(2):321–328. https://doi.org/10.1007/s10549-018-4755-5
22. van der Burg M, Poulsen TS, Hunger SP, Beverloo HB, Smit EM, Vang-Nielsen K, Langerak AW, van Dongen JJ (2004) Split-signal FISH for detection of chromosome aberrations in acute lymphoblastic leukemia. Leukemia 18(5):895–908. https://doi.org/10.1038/sj.leu.2403340
23. Baerlocher GM, Vulto I, de Jong G, Lansdorp PM (2006) Flow cytometry and FISH to measure the average length of telomeres (flow FISH). Nat Protoc 1(5):2365–2376. https://doi.org/10.1038/nprot.2006.263
24. Kraker WJ, Borell TJ, Schad CR, Pennington MJ, Karnes PS, Dewald GW, Jenkins RB (1992) Fluorescent in situ hybridization: use of whole chromosome paint probes to identify unbalanced chromosome translocations. Mayo Clin Proc 67(7):658–662
25. Rives N, Wust S, David B, Duchesne V, Joly G, Mace B (1999) Fluorescence in situ

hybridization with chromosome paint probes: a novel approach to assess aneuploidy in human sperm nuclei. J Assist Reprod Genet 16 (1):46–51
26. Rieder CL, Palazzo RE (1992) Colcemid and the mitotic cycle. J Cell Sci 102(Pt 3):387–392
27. Baksi SM, Means JC (1988) Preparation of chromosomes from early stages of fish for cytogenetic analysis. J Fish Biol 32(3):321–325. https://doi.org/10.1111/j.1095-8649.1988.tb05369.x
28. Bayani J, Squire JA (2004) Preparation of cytogenetic specimens from tissue samples. Curr Protoc Cell Biol Chapter 22:Unit 22.2. https://doi.org/10.1002/0471143030.cb2202s23
29. Deng W, Tsao SW, Lucas JN, Leung CS, Cheung AL (2003) A new method for improving metaphase chromosome spreading. Cytometry A 51(1):46–51. https://doi.org/10.1002/cyto.a.10004
30. Hemann MT, Strong MA, Hao LY, Greider CW (2001) The shortest telomere, not average telomere length, is critical for cell viability and chromosome stability. Cell 107(1):67–77
31. Tanaka H, Abe S, Huda N, Tu L, Beam MJ, Grimes B, Gilley D (2012) Telomere fusions in early human breast carcinoma. Proc Natl Acad Sci U S A 109(35):14098–14103. https://doi.org/10.1073/pnas.1120062109
32. van Batenburg AA, Kazemier KM, Peeters T, van Oosterhout MFM, van der Vis JJ, Grutters JC, Goldschmeding R, van Moorsel CHM (2018) Cell type-specific quantification of telomere length and DNA double-strand breaks in individual lung cells by fluorescence in situ hybridization and fluorescent immunohistochemistry. J Histochem Cytochem 66 (7):485–495. https://doi.org/10.1369/0022155418761351

INDEX

Lata Balakrishnan and Jason A. Stewart (eds.), *DNA Repair: Methods and Protocols*, Methods in Molecular Biology, vol. 1999, https://doi.org/10.1007/978-1-4939-9500-4, © Springer Science+Business Media, LLC, part of Springer Nature 2019

E

F

G

H

I

K

L

M

N

O

P

Q

R

S

T

U

W

X

MIX
Papier aus verantwortungsvollen Quellen
Paper from responsible sources
FSC® C105338

If you have any concerns about our products,
you can contact us on
ProductSafety@springernature.com

In case Publisher is established outside the EU,
the EU authorized representative is:
Springer Nature Customer Service Center GmbH
Europaplatz 3, 69115 Heidelberg, Germany

Printed by Libri Plureos GmbH
in Hamburg, Germany